D0762398

Networked Machinists

JOHNS HOPKINS STUDIES IN THE HISTORY OF TECHNOLOGY
Merritt Roe Smith, *Series Editor*

Networked Machinists

High-Technology Industries in Antebellum America

DAVID R. MEYER

The Johns Hopkins University Press
Baltimore

Printed in the United States of America on acid-free paper
2 4 6 8 9 7 5 3 1

The Johns Hopkins University Press
2715 North Charles Street
Baltimore, Maryland 21218-4363
www.press.jhu.edu

Library of Congress Cataloging-in-Publication Data

Meyer, David R.
Networked machinists : high-technology industries in Antebellum America / David R. Meyer.
p. cm. — (Johns Hopkins studies in the history of technology)
Includes bibliographical references and index.
ISBN 0-8018-8471-3 (hardcover : alk. paper)
1. Machinists—United States—History. 2. Metalworking industries—United States—History. I. Title. II. Series.
HD8039.M22U65 2006
331.7′68176097309034—dc22
2006009628

A catalog record for this book is available from the British Library.

Contents

Illustrations

FIGURES

MAPS

Tables

Acknowledgments

The interpretations of the machinist networks that are offered in this book rely on the many known, as well as anonymous, writers of the nineteenth century who recorded the exploits of these fascinating individuals and their firms. Numerous scholars who wrote articles and books, either directly dealing with machinists or covering broad topics in which these individuals appeared as actors, provided the many details essential to piecing together the story portrayed here. My hope is that I conveyed, in some small way, the extraordinary behavior of antebellum machinists, who set the basis for the great industrialization of the second half of the nineteenth century. Many of these machinists have been recognized in publications, but they were more than individual giants. They truly constituted communities of practitioners. Their behavior suggests that they were, to some degree, aware that they participated in a much larger enterprise—building the machinery and machine tools that were so critical to industrialization. Contemporary citizens are surrounded with equipment we consider essential to our daily lives which can be traced back to technical innovations developed by antebellum machinists and their late-nineteenth-century successors. In that sense, modern people owe an immense debt of gratitude to these machinists.

I also give thanks to the many professional colleagues in economic history, geography, history, and sociology who provided reactions to, suggestions for, and encouragement of my ideas about machinist networks. The anonymous reviewer of the manuscript offered numerous excellent suggestions for improving the manuscript. John Logan, director of the Spatial Structures in the Social Sciences initiative at Brown University generously provided support for making the computer maps; I offer special thanks to Dan Gui, who made the maps. Also, thanks are due to Geoff Surrette for making the computer graphic on Paterson's locomotive networks. Brown University's Department of Sociology, Urban Studies Program, and libraries offered support at various stages of the project, and I thank them. I owe kudos to Elizabeth Gratch for applying her editorial skills to the manuscript.

Networked Machinists

Machinists' Traces

> The vast resources of the United States are now being developed with a success that promises results whose importance it is impossible to estimate. This development . . . is by the universal application of machinery effected with a rapidity that is altogether unprecedented.
>
> *Joseph Whitworth, "New York Industrial Exhibition"*

Amid a period of seemingly dramatic technological change, contemporary observers can become euphoric and proclaim that they live in a revolutionary era. The technology boom of the late twentieth century was one such era. Manuel Castells, one of its prominent chroniclers, went so far as to write a three-volume work, *The Information Age: Economy, Society and Culture.*[1] This information technology revolution, as he called it, possessed a unifying mechanism—the network. Individual entrepreneurs, groups of actors, firms, and other units operate in networks—the means through which information is produced, exchanged, and diffused—and this interaction can generate increased innovation. Those who adhere to the view that this revolution defines the contemporary era agree that its quintessential locus is Silicon Valley, California, which is south of San Francisco between Palo Alto (Stanford University) to the north and San Jose to the south.

According to the network interpretation of this revolution, technology workers in Silicon Valley operated in social and professional networks, and individual firms and organizations such as universities and associations also possessed network links. These workers created a collaborative tradition, sometimes maintaining greater loyalty to one another than they did to organizational entities such as companies or the industries in which they worked. Technical and market knowledge flowed through the networks connecting individuals and firms. Job-hopping, often youthful workers transferred technical skills and innovations as they switched employers and even crossed over into other industrial sectors, using their networks to find new positions. Young engineers who left existing firms founded many of the new companies.

Individuals and firms in Silicon Valley used their formal and informal networks to create competitive advantages in acquiring technical skills and innovations and in accessing markets. These networks foster the innovations that buttress the valley's continual shift into new, high-technology sectors—from integrated circuits to personal computers to the Internet, for example. Networked technical communities even extend to the global scale as immigrants from Asia who worked in Silicon Valley return to their home countries. Could this effusive portrayal of the information technology revolution and praise for the network behavior of its actors be an exercise in hubris?[2]

THE NATION'S FIRST HIGH-TECHNOLOGY INDUSTRIES

From 1790 to 1860, more than 150 years before the revolution in Silicon Valley, machinists in the eastern United States created the nation's first high-technology industries (map I.1). In a set of metal manufactures termed the pivotal producer durables (metal fabricating, machinery, and machine tools) technological knowledge and skills were embedded in networks of machinists, firms, and clusters of individuals or firms; they were not the sole property of any of the constituent parts. The network behavior of machinists contributed to a substantial transformation of the pivotal producer durables by the 1850s.

When viewing such a long-ago period, it is tempting to use contemporary standards of the sophistication of high-technology industries. That backward perspective, however, dismisses too readily the underlying significance of the changes that unfolded over the seven decades after 1790. These events did not have the appearance of a radical revolution. Machinery accounted for small shares of antebellum capital investment, and the share of equipment in domestic capital stock stayed around 5 percent before 1840 (fig. I.1). Nonetheless, across a range of pivotal producer durables, mechanics developed new machines and made improvements in them. During the twenty years after 1840 equipment's share of domestic capital stock rose to about 9 percent, suggesting substantial change in the position of the pivotal producer durables.

Although most antebellum productivity improvements came from sources other than machinery, the real value of equipment per capita—a measure of the capacity to enhance the productivity of each person—managed to rise at an almost constant rate of just over 1 percent annually from 1774 to 1850 (see fig. I.1). Then the rate of increase accelerated to about 5 percent annually, and this pace continued until 1900.

Map I.1. Eastern United States

Joseph Whitworth, one of Britain's leading machinery and machine tool manufacturers, recognized the achievements of antebellum machinists. After surveying machinery works and armories in the eastern United States, he reported to the House of Commons in 1854 on what he termed the "universal application of machinery effected with a rapidity that is altogether unprecedented."

These machinists set the groundwork for the extraordinary machinery and machine tools of the late nineteenth century, when the United States moved to the forefront in making much of this equipment. The U.S. Census of 1880 displayed lithographs of sophisticated metal fabricated products, machinery, and machine tools, validating Whitworth's portrayal of the successes of antebellum machinists. By 1900 equipment accounted for 28 percent of domestic capital stock, and the machinery sector contributed 9 percent of national value added in manufacturing, ranking it as the nation's third largest industry (see fig. I.1).[3]

Antebellum machinists built networked communities in the pivotal producer durables, whose key sectors included iron foundries, steam engines, locomotives, tex-

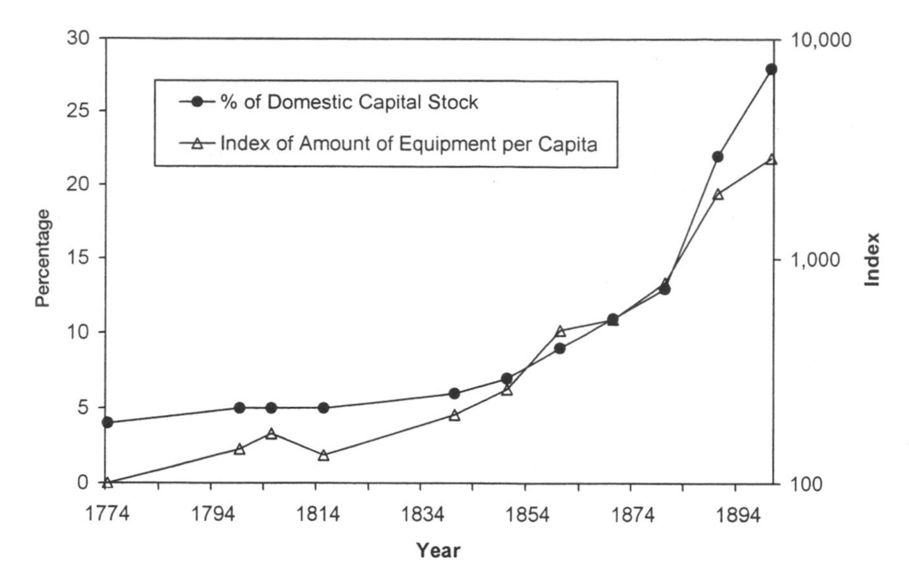

Figure I.1. Equipment as Percentage of Domestic Capital Stock and Index of Amount of Equipment per Capita (1860 Prices), 1774–1900. *Source:* Robert E. Gallman, "American Economic Growth before the Civil War: The Testimony of the Capital Stock Estimates," *American Economic Growth and Standards of Living before the Civil War*, ed. Robert E. Gallman and John J. Wallis (Chicago: University of Chicago Press, 1992), 94–95, tables 2.8–2.9

tile machinery, firearms, and machine tools. Their network approach to establishing these communities and to operating in them presaged the high-technology workers of the late twentieth and early twenty-first centuries.

MACHINISTS AS VIRTUOUS BOURGEOIS

The demand for machinists typically exceeded their supply. They underwent extensive training to acquire metalworking skills, first under a formal or informal apprenticeship and, subsequently, moving among machine shops to gain experience or they learned from coworkers, if they remained in one shop. They commanded wages that ranked among the highest of all industrial workers in the eastern United States, they received premium salaries if they were part of management, and they acquired profits from the firm if they had ownership stakes. Sometimes these machinists suffered during economic downturns, but for the most part those making the highest wages among the shop and factory workers, and especially those on salary or with ownership stakes, certainly can be termed "bourgeois." Their strategic place in the labor market of highly skilled people makes it logical to consider them prototypical *Homo economicus*; that is, they sold their talents to the highest bidders.

The visages of legendary machinists active during the antebellum years adorn the pages of many books and articles covering nineteenth-century manufacturing. They appear in lithographs produced before and after 1860 and in photographs taken after that date. These legends included, for example, the textile machinery great Samuel Slater, the locomotive giants Matthias Baldwin and Thomas Rogers, and the incomparable machine tool builders Frederick Howe, William Bement, and William Sellers. The images show them as quintessential successful bourgeois men: stern looking, bearded, wearing business suits, and shown in the later years of their life. If the lithographs appeared before 1860, they usually reveal them in their fifties or sixties. Photographs taken after that date picture them as sixty to eighty years old; some of them had retired, and a few still worked.

While the images convey the machinists' bourgeois character, a fascinating tale lurks beneath the surface which unsettles modern stereotypes of them. Certainly, they, and others like them, achieved success; they captured the financial benefits of their strategic place in the antebellum economy. Observers from the left of the political-economic spectrum critique them as bourgeois individuals who appropriated the labor surplus of others, whereas observers from the right side of the spectrum applaud their pursuit of self-interest. Both sets of observers would agree that these machinists followed an ethic of prudence; they took care of themselves. Their ethics, however, possessed more nuances than this one-dimensional portrait suggests. Observers from either side of the spectrum would agree that in their personal lives, outside the realm of business, machinists also possessed virtues such as courage, temperance, justice, faith, hope, and love. But it would be audacious to claim that they possessed "bourgeois virtue" in their business lives. While in their twenties, legendary machinists and others not quite so prominent job-hopped among shops within a metropolitan regional hinterland such as that around Boston, and some of them job-hopped up and down the East Coast from Maine to Maryland. During that time they moved along a steep upward career trajectory, reaching the position of foreman in a machine shop in their early to mid-twenties and superintendent of a shop (iron foundry, textile machinery firm, and so on) by their late twenties and early thirties; some of them acquired an equity stake of one-quarter to one-third in a shop during their thirties or early forties. Senior machinists took young men under their wings and taught them technical skills, and the mentors used their network contacts to recommend their protégés for positions.

Experienced machinists visited one another's shops to get advice about machine building, see new equipment, and even copy patterns; in their twenties, especially, they exhibited little loyalty to their firms. They changed jobs frequently without any retribution from their previous employers, even though these job-hoppers took with them ideas for new machines and skills that they learned in the shops, thus transferring them to other firms. This skill transfer occurred even though the shops might compete with one another at some time in the future. All of this career advice and assistance, without financial remuneration, cannot be interpreted simply as crass calculations that they received benefits in return which would assist their own shops. Such a view requires logical contortions to account for such a wide range of behaviors without clear, self-serving benefits.[4]

Iron foundries and steam engine shops, locomotive works, textile machinery firms, firearms factories, and machine tool firms had greater salience than being mounds of capital equipment—cupola furnaces, forges, trip hammers, lathes, and milling machines—and a labor force of skilled machinists who produced capital goods. To build the equipment, machinists needed skills that they learned from one another, beginning with an apprenticeship and continuing throughout their careers. At the broadest scope the work of mechanics, the operations of firms, and the functioning of machinist sectors required knowledge.

TYPES OF KNOWLEDGE

Several continuums structured different types of knowledge. Along one, knowledge ranged from codified to tacit forms. At the codified end, direct formats such as manuals and diagrams transmitted knowledge, or the knowledge involved, for example, simple oral explanations about how machinists used files to remove metal burrs. As the technical skills became more difficult, the machines more complex, or systems of machines were developed, knowledge approached the other end of the continuum. Machinists gained more tacit forms of knowledge through experience accumulated over time, learning by doing, or face-to-face communication as they solved metalworking problems.

Along another continuum knowledge ranged from component to architectural. At the component end, for example, knowledge related to parts of machines (such as the spindles of spinning machines or the lock mechanisms of firearms), entire machines (e.g., lathes or steam engines), or sequences of machines (e.g., stages in firearms manufacture). At the other end of the continuum architectural knowledge concerned the structure of the components, the organization of a system, including its governing rules (implicitly or explicitly understood), or the goals of the organization. The architectural knowledge of manufacturing several locomotives simultaneously, for example, became a sophisticated conceptual process that required an understanding of locomotive building as a system.

These types of knowledge continuums cross-cut one another. Component knowledge, for example, might be communicated in codified form in manuals or diagrams, such as the diagram of a steam engine, whereas the architectural knowledge of the logic of profitable firearms manufacturing, which consisted of systems of sophisticated machine tools, might only be shared as tacit knowledge within a firm. Machinists might communicate knowledge of machine-building skills through apprenticeships, or this knowledge might be obtained through visits to one another's machine shops to discuss ideas for building better machines. On the other hand, architectural knowledge of profitable firearms manufacturing might be difficult to communicate between firms, even if officials met to discuss and observe the production processes.

Clusters of machinists' firms also manifested continuums of knowledge. The conceptual and systemic nature of architectural knowledge within a firm, such as simultaneous locomotive manufacturing, may be difficult to transmit between firms in a cluster. On the other hand, component knowledge may be transmitted between firms through job mobility or meetings of machinists in one another's shops to discuss innovations in metalworking (e.g., cutting metal or shaping a bar). Over time the tacit sharing of component knowledge may build the overall skills of machine shops in the cluster, and they may share ideas on the integration of different machines in sequences, thus creating an architectural knowledge of machining. The early emergence in one cluster of a set of textile machine–building skills and the integration of these skills into a system of machines such as spinning and weaving, for example, might become architectural knowledge that remained difficult to transmit to other clusters of machine shops until these sequential processes became better understood. These barriers to the transmission of architectural knowledge among clusters, however, did not stay impermeable. Leading machinists traveled among clusters of shops or made permanent job relocations, such as from Providence, Rhode Island, to Philadelphia, Pennsylvania, and they became effective mechanisms for the transmission of architectural knowledge via tacit communication.[5]

Individual machinists communicated the different types of knowledge through their networks, and their firms possessed ties through the owners and formal ties through contractual and subcontractual relationships. Networks operated within industrial sectors, such as among textile machine shops, and sometimes networks crossed sectors, for example, between textile machine shops and locomotive shops. Lower-level machinists often switched jobs, occasionally moving considerable distances, and the premier machinists, who had changed jobs frequently during their twenties, participated in wide-ranging networks that kept them in contact with the latest ideas and innovations in their industrial sectors. The systematic behavior of machinists provided structure to their networks.

NETWORK BEHAVIOR

As with all social actors, machinists made strategic choices about their network behavior. They faced limits on the amount of time they could spend in meeting with other machinists, in adding new contacts, and in traveling to meet peers. The monetary costs of meeting also constrained them, including the opportunity cost of not working on their own projects and the travel costs involved in meeting with others located far from their shops. Machinists confronted strategic choices about how to use their structural position in networks to their own and to their firms' advantage.[6]

From the start the teenager who aimed to become a machinist faced important choices. If he lived in a subregion with a few small shops whose owners and other workers possessed limited contacts with wider machinist networks, the apprentice was at a disadvantage. He learned unsophisticated and/or obsolete skills, and the owner and other workers had few network contacts that could be tapped to help him move to more important shops. A teenager who lived in a subregion (or moved to one) with at least one top machine shop had greater career opportunities.

Network Ties within Subregions

The chief machinist (sometimes the owner) of a prominent firm served as a node in attracting other top mechanics and talented apprentices to work in the shop. These individuals came because the leading machinist taught the best technical skills in machine design and manufacturing, and, if other sophisticated mechanics also worked there, they learned from one another. The shop was a setting for tacitly sharing component knowledge. These peers maintained their network ties after they left for other shops, and they provided job referrals to help one another advance their careers. Leading machinists also possessed links with local machine shops and with other shops in their respective subregions. Employees tapped these ties when they wanted to move on to advance their skills, to find work as orders in their shops dried up, or to find new employment at higher wages. As capital goods producers, machine shops faced significant variability in orders over time; thus, when the volume of orders slackened, machinists needed to find new employment.

Within a subregion network ties took on at least two alternative types of structures, which had implications for the exchange of knowledge, transfer of technical skills, and job mobility (fig. I.2). Some subregions—such as eastern Massachusetts, Providence and vicinity, New York City and its environs, and the Philadelphia area—housed a large number of machine shops, many of them led by top machinists, and numerous ties connected the shops. These ties consisted of the personal acquaintances of each top mechanic with head machinists in the other shops, and they may have also been buttressed by subcontractual relations. Network links served as conduits to transfer tacit knowledge and technical skills and to facilitate employee mobility among the shops, and a cohesive network could result from extensive exchanges.

If the bonds exhibited multiplexity, this enhanced the exchanges. In a multiplex

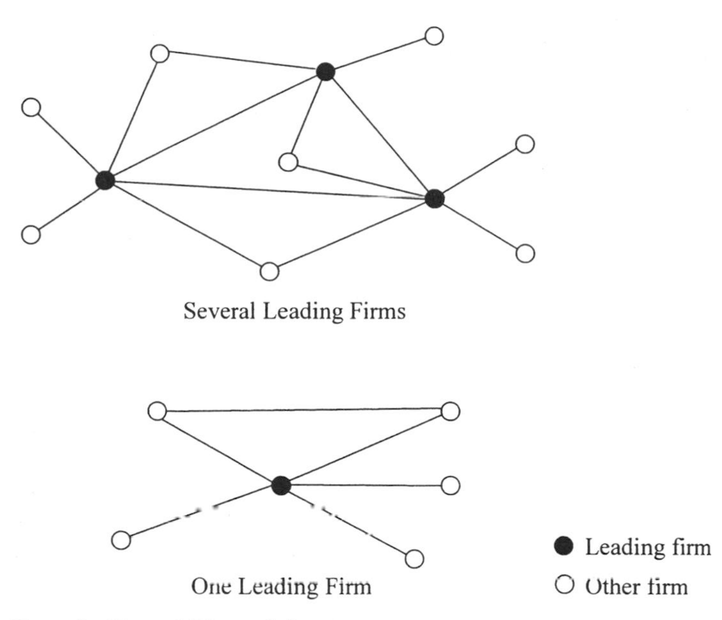

Figure I.2. Network Ties in Subregions

network machinists with different skills interacted in a range of venues (social, economic, and so on) and across various machine sectors (e.g., iron foundries and textile machine shops). In such a subregion innovation might flourish because the large pool of shops tacitly exchanged component knowledge and shared technical skills, thus creating an architectural knowledge about machine design; this became the basis for improving machines or creating new ones. These multifarious ties led to redundancies in the network because they all shared in the same knowledge and skills. If talented machinists headed different shops, such redundancies were not detrimental because over time they generated new ideas. On the other hand, in a subregion with less-talented machinists or ones who stopped innovating, technical skills failed to advance and machines became obsolete. The redundant ties reinforced the subregion's lagging character, and workers became trapped there with outmoded knowledge and skills.

Alternatively, in a subregion that housed one premier shop, the leading machinist (who typically attracted other gifted machinists) built ties to other shops based on his personal acquaintances with their heads; subcontractual relations might buttress these ties (see fig. I.2). If other shops possessed few linkages among themselves and they maintained most of their ties with the premier shop, then it acted as a hub of the subregion's networks. This shop controlled or at least significantly influenced the exchanges of knowledge and technical skills and the mobility of employees among the shops. The limited ties among the other shops produced structural holes among them. A structural hole between two shops meant that no tie existed between them, and these holes became opportunities to exert structural constraint. The premier shop and its leading machinist(s) controlled, or strongly influenced, the access of other shops to knowledge, technical skills, and workers. If the premier shop provided other shops free access, then such structural constraint may not have been detrimental. Nevertheless, such a network structure did not take advantage of the synergy that might emerge among other shops in the subregional network, which in turn could inhibit the discovery of new ideas about machining.

Network Ties between Subregions

Machinists and their firms also developed network ties between subregions of a region, for example, in southern New England, between the Boston and the Providence cores, each of which radiated about twenty-five to thirty miles from the respective city; and, in New York City and its environs, between the city's New Jersey satellites and the lower Hudson Valley. Relations between subregions introduced network behavior that was not based on frequent interactions among people within the same town or nearby towns. The time required to get together might involve a trip of several days for one-way travel. Mechanics carried out this travel for various purposes, including trips to discuss new machining ideas or to resolve problems, to observe machining methods, and to work for a short time to acquire technical skills in a new procedure or to learn to build an improved or new machine. The exchange of tacit knowledge, typically of components, required this face-to-face contact, but exchanges also might include the sharing of architectural knowledge of the shop's system of manufacturing. Longer-term exchanges might consist, for example, of employee mobility as a skilled worker left a shop in one subregion to work in a shop in another subregion for several months or even a year, thus relocating to a new residence; then the machinist would return to the original shop and share what he had learned. Sometimes machinists permanently relocated to a new job in another subregion and perhaps moved on again from there to yet another job, either locally or in another subregion.

The design of the structure of network ties that a firm (or machinist) built to other subregions arguably had an even greater impact on the success of the firm than links within a subregion. Ties between subregions required extensive time and high monetary costs to build and maintain. Strategic choices of network contacts enhanced the value of the benefits, such as the amount and quality of component and architectural knowledge which was received, of technical skills that were acquired, and of mobile machinists attracted as new employees. If a firm possessed separate ties to each of two firms in another subregion, both of which were linked to each other, the second tie

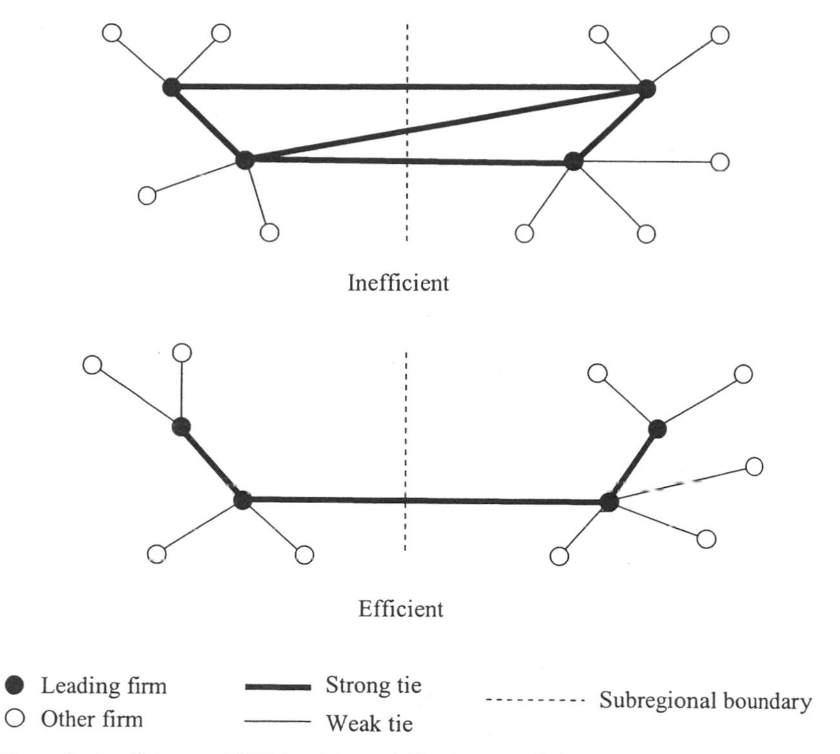

Figure I.3. Inefficient and Efficient Network Ties between Subregions

added few new benefits; one tie would be sufficient. In contrast to that inefficient network, efficient networks of firms maximized the number of nonredundant contacts, thus increasing the number of structural holes between contacts (fig. I.3). By reducing, if not eliminating, redundant contacts, the firm lowered the time and monetary costs of contact. Each additional contact contributed new knowledge, technical skills, and talented employees (attracted to work). Whenever a firm constructed a tie that was the only path between two contacts, that connection constituted a bridge. In practice such a condition was unlikely in most large networks. The concept of bridge, therefore, refers to a tie that is the shortest path between two contacts located in different local networks, and this path is better (shorter, cheaper, and so on) as the link than any other.[7]

A firm (or machinist) built an effective network between subregions of a region when it chose a primary over a secondary contact (fig. I.4). Primary contacts provided access to secondary ones, and the firm building the bridge to the primary contact did not have to invest time and money reaching other contacts. Most of the subregions with agglomerations of machine shops possessed at least one hub firm and sometimes

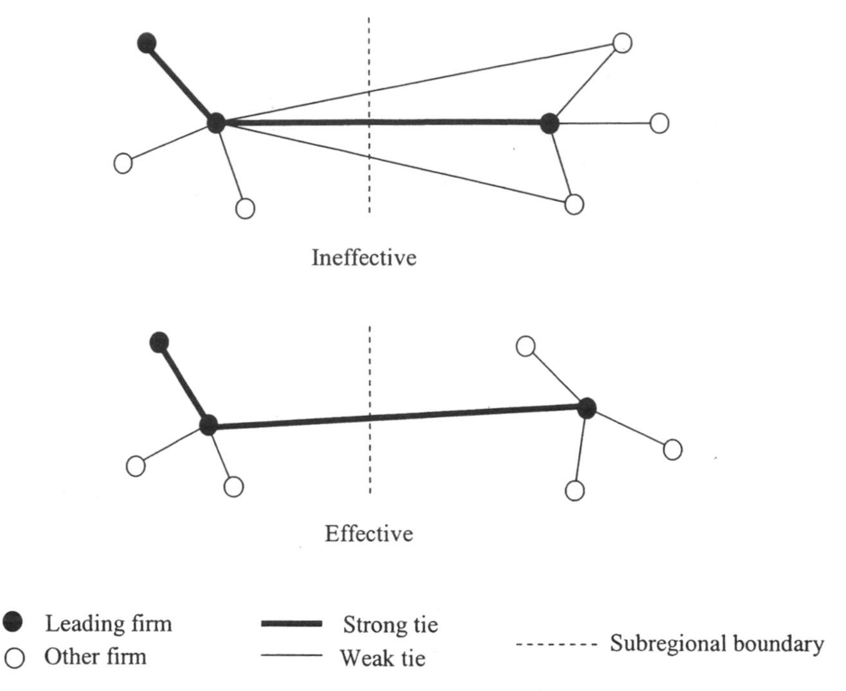

Figure I.4. Ineffective and Effective Network Ties between Subregions

several of them. Consequently, strategic firms built efficient (minimizing redundancies) and effective (ties to primary contacts) networks. Connections among subregions introduced the possibility of having greater structural constraint than within a subregion, where the short distances and multiplex ties created fewer hindrances to exchanges involving knowledge, technical skills, and workers. Hub firms created stronger filters for these exchanges between subregions because they could not be bypassed without an extensive investment of resources. Thus, their power derived from their bridging ties, which provided the only (or far better) access to other network participants.

If firms (or machinists) built bridging ties to other subregions, this enhanced the longer term viability of the subregion as a center of machinist activity. The continual infusion of knowledge, technical skills, and talented workers kept the subregion abreast of innovations elsewhere. In this setting a subregion with cohesive (close bonds) and multiplex (covering many types of ties) networks and with extensive redundancies could remain a dynamic machinist center. But, if such a subregion with these internal characteristics had few, if any, bridges to other subregions, then stagnation may be exacerbated if innovative firms faced difficulties in finding a positive reception for their ideas and products.

Interregional Networks

The same principles of bridging, of efficient and effective networks, and of structural constraint apply even more at the interregional scale. At this range the time and monetary costs of linkages are substantially higher, and fewer means exist to bypass network hubs, such as might be done between subregions of a region where other businesses (retailers or wholesalers) provide ways to reach firms (or machinists). For example, two prominent sets of interregional network links of firms and their machinists consisted of those between the Boston area and the New Jersey satellites of New York City and between the Providence and the Philadelphia areas. Network hubs within each region possessed the component and architectural knowledge, technical skills, and talented workers that firms in other regions demanded. As primary contacts, these hubs were the optimal means to access less significant firms within a region because they possessed the greatest number of, and most important, bridging ties.

In interregional networks the distinction between weak and strong ties as vehicles for the exchange of resources, such as knowledge, technical skills, and talented workers, can be seen in greater relief than both within subregions and between subregions within a region (fig. I.5). Weak ties formed bridges that connected two firms (or machinists), but they engaged in such infrequent and perhaps impersonal contacts that the bonds between them remained circumscribed. Nonetheless, these types of bridges still may have been sufficient for resource exchanges. Firms with proven reputations constituted valuable exchange partners, and hub firms had a large number of contacts with whom they had weak ties; this provided widespread access to resources. On the other hand, hub firms may have had strong ties that also served as conduits for interregional exchanges, and they could possess these links even if they had few, if any, other contacts in common. Deep family bonds might have established these connections, yet the machinists created hub positions in different regions. Alternatively, these strong ties also might have been rooted in close relations established when machinists worked in the same machine shop earlier in their careers.

Regardless of whether interregional bridging ties between hub firms (or machinists) were weak or strong ties, hubs possessed a powerful capacity to exert structural constraint over other firms within their regions (and subregions). Firms had greater difficulty establishing interregional contacts than creating links between subregions within the same region because long-distance connections involved extensive time and monetary costs. They needed to rely on the hubs for their knowledge, technical skills, and talented workers, and these non-hub firms located in different regions had structural holes between them. Nonetheless, structural constraint within machinist networks remained circumscribed, and, ultimately, it did not fully determine net-

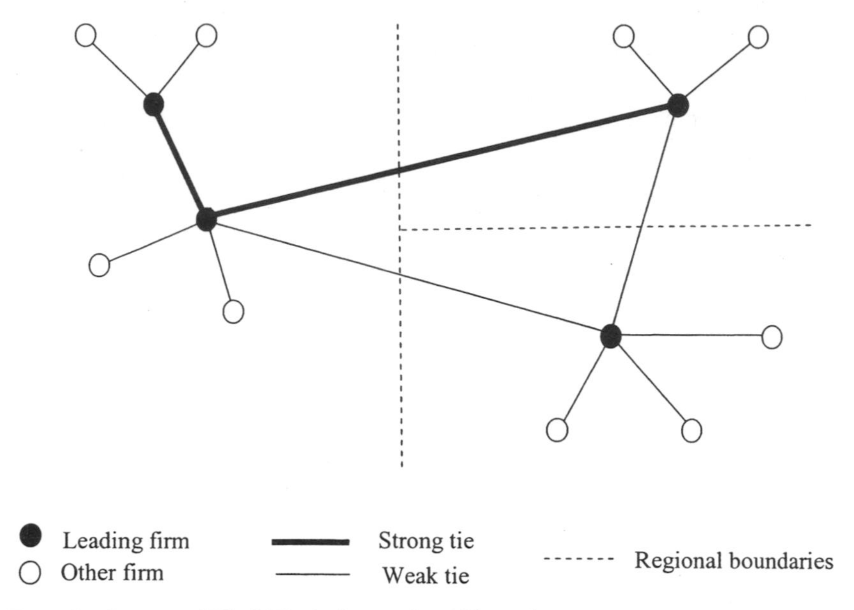

Figure I.5. Strong and Weak Ties in Interregional Networks

work relations. Talented machinists and innovative firms possessed many opportunities in antebellum machinist work because demand for machinists exceeded supply. Furthermore, the rapid pace of technological change across so many machinist sectors—iron foundries, steam engines, locomotives, textile machinery, firearms, and machine tools—provided numerous alternative avenues for network exchanges.

Interregional bridging ties, which became increasingly significant over time, channeled the diffusion of knowledge, technical skills, and talented workers by the early part of the nineteenth century. These bridges were effective in spite of the fact that overland long-distance travel remained arduous. Railroad travel did not commence until the mid-1830s, and most of the early lines radiated less than fifty miles from major cities; interregional rail travel did not become consequential until the late 1840s. Telecommunication via the telegraph did not exist before the mid-1840s, when it first became operational, and commercial messages dominated for the next several decades.

Before 1850 most interregional industrial shipments consisted of processed raw materials, such as flour and sugar, or of light, high-value (relative to their weight) goods, such as cotton textiles, shoes, buttons, and tinware. Interregional shipments of machinery and other heavy metal fabricated goods remained small for much of the antebellum. Firms in the Philadelphia region, one of the East's most important producers of machinery and fabricated metals, did not ship much of these goods long distance until at least the late 1840s. Hence, before that time interregional flows of pivotal producer durables did little to encourage machinists to build networks that reached outside metropolitan regional hinterlands.[8]

Nevertheless, interregional networks of machinists transcended these limitations of transportation, telecommunication, and shipments of manufactures prior to 1840. This testifies to the power of their networks as conduits for the exchange of knowledge, technical skills, and talented workers throughout the antebellum East. These networks created "communities of practice," and the machinists' behavior followed a template that can be recognized among many types of occupations that deal with technology. While the antebellum machinist communities might have had antecedents, we can say without exaggeration that they were forerunners of modern technology communities.

COMMUNITIES OF PRACTICE
The Artifact-Activity Couple

Machinists employed visual and tactile skills to work with artifacts such as metals, tools, and machines. They acquired these skills through direct, extensive training because, prior to 1860, science and engineering theory had little impact on manufacturing and many mechanical principles were not well understood. Machinists entered into formal or informal apprenticeships, moving among shops to improve their skills and gain experience. The concreteness of visual and tactile skills obscures the critical place of knowledge in machinists' work. A machine embodied the accumulation of prior skills and knowledge; machinists thus used their intellectual abilities to understand the construction of, and working of, the machine. They employed this knowledge to design, build, repair, and modify machines and to innovate improvements or new designs.

Verbal descriptions or diagrams, which often possessed ambiguity even when they utilized technical language, did not suffice for communicating the character of machines. Machinists needed to observe and study a physical machine, focusing on, as the anthropologist-historian Anthony Wallace termed it, the "grammar of the machine." This mechanical system represented the transmission of power from a source (such as waterpower) through belts and gears to the moving parts of a machine. During the period from 1790 to the early 1850s machinists increasingly acquired a technical literacy for understanding machines. They had to learn notational systems and associated vocabularies and grammars dealing with alphabetic expressions, scientific and mathematical notations, and spatial-graphical representations. They did this through self-teaching on the job or instruction in schools such as mechanics institutes. Nevertheless, technical literacy did not remove the communication problem—

machinists needed to participate in tacit exchanges of component knowledge with other machinists while observing and working on actual machines. These exchanges probably constituted the most important kind of communicated knowledge because they embodied the skills and conceptual learning that machinists had acquired.

The concept of the "artifact-activity couple" captures the bond between the artifacts with which machinists worked and the skills and knowledge that they brought to bear as they worked on machines and communicated about equipment with other machinists. Technological change within a firm and the transfer of technology among firms (and their machinists) worked through the mechanism of this artifact-activity couple. Even the successful replication of a machine that was purchased or borrowed from another shop required this mechanism because most machines embodied skills and knowledge that could not be fully grasped only through visual examination. During the antebellum years the mechanism of the artifact-activity couple generated a gradual accumulation of know-how about machines, and machinists and firms collaborated in this effort. Their networks and communities of practice shaped this accumulation process.[9]

Practicing in Communities

Machinists operated in various institutional environments. During the antebellum they had minimal direct impact on the growing capital investment in machines, yet this investment generated two consequences for their behavior (see fig. I.1). First, the growing investment in equipment reflected greater demand for machines to power waterworks, factories, and transportation vehicles (e.g., steamboats and railroads) and for machines to produce goods and manufacture machines. This demand stimulated greater innovative and inventive activity. Nonetheless, demand by itself cannot necessarily frame the direction in which innovations and inventions should proceed. Machinists could not always predict which innovations actually improved the performance of a machine, and consumers may have no conception of a product that does not exist. Thus, the second consequence may be more important: the growth of capital investment, implying an expanding output of machines, lowered the risks and costs of making improvements to existing machines and of experimenting with new types. This supply response directed the behavior of machinists as they individually and cooperatively built machines. Collaboration also extended across firms as machinists often shared skills and knowledge with one another, thus enhancing innovation.[10]

To present-day observers of the culture of high-technology firms, which are viewed as attracting young, risk-taking workers to the exciting life of the information technology revolution, it is difficult to grasp that young mechanics saw antebellum machine shops as places to achieve fame and fortune. Every machine shop did not fit this model, because many were tiny businesses that offered well-paying jobs but not distinction or wealth. Nevertheless, larger shops attracted not only the young, upwardly mobile residents of farms, villages, and small towns but also young men from the upper socioeconomic groups and from the professional classes of American society.

The reason is clear—the antebellum manufacturing expansion posed enormous intellectual and technical challenges. Big-city waterworks (in New York City and Philadelphia), steamboats, and railroads needed steam engines, and their construction supported the development of iron foundries, the heavy industrial firms that contained cupola furnaces, forges with heavy hammers, and machine shops. The rich Boston merchants who funded large cotton mills in New England gave cache to the big textile machinery shops. The federal government's effort to improve arms manufacturing generated sizable sums of money to support innovative firearms production. As railroads became the rage and the railroad company took on the character of a large, complex organization—the country's first modern business enterprise, in Alfred Chandler's terms—locomotive manufacturing attracted capital to construct some of the most expensive, complex machines of the era.

The ranks of machinists included a cadre of the elite of American society, which for the 1820–40 cohort probably numbered at least several hundred machinists in the East. The institutional environment created by these elite framed machinists' networks. They knew one another directly through personal and family connections, or their contacts were only one level removed as friends of friends or friends of relatives. They communicated with one another about their work, and this created ongoing exchanges of technical knowledge; often these networks were cohesive. This elite network embraced the entire East Coast of the United States, and, through family members, its tendrils extended to Britain. Transatlantic visits of machinists, especially Americans traveling to Britain to observe machine shops and discuss mechanical technology, also created these contacts. Likewise, British immigrant machinists shaped transatlantic networks, and, later in the antebellum, visits of leading British machinists buttressed them. Anthony Wallace termed these elite the "international fraternity of mechanicians."

The emergence of a technical literacy, which became a shared language of communication, acculturated the elite into a community of machinists, reinforcing their self-conscious identity. The young, upwardly mobile residents of farms, villages, and small towns who demonstrated technical skills also were drawn into this community. Using a rough estimate of about twenty journeymen and apprentices for each elite machinist suggests that this larger community in the East numbered perhaps two thousand to three thousand machinists between 1820 and 1840. It is useful to conceive of antebellum machine shops as possessing a "shop culture," as the engineering historian Monte Calvert terms it. This culture consisted of an orientation to technical work and problem solving, of emergent traditions, and of an institutional framework of a machinist community which united them and which bonded them with their customers. Machine shops constituted one of the key foundations, along with railway repair shops (essentially big machine shops) and naval engineering, of the mechanical engineering profession in the second half of the nineteenth century.

Social class distinctions entered the machine shop, but the effectiveness of the shop required close working relations between the top machinists and the ranks of the mechanic labor force. The leading machinists, either from the elite of society or from lower in the social hierarchy, typically commenced their skill acquisition working with the regular mechanic workforce, often as apprentices. Once the top machinists had attained their senior positions, they often led efforts to improve machines and solve technical problems. Yet, in order to implement their ideas, they needed to collaborate with machinists on the shop floor—a consequence of the tightness of the artifact-activity couple. This produced considerable sharing of know-how (tacit knowledge) within the shop. That collaboration extended to the firm's customers, who raised technical problems that required solutions, because much of the machine shop's work, with the exception of some textile machinery firms and firearms factories, consisted of custom work.

Machinists in each shop also were members of larger informal communities of practice that included other shops. These communities formed around each sector such as iron foundries, steam engine shops, locomotive shops, textile machine works, and firearms factories. At the same time, informal communities connected these sectors because they shared design, pattern-making, metalworking, and machine-building skills. The artifact-activity couple, signifying the sharing of technical skills and the tacit communication of component knowledge while mechanics worked together on equipment, supported the creation and continued functioning of these communities. Machinists maintained their access to them by bridging ties, and, in order to sustain a stable machine-building capacity in a firm, owners had to participate in informal communities to replace machinists who left the firm. During the technological ferment of the antebellum, with so many sectors emerging and changing rapidly, the stakes were high. Machinists and their firms needed to participate actively in communities of practice in order to stay abreast of developments; any lesser approach made a machinist or a firm uncompetitive.[11]

No governing organizations or groups deliberately formed these communities of practice. Machinists did not abide by a set of rules, and they succeeded precisely because they followed informal, noncanonical practices. Certainly, they shared accepted practices, which over time became part of the tradition; that mechanism helped them accumulate skills and tacit knowledge. Nonetheless, they also reconsidered these skills and this knowledge. Machinists competed to solve technical problems and to gain recognition for their success, and they jostled for positions within and among firms. This competition motivated them to tackle difficult technical problems. At the same time, they collaborated within and across firms to solve technical problems, transfer skills, and exchange component, and even, architectural (organizational and systemic), knowledge. Their job mobility, especially during their twenties and early thirties, enhanced know-how trading as they brought new practices to a firm. The network structure of this mobility framed the transfer of technical skills and knowledge. Because these mobility networks reached between subregions and interregionally, the tacit exchanges of component and architectural knowledge operated among widely separated communities of machinists.

This competitive environment, coupled with a dynamic exchange of collaborative know-how, constituted a potent mechanism for enhancing innovation because incremental improvements passed among the practitioners. During the first several decades following the mid-1830s, for instance, railroads often exchanged component (e.g., wheel and axle machining) and architectural (e.g., locomotive repair strategies) knowledge through their top machinists, who supervised the equipment and ran the railroad repair shops—basically, big machine shops. These machinists also exchanged tacit knowledge with railroad suppliers such as the locomotive shops. Machinists and firms that did not employ network strategies to participate in informal communities of practice cut themselves off from innovation; this approach became a recipe for reduced competitiveness and stagnation. Because multiple communities existed, technological change advanced on various fronts, thus contributing to the success of so many antebellum machine sectors. Besides the networks that linked these communities, machinists and their firms exchanged knowledge with any industry that mechanized, because these actors possessed the critical skills required to build the machines. The communities of practice learned about new machine problems that needed resolution, and this information impelled their members to enhance their existing network ties or to build new ones.[12]

THE ANTEBELLUM TRANSFORMATION

From 1790 to 1860 machinists in the eastern United States built sophisticated networked communities of practitioners across a range of metalworking industries. These communities operated at local, such as a city and its immediate environs (e.g., Providence and adjacent towns), and at subregional scales, such as a metropolis and its inner hinterland (e.g., New York City and its New Jersey satellites and the lower Hudson Valley). They also extended to regional territories such as a metropolis and its hinterland (e.g., Philadelphia and much of Pennsylvania), and they were integral constituents of the areas of rising agricultural prosperity in the East. Localized learning went on within some of these territorial communities, and the sharing of technical skills, knowledge, and workers could generate competitive advantages for firms that other territorial communities had difficulty duplicating. Nonetheless, territory neither determined all, nor necessarily the most important, features of these networks. Communities spanned a manufacturing sector, such as locomotives, or several sectors, such as steam engines and locomotives. Machinists within and among sectors engaged in network behavior that bridged territories, such as corresponding by mail, visiting one another's shops, sending skilled workers for training, and moving from one job to another. Their system precluded strict territorial bases of these communities.

Within the first several decades after 1790 some communities of practitioners encompassed the East Coast, and several even extended into the distant areas of agricultural prosperity in central New York state. A set of networked communities emerged out of the relationships formed by the earliest civil engineers who worked on the design and construction of bridges, roads, and canals during the 1810s and 1820s. Although they did not constitute national networks, from the start of the new republic the United States possessed communities of machinists (and civil engineers) which spanned extensive territories. This is perhaps two or three decades earlier than has been suggested for the first appearance of national communities of technologists, such as those in geology and mining, which started in the 1840s. Almost from the start, American machinist communities had an international dimension. The early emergence of some communities that spanned much of the East Coast—well before the existence of large-scale industries, effective interregional transportation, and substantial interregional trade in manufactures—suggests that the network behavior of machinists through exchanges of technical skills, knowledge, and workers overcame some of the barriers of distance.

The machine tool industry has been touted as the pivotal capital goods sector, and from 1840 to 1880, according to this argument, it underwent a significant transformation. This industry supported the emergence of entirely new industries such as typewriters, bicycles, and automobiles. Because technical issues possessed commonalities across so many machinery and metal-using sectors, this technological convergence constituted a milieu in which machinists directed their efforts toward improving a core set of metalworking processes and machine tools. Initially, firms that used these tools also built them. As workers mastered the skills and techniques required to manufacture this equipment and as it became employed in more metalworking sectors, specialized firms began to build machine tools. This argument has appeal, but it dismisses too readily the larger set of pivotal producer durables and the broader knowledge and skills applicable to building machine tools during the antebellum.[13]

Instead, machinists built networked communities after 1790 in the pivotal producer durables, whose key sectors included iron foundries, steam engines, locomotives, textile machinery, firearms, and machine tools, and this development contributed to a material transformation of these sectors by the 1850s. For at least three to four decades after 1860, machine tools, for the most part, formed a business line and a set of skills within many types of firms; they were not a product of specialized machine tool firms. Technological knowledge and skills were embedded in networks of machinists, firms, and clusters of individuals or firms; they were not the sole property of the constituent parts. The dynamic communities of practitioners were open to, and promoted, technological change, and the failure by mechanics and their firms to participate actively in these networks condemned them to an uncompetitive status.

From 1790 to 1820 the tendrils of the pivotal producer durables—urban iron foundries and steam engine works, textile machinery shops, and federal armories and private firearms firms—formed within the East. Interrelations among them increased over time, and after 1820 they spun off a widening array of pivotal producer durables, including the machine tool industry.

PART ONE

The Formation of the Networks, 1790–1820

CHAPTER ONE

Iron Foundries Become Early Hubs of Machinist Networks

> THE subscriber has established . . . a set of works, consisting of an IRON FOUNDRY, STEAM ENGINEER'S SHOP, MOULDMAKER'S SHOP, . . . BLACKSMITH'S SHOP, MILLSTONE FACTORY, and a STEAM MILL for grinding plaister and turning heavy cast and wrought Iron. And he is prepared . . . to execute such orders as he may receive, either for steam engines or mill work; such as cast iron cog wheels, gudgeons, cranks, spindles, rynes, &c. sugar and soap boilers' kettles or kerbs; rollers for sugar mills . . . ; forge and rolling mill work, &c, In short all that may be wanted either of cast or wrought iron for machinery. The Steam Engines are of his own invention, and are patented; they will serve as a substitute for water falls to produce power much cheaper than by cattle, to move machinery.
>
> *Oliver Evans, Mars Works*

Upon first reflection, to claim that one of the earliest hubs of machinists' networks arose in the iron industry seems farfetched. Certainly, the manufacture of basic iron products such as pig iron, wrought iron, simple castings, plates, and rails required metalworking abilities. Their production, however, consisted chiefly of technical knowledge of metallurgy and the possession of skills in heating ores, pouring molten metal into molds, and beating metal with hammers or rolling it into standard shapes. This heavy metal work does not convey images of skilled machinists using tools or equipment to shape metal into complex forms, but the subsector of the iron industry which consisted of iron foundries stood apart. These firms and their mechanics played central roles in the development of machinist networks in the early antebellum. To see how this came about, iron foundries must be positioned within the larger iron industry.

INPUT-OUTPUT CHAINS

The iron industry consisted of input-output chains of manufacturing, starting with blast furnaces, whose initial output, molten iron, either was turned into pig iron bars or was poured into molds to make castings. Over time separate firms that specialized in the manufacture of castings from pig iron increasingly dominated this production, and sometimes iron foundries, machine shops, or machinery firms made castings. Forges acquired pig iron from blast furnaces and employed waterpowered trip hammers or skilled workers with hand hammers to transform pig iron into wrought-iron bars. Subsequent stages of production met various demands of households, farms, and businesses. Blacksmith shops acquired wrought-iron bars from forges and transformed them into innumerable products. Iron foundries acquired pig iron from blast furnaces and transformed it into manifold castings, or they acquired wrought iron from forges and shaped it into diverse products. Each of these venues—blast furnaces, forges, blacksmith shops, and iron foundries—housed skilled metal fabricators.[1]

During the early antebellum the prohibitive transport cost of bulky raw materials kept most blast furnaces confined to rural sites that possessed iron ore and timberland. The ore contained mostly waste material, and eliminating it by operating blast furnaces near ore sites reduced total transport costs. The furnaces' large consumption of charcoal fuel required them to have immediate access to sizable timberland acreage, ranging from three thousand to about eight thousand acres per furnace. They turned most, if not all, of the molten iron into pig iron bars, but, if they made castings, these took many forms, such as kettles, stove plates, hollow ware, and sash weights, depending on the mold. Furnaces sold pig iron to nearby forges, and they sold the castings to retailers within a radius of about thirty miles of the furnace. They also marketed pig iron and castings through wholesale merchants in East Coast metropolises including Boston, New York, Philadelphia, and Baltimore. Often these merchants owned iron furnaces, and many of the facilities operated within fifty miles of their metropolitan offices. By 1800 furnaces commenced production in western Pennsylvania, and some of them looked to the Pittsburgh market.

Blast furnaces required few machinists. The iron founder who served as manager of the furnace and evaluated the quality of the molten iron and the molder who produced molds for castings constituted the main exceptions; however, neither of them reshaped metal. Because furnace pig iron retained considerable carbon impurities, it was unsuitable for high-quality castings; consequently, castings made at blast furnaces mostly consisted of simple products or parts. The typical furnace produced enough pig iron to supply several forges; therefore, furnaces rarely integrated forward to include a forge, which required some machinist skills to work pig iron into wrought iron. Restrictions on the hearth size required to reheat pig iron, on the scale of individual waterpowered trip hammers, and on the effort of a worker with a hammer limited the economies of scale of production in forges. Greater output resulted from adding more hearths, trip hammers, and workers, and this approach generated little or no reduction in cost per unit of output. Because minimal weight loss occurred when the forge processed pig iron into wrought iron, locating forges next to blast furnaces offered few savings in total transport cost (blast furnace to forge to market). Forges required large amounts of charcoal, and they utilized the same timberlands as blast furnaces. Accordingly, most forges agglomerated near furnaces to access fuel and pig iron supplies. Furnaces that internalized forges gained few savings in transaction costs, and they needed much more capital and added the risk of using their own pig iron in their forges, rather than spreading that risk by selling the pig iron to numerous consumers.

Likewise, furnaces rarely added the iron foundry stage because a furnace produced more pig iron than a foundry could consume. Little transport cost savings accrued from close proximity of foundries to pig iron or to wrought iron suppliers because minimal weight loss resulted from foundry processes. The coordination costs between foundries and the firms that consumed their output exceeded the minor transport cost savings from eliminating extra metal in foundry processing at a blast furnace. Because a blast furnace required substantial capital, the extra amount needed to integrate forward into forges or foundries added material risk to the firm. Thus, blast furnaces integrated backward to raw materials (iron ore and charcoal), and they sold most of their pig iron to forges and to iron foundries.

Although most forges agglomerated near blast furnaces and large supplies of cheap timber (for charcoal), few rural forges integrated forward into iron foundries. Because the transformation of pig iron into wrought iron at forges generated little weight loss, the wrought iron's transport cost embedded in the foundry's final product sold at its market (dense agricultural areas, cities, and factories) remained the same, regardless of the location of the foundry processes—in timber areas or near markets. On the other hand, foundries often needed close coordination with buyers and access to market information. Metal-fabricating skills at forges chiefly consisted of employing trip hammers and hand hammers to remove carbon impurities from pig iron in order to transform it into wrought iron, but these skills spanned a limited range. Consequently, forges, similar to blast furnaces, did not become the early hubs of machinist networks in the heavy capital equipment industry; instead, foundries occupied that position.

IRON FOUNDRIES
Production and Markets

Foundries purchased pig iron from blast furnaces and wrought iron from forges, and foundries bought both products from merchant wholesalers in Boston, New York, Philadelphia, and Baltimore. After 1815 specialized iron merchants in these metropolises also served as suppliers. An indeterminate share of early antebellum wrought iron came from foreign sources. Estimates for around 1810 suggest that imports amounted to about ten thousand tons, whereas domestic sources supplied between twenty-five thousand and forty thousand tons. Imports may have reached about twenty-six thousand tons annually by the early 1820s. Blacksmith and machine shops purchased some of this imported wrought iron, but iron foundries must have bought sizable amounts, and wholesalers or specialized iron merchants supplied most of that.

Local economies that generated few surpluses of agricultural or other goods did not require iron foundries. In these areas distant from major cities one or two blacksmith shops supplied enough metal products to meet demand, and their market information came from local consumers and retailers. Because blacksmiths faced little competition, they maintained simple techniques. Rising agricultural prosperity and greater local and external trade, however, boosted demand for metal goods. This encouraged some blacksmiths to expand output, to improve techniques and raise productivity, and to invent new products. They captured a larger share of the local market and competed with blacksmith shops in nearby local economies. Some entrepreneurs expanded forging operations from simple blacksmithing to a full-fledged forge shop, and a few invested in a cupola furnace to produce castings, thus operating a small foundry.

Before 1820 most foundries operated in prosperous farming areas in the East. Household consumers wanted stoves and utensils, farms used plows, raw material processing required mill equipment, factories employed machinery, transport modes were powered by steam engines, and urban infrastructure needed sash weights and pipes. Foundries located in villages, towns, or small cities within the inner hinterlands of Boston, New York, Philadelphia, and Baltimore. During 1810–20, for example, foundries near Boston included, Shepard Leach's at Easton and Charles Leonard's at Canton. Nonetheless, the metropolises, their nearest satellites, or subregional metropolises such as Providence and Albany housed most of the major foundries.

Foundry goods could be shipped moderately long distances, but most production consisted of custom work for clients or it required access to market information, such as the types of mill equipment and machinery which processing firms and factories needed. This information came through social networks, and merchant wholesalers in the large metropolises supplied the broadest array. Their networks connected these cities and also reached to businesses in their inner hinterlands and in subregional metropolises. Occasionally, machinist networks that reached beyond local areas provided market information, but their primary long-distance information concerned technology and supplies of talented mechanics.

The centrality of urban iron foundries in fabricating heavy metal components and in manufacturing heavy industrial machinery during much of the nineteenth century has faded from view. Few freestanding iron, steel, or aluminum foundries remain, whereas numerous small and large foundries stay hidden as a component of a large factory such as an auto or aerospace plant. Urban foundries did not face effective competition either from rural blast furnaces, which converted their pig iron directly into castings (such as kettles or plows), or from rural forges, which converted their wrought iron into diverse products (such as wagon rims or building hardware). These rural firms sold most of their goods in nearby markets because heavy, bulky, low-value products incurred high transport costs for sales in distant markets.

Consequently, the urban foundry purchased pig iron or wrought iron from a rural furnace or forge as well as perhaps from a foreign importer. The final product made out of pig iron or wrought iron lost little weight during manufacture, but the transport cost per ton of the bulky, low-value finished products exceeded the transport cost per ton of the compact pig iron or wrought iron. The final local sale prices of castings or wrought-iron goods made in the city, therefore, typically fell below their rural competitors. Urban foundries that made cast- and wrought-iron products possessed another advantage: local manufacturers gave foundries custom orders for many goods. This required close communication among firms and sometimes cooperation among several foundries, but distant manufacturers of cast- and wrought-iron products remained uncompetitive in that environment.[2]

Inside the Foundry

Urban foundries established after 1790 awed contemporaries. Even the smallest of them needed plenty of capital and many skilled and unskilled workers to manufacture their products. The foundry, the term for the overall enterprise as well as one of its parts, typically consisted of at least four distinct components—the foundry, the pattern shop, the forge shop, and the machine shop—housed in different buildings or shops. The cupola furnace loomed over all activities in the foundry building, and this vertical blast furnace used charcoal and, later, mineral coal to fire the mixture of pig iron and scrap iron. Workers ladled molten metal into molds of damp sand on the floor of the foundry or into freestanding molds. After the castings cooled and solidified, workers cleaned the sand off and perhaps filed or chiseled the castings to remove irregularities. These castings were either sold or moved to the machine shop for more work. Urban foundries had the cheapest access to scrap iron because consumers (such as manufacturers and households) in the large cities and in the densely populated inner hinterlands of large metropolises generated much of it. The addition of scrap iron to the charge of the cupola furnace gave the castings a more uniform, less brittle character, making them stronger and less prone to breakage, thus affording urban foundries a competitive advantage in the production of high-quality metal components and machinery.

In the pattern shop, a critical adjunct to the foundry, skilled craft workers created wooden patterns that molders used to make forms in the damp foundry sand or to set up as freestanding molds to hold molten metal. Some patterns took standard shapes, such as kettles or pipes, whereas others met specific needs of customers such as for metal parts. As the foundry increasingly served the diverse needs of city public works and of manufacturers for parts or entire machines, the inventory of patterns in the shop grew and represented a sizable capital investment. Skilled molders developed the capacity to replicate patterns, which enabled the production of multiple copies of a casting (e.g., kettles). Some foundries made only castings, but this restricted product line limited their capacity to respond to the needs of manufacturers for metal components and machines. Urban foundries that followed narrow lines typically developed specialization in a few products such as stoves or plows; in effect, they became manufacturers of these goods rather than full-line foundries.

The forge shop of the foundry worked wrought-iron bars, rods, and sheets into desired shapes, which gave the foundry a broad product line to respond to diverse demands. Forge shops housed multiple "fires" for heating wrought iron, which workers then hammered on anvils. Foundries called these workers blacksmiths, and sometimes that name was applied to the forge shop, but it stood far removed from village blacksmith shops that made small items for homes, farms, and gristmills. The foundry's forge shop focused on metal components for machines and factories, and it contained large anvils, hoisting devices, and trip hammers. Its power sources consisted of horses, water, or steam for shaping large items, and workers used small anvils and hand hammers for shaping little pieces. Castings and wrought-iron components sometimes made a final stop in the machine shop. Workers used hand tools to file, chip, or plane the castings or wrought-iron pieces—thus, the label "file shop" fit. This handwork remained important long after 1820, even as foundries increasingly used steam-powered machine tools.[3]

Because iron foundries integrated the foundry (casting in a cupola furnace), pattern shop, forge shop, and machine shop, they brought together mechanics with diverse talents to construct sophisticated components and machinery. Thus, urban foundries became the early hubs of machinist networks in the heavy capital equipment industry, and they continued as key members of the pivotal producer durables for the rest of the nineteenth century. Factories needed their goods to set up production (e.g., machinery), maintain equipment, and supply parts that they could not manufacture themselves, especially large castings and wrought-iron pieces that required trip hammers to shape. Following 1790, developments in steam engines caused a transformation of some urban foundries, setting the stage for the emergence of heavy industrial machinery firms in several eastern cities.

STEAM ENGINES FROM THE ELITE

For three decades after 1790 social, political, and economic elite in the United States furnished potent support for the insertion of steam engines into urban and industrial life. To be sure, the elite constituted a small share both of the numerous individuals working on steam engine innovations and inventions and of the many machinists making components or entire engines. Nonetheless, the elite provided critical political contacts and financial support that conferred competitive advantages on a select number of steam engine builders. On the one hand, their support devolved from the early use of steam engines to pump water for large urban waterworks, which constituted dramatic symbols of the growing cities that they governed; the innovations in steamboats, which provided improved locomotion for the elite on inland waterways; and the application of steam engines for power in factories, which they invested in.

On the other hand, firms required much more capital to engage in routine steam engine design and construction than they needed to build experimental prototypes and a few customized engines for local firms. Only the elite could muster this capital through their personal investment or by extracting financing and other support from government. Diversified iron foundries built most of the steam engines, though only a small share of firms participated, and a few specialized steam engine companies—foundries with a single business line—also contributed to the supply. These firms had the extensive fixed capital of buildings (foundry and forge shop) and equipment (cupola furnace and trip hammers). Furthermore, they employed the highly paid engineers, pattern makers, and metalworkers who had the skills to design and build engines. Foundries also needed abundant working capital to fund wages and inventories of materials during the complex, time-consuming process of building engines. Boston, New York, Philadelphia, and Baltimore, as well as their immediate vicinities, possessed advantages as bases for the early leading steam engine firms. They had a sizable elite group, a large and concentrated demand for steam engines (for waterworks, steamboats, and factories), and big foundries capable of manufacturing engines. Their local firms controlled the resources to attract the top mechanics from Europe and the eastern United States.

The Transfer of Technology to the United States

In Britain the flooding of deep ore and coal mines intrigued the iron monger Thomas Newcomen, and by 1712 he had built one of the first successful steam engines to pump water from a mine. The Newcomen engine worked on the principle of creating a vacuum by heating steam and condensing it in a chamber; atmospheric pressure then forced a piston and rod down on the vacuum. As the vacuum dissipated, weights on the opposite end of the rod raised the piston. Over the next several decades following its invention steam engines built on this principle were adopted throughout Europe. Their pumping action suited the needs of miners, and soon major cities used them to pump water for waterworks. American colonial elite—such as John Adams, Thomas Jefferson, Benjamin Franklin, and William Small, the influential professor at the College of William and Mary in Virginia—stayed abreast of these European developments. The large, expensive Newcomen engines, however, were not suitable for colonial markets because few mines reached depths at which flooding occurred and most cities remained too small to need waterworks. By the end of the Revolutionary War only three Newcomen engines operated in the country—at the copper mine of John Schuyler near Passaic, New Jersey; at the New York City waterworks; and at the iron ore mine of the Hope Furnace in Rhode Island, owned by the Brown merchant family in Providence. The Schuyler mine employed an imported engine that an English engineer, Josiah Hornblower, set up and ran, whereas Peter Curtenius, owner of an iron foundry in New York City, cast the cylinders of the other engines.

For several decades after the Revolutionary War the United States diverged from the European focus on steam engines for mines, waterworks, and factories. The demand for pumping water from mines and through urban waterworks remained subdued, and the East Coast, where the first factories emerged, possessed ample supplies of low-cost waterpower. Rapid population growth and frontier migration escalated demands for improved inland waterway transportation, the only feasible, efficient means of long-distance movement before the railroad; thus, dreams of harnessing steam engines to power boats energized entrepreneurs. They targeted markets on the highly traveled waterways serving New York City—the Hudson River to Albany, Long Island Sound to Connecticut, and New York harbor to the New Jersey ports—and serving Philadelphia on the Delaware River. Nevertheless, building steamboats to run on the great western rivers—the Ohio and the Mississippi—remained a grand dream.

The heavy, bulky Newcomen engine did not work well for steam navigation. Fortuitously, Matthew Boulton, successful manufacturer of small metal goods such as buttons and plated ware, and James Watt, a steam engine inventor, established their famous partnership in 1775. By the early 1780s Watt perfected the low-pressure, rotative steam engine, with parallel motion, at the Soho Manufactory of Boulton, near Birmingham, England. This compact, elegantly designed engine used both steam pressure and the vacuum principle to give motion to the piston in a cylinder. A separate condenser cooled the steam, and a set of valves permitted alternate injection of steam above and below the piston, thus the appellation *double-acting engine*. By 1800 Boulton and Watt had finished 496 engines, with 164 of them for pumping and most of the remainder for driving machinery in textile mills. Americans had to devise their own application to steamboats.

After 1785 information about the Boulton and Watt engine slowly percolated to the United States, initially through traveling political elite, such as Thomas Jefferson and John Adams, and economic elite, such as Silas Deane, onetime Connecticut merchant. Before 1800 at least a dozen individuals from New Hampshire to Georgia and as far west as Kentucky struggled to build engines and steamboats based on their rudimentary knowledge of steam engine technology and of the engineering principles of gearing and propulsion. During the 1790s as many as eleven people took out a total of fourteen steam-related patents (table 1.1). Over two-thirds of these inventors lived in Boston, New York, Philadelphia, and Baltimore or in their satellites. This suggests that successful steam-related work required access to metropolitan-regional knowledge networks and to their infrastructure of machinists and machine shops. John Fitch achieved the most notoriety through his contacts both with politicians, which garnered him monopolies from several states to operate steamboats, and with leading builders of steam engines. Given the meager industrial infrastructure of urban foundries, skilled engine builders, and engineers, the best of these early efforts resulted in steamboat curiosities that moved short distances.

Machinists faced challenges operating through the artifact-activity couple to construct, assemble, and maintain the Boulton and Watt engine. It was sufficiently complex that mechanics could not duplicate it from secondhand information; they needed to work with others in order to learn to build it. Furthermore, before 1795 Boulton and Watt did not supply assembled engines from their shop. They made regulators and valves, while the Wilkinson foundry at Bersham, England, supplied cast-iron parts, including the cylinder, which required accurate boring. Boulton and Watt shipped parts, never before assembled, to the site, and their skilled engine erectors, not local mechanics, set it up. Engines needed extensive adjustments, and skilled workers had to maintain them. In 1795 Boulton and Watt established the Soho Works, a complete engine factory that included a foundry, forge, boring mill, blacksmith shop, and fitting and carpenter shops. For the first time customers could purchase a complete assembled engine or a working one that had been disassembled for shipment.

TABLE 1.1
Steam-Related Patents, 1791–1820

Boston and Its Hinterland			
Metropolis	Patentee and Year	Satellites	Patentee and Year
Boston	A. Quincy (B), 1812	Salem, Mass.	N. Read (B), 1791
	S. Morey (E), 1815		
	N. Read (E), 1817		
	J. Sullivan (E), 1817		
	J. Sullivan (3 B), 1818		
	J. Sullivan (E), 1818		
Inner Hinterland	Patentee and Year	Outer Hinterland	Patentee and Year
Massachusetts	S. Morey, R. Graves, and G. Richards (E), 1803	Orford, N.H.	S. Morey (A), 1795
Portland, Maine	E. Jenks (B), 1805		S. Morey (A), 1799
Westfield, Mass.	R. Yeaman (E), 1812		S. Morey (A), 1800
		Somerset, Maine	J. Hilton (E), 1816
		Orford, N.H.	S. Morey (B), 1817
			S. Morey (B), 1818

New York City and Its Hinterland			
Metropolis	Patentee and Year	Satellites	Patentee and Year
New York City	J. Martin (P), 1797	Bergen County, N.J.	J. Stevens Jr. (A), 1791
	D. French (E), 1809		J. Stevens Jr. (B), 1791
	J. Stevens (E), 1810		J. Stevens (E), 1791
	D. Dodd (E), 1811		J. Smallman and N. Roosevelt (E), 1798
	R. Fulton (E), 1811	Elizabethtown, N.J.	D. Dodd (E), 1812
	J. Stevens (E), 1811	Essex, N.J.	F. Ogden (E), 1813
	H. Ricketson (B), 1816	Mendham, N. J.	D. Dodd (S), 1814
	W. Schultz (B), 1816	New Jersey	N. Roosevelt (P), 1814
	I. Jennings (E), 1817	Elizabethtown, N.J.	E. Burt (2 E), 1816
	A. Batby (E), 1819	Hoboken, N.J.	J. Stevens (E), 1816
Inner Hinterland	Patentee and Year	Outer Hinterland	Patentee and Year
Connecticut	J. Starr (P), 1797	Herkimer, N. Y.	M. Battle (E), 1812
	B. Platt (B), 1803	Plymouth, N.Y.	J. Curtis (E), 1812
New Haven, Conn.	E. Gunn (B), 1808	Herkimer, N. Y.	P. Hackley (B), 1815
Albany, N.Y.	R. Letton (A), 1810	Cazenovia, N. Y.	L. Richards (B), 1815
	J. Rodgers (A), 1811	Whitestown, N.Y.	B. Wilbor (E), 1815
New London, Conn.	J. Sizer (B), 1811	Geneva, N.Y.	I. Richardson (E), 1817
New Haven, Conn.	J. Morris (B), 1812	Springfield, N.Y.	B. Hendry and H. Price (E), 1818
Newtown, Conn.	M. Parmalee (A), 1812	Hartwick, N.Y.	T. Pierce (B), 1819
Hudson, N.Y.	P. Hackley (B), 1813		
Goshen, N.Y.	G. Philips (B), 1814		
East Windsor, Conn.	C. Reynolds (E), 1814		
Greene County, N.Y.	T. Barker (B), 1819		

TABLE 1.1
continued

Philadelphia and Its Hinterland			
Metropolis	Patentee and Year	Satellites	Patentee and Year
Philadelphia, Pa.	J. Fitch (P), 1791 O. Evans (E), 1804 S. Bolton (B), 1809 A. Anderson (E), 1812 D. Large and F. Grice (S), 1812 J. Dowers Jr. (E), 1814	Delaware Burlington, N.J.	W. Thornton (B), 1803 J. Sexton (E), 1812
Inner Hinterland	Patentee and Year	Outer Hinterland	Patentee and Year
Gettysburg, Pa.	M. Longwell (B), 1807	Connellsville, Pa.	T. Gregg (P), 1813
Baltimore and Its Hinterland			
Metropolis	Patentee and Year		
Baltimore, Md.	E. Cruse (E), 1791 W. Eaton (B), 1810 J. Curtis (E), 1815 R. Yeaman (B), 1815 R. Yeaman (B), 1817 W. Church (E), 1818		
Inner Hinterland	Patentee and Year	Outer Hinterland	Patentee and Year
Washington, D.C.	W. Thornton (B), 1809 J. Russell (E), 1813 W. Tatham (P), 1816	Berkeley County, W. Va.	J. Rumsey (2 E), 1791
South	Patentee and Year	Midwest	Patentee and Year
Virginia	E. Bachus (B), 1796	Kentucky	E. West (S), 1802
Georgia	S. Briggs (E), 1802	Chillicothe, Ohio	J. Stubbs and N. Parsons (E), 1811
Richmond, Va.	J. Staples (E), 1811	Marietta, Ohio	S. Fairlamb (E), 1816
Baton Rouge, La.	S. Briggs and A. Steele (E), 1813	Kentucky	E. (for J.) Rumsey (B), 1816
Norfolk, Va.	M. Cluff (E), 1815	Kentucky	E. Rumsey (E), 1816
New Orleans, La.	S. Briggs (A), 1817	Ross County, Ohio	C. Leeke (E), 1820
Virginia	J. Heavin (E), 1817		
Augusta, Ga.	J. Eve (E), 1818		
Richmond County, N.C.	R. Thomas (E), 1819		

SOURCE: Martin Van Buren, "Patents Granted by the United States," Secretary of State, Communicated to the House of Representatives, January 13, 1831, 21st Cong. 2nd sess., Doc. No. 50, *New American State Papers, Science and Technology*, vol. 4: *Patents* (Wilmington, Del.: Scholarly Resources, 1972), 284, 289, 290, 417–18, 436–38, 440.

NOTE: Abbreviations for patents and their number are enclosed in parentheses, if more than one was awarded to a patentee in the same year: P, steam propulsion; S, steamboat; A, steam applications; B, boilers for steam engines; E, steam engines. J. Curtis (Plymouth, N.Y.) and R. Yeaman (Westfield, Mass.) jointly received a steam engine patent in 1812.

Before 1795 the prohibitive cost of sending parts and engine erectors to the United States precluded Fitch and other steamboat builders from buying Boulton and Watt engines. Even after that date, steamboat enthusiasts required substantial capital to purchase the engines, and, if they purchased disassembled engines, they needed skilled erectors; in either case skilled mechanics had to maintain the engines. If entrepreneurs aimed to move beyond experimental prototypes to the routine construction of steam engines and steamboats, they needed a package consisting of large amounts of capital, sophisticated mechanics, management skills, and an industrial infrastructure of the iron foundry, but that package remained in short supply when so few successes existed as models.[4]

The Concentration of Steam-Related Invention

During the formative thirty-year period of steam engine development following 1790, numerous individuals contributed steam-related inventions (see table 1.1). As many as sixteen inventors took out patents for steam propulsion, steamboats, and steam applications. The large metalworking components, however, dominated inventors' efforts; twenty-six individuals took out thirty patents for boilers, and forty-four people acquired forty-seven patents for steam engines. Several leading inventors—Nicholas Roosevelt, Daniel Dodd, John Stevens Jr., Samuel Morey, and John Sullivan—operated across more than one category of invention. Nevertheless, the large number of participants in invention constituted the most striking feature of technological developments in steam engines and their use in steamboats. They resided in widely separated locations, mostly in the East. Few inventors operated in the South, an omen of its weak participation in nineteenth-century industrialization. Inventors in the Ohio Valley commenced efforts by 1802, foreshadowing the agricultural and industrialization of the Midwest which gained momentum after 1830.

Yet the dispersal of inventors across the East from Maine to Maryland and the reach to central New York state and western Pennsylvania gives a misleading view of invention. The greatest share of inventors, almost one-third of them, operated in Boston, New York, Philadelphia, and Baltimore (see table 1.1). New York City's satellites of Bergen County, Elizabethtown, Essex, and Hoboken were beehives of invention, whereas satellites of Philadelphia and Boston were minor centers, and Baltimore's satellites had none. Inventors worked beyond these satellites in the metropolises' inner hinterlands, which extended to a distance of about 100 to 150 miles. These areas around New York City included Connecticut (New Haven, New London, Newtown, and East Windsor) and the Hudson Valley and vicinity (Greene County, Goshen, Hudson, and Albany), and around Boston the areas included Portland, Maine, and counties across Massachusetts. Each of these areas housed prosperous farmers, a budding village and small-town urban infrastructure, and bustling workshops and small factories. Inventors also worked farther away in what can be termed the "outer hinterland." The large number active in boiler and steam engine patenting after 1811 in central New York state (Herkimer, Plymouth, Cazenovia, Hartwick, Geneva, Whitestown, and Springfield) resided in one of the richest agricultural areas of the East. Its farmers had shifted into specialized agriculture such as cheese and butter; thus, local markets provided opportunities for eager manufacturing entrepreneurs, including cotton mill owners.

The concentration of steam-related invention, especially in the East's metropolises and in their satellites and inner hinterlands, confirms that inventors responded to swelling transportation and industrial markets. Transport firms operating on the inland and coastal waterways that served these areas offered the first profitable steamboat markets, and the same areas housed the largest number of factories that demanded stationary steam engines. Furthermore, a sizable supply network of machine shops and foundries in these places brought the necessary skills and infrastructure to bear on building boilers and engines. Collectively, this territory accounted for 70 percent of steam-related patents (see table 1.1).

Information circulated rapidly among the metropolises and to their satellites and inner hinterlands. Some of the inventors active in steam-related patenting, especially in the core sectors of boilers and steam engines, may have operated autonomously and limited their contacts to occasional written and face-to-face communication. The concentration of patentees in the vigorous metropolitan and agricultural-industrial areas of the East, which had extensive channels of knowledge exchanges, implies, however, that networks of inventors, industrial entrepreneurs, and the social-political-economic elite provided the milieu for this flowering of invention. Communities of practice in steam engine building probably emerged within each of the metropolitan satellite territories, in the Hudson Valley, and in central New York state, where the prosperous, highly educated populations kept in close contact. Moreover, the substantial inter-metropolitan communication among the business elite in the East supported the rise of multiregional communities of practice in steam engine building.[5]

With almost half (46 percent) of the East's patents, the New York metropolis and its satellites, its inner hinterland, and the distant rich agricultural-industrial area in central New York state, dominated steam-related invention, especially in boilers and steam engines (see table 1.1). Nevertheless, the tight agglomeration of inventors in New York City and its New Jersey satellites, which accounted for 23 percent of the East's patents, remained unsurpassed as a seedbed for early steam engine development.

New York's First Engines

The construction of the "Soho Works" at the site of the old Schuyler copper mine near Passaic signaled that Americans had commenced serious steam engine work. Nicholas Roosevelt and his associates purchased the inactive copper mine in 1793 and hired Josiah Hornblower, the original assembler of the mine's Newcomen engine, to restart it. Roosevelt had big ambitions; after all, he had audaciously named his foundry, machine shop, and stamping mill after the Boulton and Watt Works in England. Within a few years he hired outstanding engineers and mechanics trained in Europe, such as Frederick Rhode and Charles Stoudinger from Germany and John Hewitt and James Smallman from the Soho Works of Boulton and Watt. This move immediately inserted Passaic's and New York City's local mechanics into an international network of elite machinists. By 1798 Roosevelt and his workers joined forces with Robert Livingston, a wealthy Hudson Valley landowner and a member of the New York state and the national political elite, and with John Stevens, rich landowner and steamboat and engine designer in Hoboken, New Jersey. They built a steamboat that used a Boulton and Watt type of engine which Roosevelt's Soho Works constructed. Trials on the Passaic River, however, revealed flaws in the boat and engine design.[6]

Philadelphia's First Engines

During the 1790s Philadelphia did not have an equivalent critical mass of engineers and mechanics with skills to build steam engines. No one would have predicted that Oliver Evans's move to Philadelphia in 1792 would help catalyze its emergence as a center of steam engine technology. He leveraged his invention and introduction of automated flour-milling machinery at mills in the Brandywine region near Wilmington, Delaware, and established a firm to sell milling supplies, cut millstones, and build flour mills. Philadelphia's business leaders possessed wide-ranging wholesale and financial contacts; thus, the city offered an ideal base for marketing Evans's milling business throughout the nation. He also had another project in mind which had germinated the previous decade. In 1786 he had applied to the Pennsylvania legislature for a patent on a steam engine to power mills and to drive a steam carriage, but it refused his request. He deployed profits from his successful milling business and from occasional forays into mercantile trade in Philadelphia to experiment with a radical alternative to the low-pressure engine of Boulton and Watt. This high-pressure engine rested on the principle of variable steam pressure, rather than the combination of small pressure and a vacuum created by condensation. Nevertheless, Evans did not build a workable engine until 1802; his steam engine contributions lay ahead.

Benjamin Latrobe's arrival in Philadelphia provided another spark to the city's emergence as a steam engine center. Following the death of his wife, he had migrated from England to the United States in 1796 and set up residence in Virginia, where he practiced his architectural profession. After receiving the commission to serve as architect for the new headquarters of the Bank of Pennsylvania, he relocated to Philadelphia in December 1798 and established contact with his mother's family, who moved among the state's social elite, and an uncle who lived in the city. Shortly after his arrival, he submitted a report to the city council dealing with improving the water supply, and by the following March the council approved his plans and appointed him chief engineer for constructing Philadelphia's waterworks.

Latrobe possessed substantial assets: a classical education in England and Germany in languages, mathematics, and philosophy; experience moving among London's social elite; contacts with the pinnacle of Philadelphia society through his uncle and Samuel Fox, president of the Bank of Pennsylvania, who chose Latrobe as architect of the impressive new bank headquarters; and a grand waterworks plan that appealed to the social elite. This plan included two large steam engines and could not succeed without the elite's political and financial support. Only they could afford to buy the one hundred dollar bonds for the first offering; they paid the preponderant share of the subsequent special tax of fifty thousand dollars on estates (real and personal), and they cosigned notes to guarantee repayment of loans from the Bank of the United States.

The decision to use two steam engines—the lower engine to pump water from the Schuylkill River to a basin that connected through a tunnel to Centre Square and the upper engine to pump water from the square to a reservoir from which pipes ran throughout the city—did not follow unassailable logic. In 1822, after years of struggling with the engines and sinking vast sums of money into them, the city switched to a dam and waterpower system that provided efficient, low-cost pumping. The initial choice of steam engines to pump water, therefore, probably represented an experiment of the social elite, stimulated by the persuasive abilities of Latrobe, a brilliant recruit to their group. The choice of engine builder was obvious. Only Roosevelt's Soho Works, located on the Passaic River near New York City, possessed the foundry skills, equipment, and engineers and mechanics trained to build Boulton and Watt low-pressure steam engines. Latrobe and Roosevelt became close friends and, later, partners; Roosevelt also married one of Latrobe's daughters, cementing their professional ties.

Problems with boring the large cylinders caused lengthy delays, but finally, in early 1801, the Soho Works had the steam engines ready for installation. Roosevelt sent James Smallman, one of his most experienced workers who had trained at the Boulton and Watt works in England, to install and operate the engines. Because they were built to handle growing demands for pumping, Philadelphia agreed to a contract with Roosevelt, who later added Latrobe as a partner, to use surplus power of the lower engine on the Schuylkill River to run a rolling mill for iron and copper sheets. Smallman stayed to run this mill, but it faced continual financial problems. In 1804 he started his own firm in Philadelphia to build low-pressure stationary steam engines and completed several of them. He enhanced his reputation six years later by building the engine to run the scrap metal stamper, forge hammer, stone grinder, bellows, block mill, and sawmill at the Navy Yard in Washington, D.C. That order came courtesy of Latrobe, who at the time served as engineer for the Navy Department; over the years he recommended Smallman to build other engines. Latrobe's support of Smallman's career exemplified the bourgeois virtue driving the behavior of many leaders in the emerging community of practice of steam engine building.

The construction of steam engines for Philadelphia's waterworks boosted the Soho Works' reputation and experience, thus benefiting New York as a steam engine center, but Philadelphia gained more. Because Smallman was one of the nation's leading builders of low-pressure engines, his move to Philadelphia strengthened it as a hub of machinist networks. A local community of practice emerged which rested on the well-defined artifact-activity couple of steam engine construction and maintenance, on the one hand, and of the training and tacit sharing of component knowledge among the skilled machinists, on the other. Its growth contributed to steam-related patenting in Philadelphia (see table 1.1). Philadelphia operated the premier ongoing experiment in urban waterworks, and numerous visitors came to see the great engines and pumps and elegant architecture of the buildings that housed them. This enhanced the city's reputation as a center of architectural knowledge of steam engineering, attracting mechanics such as Daniel Large, who arrived in 1807. Soon he started a firm to build stationary steam engines and, later, steamboat engines. Likewise, Oliver Evans boosted Philadelphia's steam engineering capabilities; in 1802 he completed a working prototype of the high-pressure Columbian steam engine that he used to grind plaster at his factory. Its boiler withstood high pressure, and the engine achieved much greater horsepower over a larger range. This enhanced operating flexibility at less cost than a low-pressure Boulton and Watt engine. Over the next few years Evans started to advertise his engines and acquired several orders.

From 1795, with the founding of the Soho Works near New York City, to 1805, with the start of Evans's high-pressure steam engine business, New York and Philadelphia embarked on a shared destiny as leaders in manufacturing steam engines and other heavy industrial machinery. Top machinists in each center knew one another as friends or competitors, and they engaged in know-how trading about steam engine technology. This collective approach probably contributed to the concentration of so much steam-related invention in New York City and its New Jersey satellites and in Philadelphia (see table 1.1). A larger pattern of knowledge exchange between New York and Philadelphia overlay this exchange of technological knowledge and buttressed it. Codified knowledge that flowed via newspapers, mail, commodity shipments, and passenger travel moved more rapidly and in greater volume between New York and Philadelphia than between any other cities.[7]

The Steamboat Push

In contrast to the bustle of steam engine work in Philadelphia, New York became quiescent after the Soho Works finished the engines for the Philadelphia waterworks in 1801. Between 1799 and 1808 no one acquired steam-related patents in New York City and its New Jersey satellites (see table 1.1). By 1804 a steam engine, possibly built by the Soho Works, operated a sawmill on Rhinelander's Dock, and the Manhattan Water Company installed an engine imported from Boulton and Watt in England to pump water, but they did not lead to other engine projects. In Hoboken, New Jersey, John Stevens continued to experiment with steamboat and engine designs, producing some working models, including a small propeller-driven steamboat in 1804. He began experiments with high-pressure steam engines, convinced that only these engines generated enough power for steamboats to achieve high speeds. His lack of trained mechanics and of a fully equipped engine shop at this time, however, hindered his efforts.

The trajectory of steamboat innovation changed from an unexpected direction when Robert Fulton met Robert Livingston, the new minister to France who arrived in Paris in 1801. They had a potent chemistry to build a partnership. Livingston, the New York patrician, had acquired wide knowledge of steamboat projects in the United States and possessed social and political influence in New York to renew his monopoly on steamboat navigation in state waters. He obtained the monopoly from the legislature in 1798, when he associated with Nicholas Roosevelt and John Stevens in their abortive steamboat venture. Fulton had acquired a background in civil and mechanical engineering, and between 1786 and 1797 he used contacts with the British nobility to become active in canal engineering. During his residence of over a year among Manchester's leading engineers and scientists, he acquired a first-rate education, including awareness of Boulton's and Watt's work on the low-pressure steam engine. In 1797 Fulton moved to France and shifted focus to the design and construction of a working submarine armed with mines and torpedoes, a task at which he succeeded. Fulton and Livingston met in Paris and discussed steamboats; in 1802 they signed an agreement to build a steamboat to run on the Hudson River.

Like Latrobe, Fulton did not have expertise in the component details of engines and shipbuilding, but he possessed an architectural knowledge of the issues which permitted him to break the logjam in steamboat design and construction. He focused on the overall engineering of the boat, and in 1803 his steamboat successfully ran on the Seine River through Paris. The next year Fulton ordered a twenty-four-horsepower engine from Boulton and Watt, but the export permit was not approved until 1805, because it became hostage to Fulton's negotiations with the British government on building torpedoes to sink the French fleet. Finally, the government awarded him fifteen thousand pounds sterling as the reward for demonstrating practical torpedoes. He invested most of this money to provide a comfortable income and in 1806 returned to the United States. The following year his steamboat, the *Clermont,* built in the shipyard of Charles Brown in New York City, made its maiden trip of 150 miles on the Hudson River between New York and Albany. The boat demonstrated, unequivocally, the feasibility of commercial steamboat navigation, and the race to build these boats commenced.

The patronage of the social, political, and economic elite which powered the rise of Philadelphia as a steam engine center also operated in New York City. In 1803 Livingston used his influence in the state legislature to renew his monopoly on steam navigation in New York waters. Now he shared that monopoly with Fulton, and in 1807 the legislature extended it for thirty years. This monopoly covered both intrastate travel, such as on the Hudson River, and interstate travel from New York City to New Jersey, to Connecticut shore towns, and to Providence. The monopoly on interstate steamboats continued until 1824, when the United States Supreme Court ruled in *Gibbons v. Ogden* that only Congress possessed authority to regulate interstate commerce; the following year the New York legislature repealed the intrastate monopoly. Steamboat construction required substantial capital to cover the steam engine, gearing, paddle wheels, and hull. By 1813 Latrobe and Fulton used a cost estimate, excluding overhead, of roughly twenty-five thousand dollars per boat, though it could reach over forty thousand dollars. A single steamboat represented an investment equal to one-quarter to one-half of a large cotton mill. Only highly capitalized iron foundries, steam engine firms, and steamboat construction companies could build more than a few boats or their major components such as engines, hulls, and superstructures.

Under the protection of the monopoly, the Fulton-Livingston group embarked on a binge of steamboat construction, accessing one of the East's greatest concentrations of skilled foundry machinists in New York City and its New Jersey satellites. Charles Stoudinger, one of the expert builders of Boulton and Watt low-pressure engines at the Soho Works, supervised engine building, and foundries in New York, such as those owned by Robert McQueen and James Allaire, supplied castings. Because they needed to collaborate on building the parts for steam engines, these foundries must have engaged in extensive tacit exchanges of component knowledge.

By 1811 the Fulton group had built three more steamboats, and their dimensions quickly doubled. Within a year it turned out steam ferryboats to connect New York and its satellite cities around the harbor. In 1813, the year the Fulton group moved into its Jersey City Works, it built five steamboats. The facility covered three acres, cost forty thousand dollars to build, and housed a complete set of shops—a mill to bore cylinders, a forge shop, a boiler shop, and a dry dock to build and repair steamboats. It did not have an iron foundry because McQueen and Allaire continued to supply castings. By 1815, the year Fulton died, the group had finished fifteen steamboats. The Fulton group and other associates, including Roosevelt, Stoudinger, and Latrobe, set up temporary steamboat shops in Pittsburgh, but they were not as successful as the Jersey City Works. Roosevelt built the first steamboat, the *New Orleans*, and made the run to that city in 1811 to demonstrate the feasibility of using large steamboats on the Ohio-Mississippi river system. Their shops completed two other boats, but in 1814 they backed out of a fourth boat started by Latrobe.

The Fulton-Livingston monopoly hindered, but did not prevent, the expansion of other steam engine manufacturers in New York. John Stevens, brother-in-law of Robert Livingston, harbored bitterness at the restrictions that the monopoly posed for his Hoboken engine shop. He excelled at high-pressure steam engine and boiler innovations, far surpassing the efforts of the Fulton group. Stevens's paddle wheel steamboat, the *Phoenix*, had its trial run in 1808, but he capitulated to the monopoly the following year and sent it to run on the Delaware River between Philadelphia and Trenton. John and his sons, especially Robert, kept busy building steamboats and high-pressure engines, but their bigger success came after 1824, when the Supreme Court and the New York legislature terminated the Fulton-Livingston monopoly. The Stevens family became the foremost manufacturers of high-pressure steam engines for eastern steamboats, contributing to the demise of the low-pressure Boulton and Watt type engine that the Fulton-Livingston group favored.

Stoudinger joined with Allaire to lease the Jersey City Works of Fulton after he died, but Stoudinger also died within a year. Allaire moved the equipment to his foundry in Manhattan, thus acquiring the capacity to build engines alongside his regular foundry work. Similarly, by 1811 Robert McQueen, who made iron castings for Fulton, turned to building engines in New York, and, shortly afterward, he built a full-fledged foundry and engine works. During the decade after the first run of the *Clermont* in 1807, New York foundries and engine works had elaborated a community of practice which made the city a center of steam engine manufacturing rivaling Philadelphia, but machinists in that city did not rest on their laurels.[8]

THE INDUSTRIAL MACHINERY WORKS

In 1807 Oliver Evans proudly advertised the opening of the Mars Works in Philadelphia. The choice of the Roman god of war for the name may have symbolized Evans's challenge to the Soho Works and to others who built Boulton and Watt low-pressure steam engines.[9] Evans's fully equipped engineering works housed an iron foundry, pattern shop, forge shop, steam engineer shop to assemble engines, millstone factory, and a steam mill for grinding plaster and turning heavy cast- and wrought-iron pieces. He made improvements in the high-pressure Columbian engine, but the small demand for steam engines could not support specialized builders. Along with other successful foundries and machine shops, the Mars Works sold a full range of castings and machinery. By 1811 it employed thirty-five workers, melted about ten tons of pig and scrap iron in the cupola furnace every week, and sold products valued at fifty thousand dollars annually.

Like the Fulton-Livingston group, Evans recognized that the Midwest offered a lucrative market for steam engines, but he diverged from his New York competitors, who only built temporary shops in Pittsburgh to supply engines for steamboats. In 1811 Evans founded the Pittsburgh Steam Engine Company, a replica of the Mars Works, and placed his son, George, in charge; he set the Pittsburgh firm on a foundation of diversified foundry work. With far fewer good waterpower sites than the East, the Midwest demanded engines for factories as well as for steamboats. By 1812 Evans's Mars Works had completed ten engines that operated in the East or the Ohio-Mississippi valleys, and he had ten more in process or on order and two engine works to meet them. Within two years he had finished twenty-eight engines (eighteen in the Midwest and ten in the East), most of them employed in urban factories. In 1816 the Pittsburgh branch employed one hundred workers, testimony to the growing significance of the midwestern market.

In 1820 New York and Philadelphia were the leading producers of heavy industrial machinery, and their status rested on substantial strengths. They generated the greatest demand for steam engines, they housed the largest agglomeration of iron foundries and machine shops, their leading-edge technology firms attracted top machinists, and their sizable elite group provided financial support for the new technology. The steamboat's glamour gave an impetus to steam engine construction, especially in New York, but stationary steam engines also became more important as factories in the metropolises expanded and as factory owners began more fully exploiting large water-power sites away from the coast. The focus on marine engines at the Stevens works in Hoboken and at Fulton's steamboat works in Jersey City remained aberrations before 1820. Fulton's specialization lasted only until 1816, when Allaire moved the equipment to his foundry in New York City. Firms owned by Evans, Smallman, and Large in Philadelphia and by McQueen, Allaire, and Roosevelt in New York and vicinity built engines for steamboats and for factories and waterworks.

The greatest firms typically combined diversified iron foundry work with steam engine building because the demand for engines remained small. Foundry work provided steady employment of capital equipment and generated profits to support steam engine building. In New York City the Allaire Works employed sixty men and ten boys and had a capital investment of one hundred thousand dollars in 1820. At that time Robert McQueen's Columbian Iron Foundry and Steam Manufactory employed seventy men and twelve boys and had an investment twice the size of Allaire's. The West Point Foundry, established by Governor Kemble and others in 1817 and capitalized at ninety thousand dollars, operated at Cold Spring on the Hudson River. It manufactured cannons for the U.S. Navy and became a top producer of large castings, steam engines, and heavy machinery. By 1817 Evans claimed to have built seventy to eighty steam engines, most of them stationary, and the Pittsburgh branch probably made the majority of them. After Evans's death in 1819, his sons-in-law, Irwin Rush and Peter Muhlenberg, took over the Mars Works and maintained the Philadelphia firm as a leading engine builder and foundry.

The social, political, and economic elite, therefore, allied their capital and influence with talented engineers and mechanics, some of whom also came from the elite, to propel steam engine manufacturing from feeble efforts in the 1790s to the great industrial machinery works of the New York City area and of Philadelphia in 1820. On the one hand, pursuit of fame and fortune energized prominent participants such as Latrobe, Evans, Stevens, and Fulton, and, on the other hand, elite norms governed their behavior. Carried to an extreme, these perspectives view prominent actors as atomized individuals, undersocialized in the former case and oversocialized in the latter. The prominent participants, however, did not always operate in a narrow pursuit of self-interest to the pinnacle of the industry, and the elite did not follow such programmed behaviors that their ongoing social relations had little impact on their actions.[10]

ELITE TIES POWER STEAM ENGINES

Neither the absence of ties nor overly cohesive ties among participants channeled their rapid and successful launch of the steam engine industry beyond the scattered efforts of early inventors, thus presenting a paradox. Enthusiasts among the elite and their allies in engineering and mechanics recognized that they could not operate alone because engine and steamboat construction consumed substantial capital, posed great risks, and required sophisticated skills in engineering and mechanics.

Overly cohesive ties within a small group, however, posed dangers to that effort because they restricted the range of allies and of knowledge available to investors in, and builders of, steam engines and steamboats. That narrowness could arise because individuals possessing these types of social ties invested tangible effort and resources in maintaining them, limiting the number and variety of people in their network. These individuals more likely shared friends, thus exchanging similar knowledge about steam engine technology. Likewise, if hub individuals of social groups forged bridges to one another, all members of the groups shared in the same pool of knowledge. In contrast, network ties reached optimal efficiency when individuals within each social group of engine builders spent only part of their time on cohesive ties within the group and hub persons created bridges to other social groups that had few linkages among them (see fig. I.3). This system maximized the exchange of knowledge about innovations in steam engine technology, provided alternative checks on ideas, and enhanced chances for individuals to form alliances to underwrite and develop steam engines.

Latrobe occupied a hub position. From 1799, his first contact with Roosevelt, owner of the Soho Works near Passaic, almost to his death in 1820 in New Orleans, where he was finishing its waterworks, Latrobe forged a staggering array of social ties that spread knowledge about low-pressure Boulton and Watt steam engines. Roosevelt became a friend and son-in-law as well as business partner in a variety of ventures, but this cohesive tie never inhibited Latrobe or Roosevelt from maintaining independent links to other engine builders. The order given to the Soho Works for the two Philadelphia waterworks engines and the subsequent move of Smallman from Soho to Philadelphia constituted a direct channel through which to exchange steam engine knowledge between New York and Philadelphia. Fulton also became a friend and partner of Latrobe, thus inserting other bridges, besides the one through Roosevelt, between Latrobe and steam engine and steamboat builders in New York. He used wide-ranging professional contacts to recommend steam engine builders, such as Smallman, builder of the engine for the Washington Navy Yard. Architectural competitions in which Latrobe participated and commissions he received in New York, Philadelphia, Baltimore, and Washington provided many contacts, thus easing his movement among the social, political, and economic elite of the East Coast. In 1803 his appointment by Thomas Jefferson as surveyor of the Public Buildings of the United States positioned him among the national political elite. Latrobe served as a bridge across many social groups of steam engine and steamboat builders which did not necessarily possess links among themselves.

Initially, New York's steam engine and steamboat builders created a maze of ties which did not tightly link each person to every other. From his position at the Soho Works near Passaic, Roosevelt acted as a hub through his employment of leading engineers such as Stoudinger, Rhode, Hewitt, and Smallman, and during periods of overlap in their employment these talented individuals must have shared tacit knowledge about steam engine building. They developed their own bridges over which they exchanged this tacit knowledge with others: Stoudinger worked with Stevens in Hoboken and traveled to Pittsburgh to build a steamboat; Rhode or a relative trained Allaire, who also worked with Stoudinger for a year; and Smallman moved to Philadelphia. Roosevelt and Stevens joined with Livingston to build a steamboat in 1798. Stevens acquired social, political, and economic allies in New Jersey. A grandfather on his mother's side was a wealthy landowner, lawyer, and politician, and his father-in-law was a prosperous iron founder. These contacts served Stevens when he challenged the extension of the Fulton-Livingston monopoly into New Jersey.

During the period of this monopoly, however, cohesive ties revolving around participating foundries and engine builders may have restricted the exchange of component knowledge about steam engines and hindered cooperative ventures with those outside the monopoly. Livingston's powerful links to the New York political elite kept the Fulton-Livingston monopoly on steamboat service in New York waters secure until the United States Supreme Court overturned it in 1824. Stevens, the brother-in-law of Livingston, never translated that family bond into a cooperative business relation after the formation of the Fulton-Livingston partnership. The monopoly hindered the steam engine projects of Stevens and reduced his ties to other New York engine builders. Fulton became a local hub through the power of the monopoly and forged ties to Allaire and McQueen, owners of foundries, who supplied component knowledge about castings and other metal parts of steam engines. After Fulton's death, Allaire and McQueen could more freely exchange component knowledge with other foundries, as they had done before their participation in the monopoly.

In Philadelphia, Smallman and Large, builders of low-pressure engines, maintained some contacts, but they never flowered into full cooperation. Evans, the leader in building high-pressure engines, achieved some autonomy and served as a hub for these builders' network. The Mars Works in Philadelphia and the Pittsburgh Steam Engine Company gave Evans the freedom to institute a wide array of network bridges to other groups working on steam engines. His ongoing patent disputes could consist of tacit exchanges of component knowledge of engines and even architectural knowledge of the engine as a system, especially when these disputes landed in courts in which experts testified. Memorials to the United States Congress, correspondence with national political leaders such as Thomas Jefferson, and publication of *The Abortion of the Young Steam Engineer's Guide* in 1805 constituted forums for exchanges of codified knowledge. Stevens and Evans maintained links through Dr. John Coxe of Philadelphia, brother-in-law of Stevens, but this was likely a weak tie, because they never formed an alliance to challenge the low-pressure engine group. William

Thornton, head of the federal patent office in Washington, filtered some exchanges of knowledge about steam engines.

Networks of the social, political, and economic elite and their allies in engineering and mechanics offered multiple channels for spreading knowledge about steam engine technology and for know-how trading of technical skills. These ties nurtured low- and high-pressure approaches to steam engines and permitted cross-fertilization between them. By 1820, however, the high-pressure engine surged past the low-pressure competitor as the strength of the Fulton-Livingston group declined and the superiority of the high-pressure engine for American steamboats, mills, and public works became evident, especially its lower initial cost, simpler construction, and wider range of power.

Nevertheless, during the first decade of the nineteenth century the success of the high-pressure engine did not appear assured. The support of the elite remained crucial to development of the steam engine, and most of them threw their support behind the low-pressure Boulton and Watt type of engine. The low- and high-pressure engine advocates fought wide-ranging battles in the courts, the federal patent office, and the state and national legislatures in search of approvals for monopolies, and both groups lobbied high governmental officials, including state governors and United States presidents. The refusal of most midwestern states to support monopolies—the states rejected eastern interlopers—and the inability of the elite to convince the United States Congress to support national monopolies contributed to the stalemate.

The 1824 Supreme Court decision in *Gibbons v. Ogden* became the final nail in the coffin of elite battles. Free-ranging competition among firms for dominance came after the political and legal conflicts subsided, without leaving clear winners. Had the federal government engaged in a national industrial policy, the low-pressure engine group might have prevailed because it had the most formidable lineup of elite. This situation might have retarded regional economic development, especially in the Midwest, which overwhelmingly favored the low-cost, high-pressure engine for steamboats and factories; this equipment constituted an advantage in the early stages of settlement, with limited supplies of capital, and steamboats needed powerful engines on the area's fast-flowing rivers, such as the Ohio and Mississippi. The expansion of steam engine manufacturing in Pittsburgh before 1820 signaled that New York and Philadelphia would not monopolize engine manufacturing; nevertheless, they had a secure position as leading centers of industrial machinery.[11]

THE GLORIOUS DAYS OF THE ELITE

The nation's social, political, and economic elite contributed materially to the penetration of steam engines into urban-industrial life during the early antebellum years. Because these engines required large amounts of capital, the elite's involvement proved crucial to spurring technological change. They possessed substantial personal capital and could collectively marshal larger sums to fund iron foundries and steam engine firms, and they astutely used their connections on occasion to tap government resources or influence policies. Within Boston, New York, Philadelphia, and Baltimore, multiple sets of cohesive networks existed, with bridges between them structured around hub individuals. The elite's networks also connected metropolises, sometimes through family ties but mostly through social, political, and economic connections. These ties facilitated the exchange of knowledge about technological change and know-how trading of technical skills, and the elite used their networks to arrange for leading mechanics to move among firms. The direct involvement of the elite in advancing technological change through inventive activity quickly receded after 1820, but their indirect involvement through provision of capital and governmental largesse continued.

The elite participated in underwriting the major iron foundries that built steam engines, some of them as engineers who worked on the machinery, thus inserting them directly into the machinist networks of foundries and machine shops. Individual mechanics who built steam engines exhibited highly networked behavior. Their communities of practice had local components as well as interregional ones as early as the first decade of the nineteenth century. The foundries, including those that also built steam engines, served as one tendril of the pivotal producer durables. They spawned heavy industrial machinery firms, and foundries would serve as one of the seedbeds of the machine tool industry. By the early decades of the nineteenth century most foundries and engine shops met custom orders, and castings, which only needed some filing, constituted the majority of the products made in substantial quantities. These firms did not engage in a large enough number of repetitive metalworking activities on identical products to justify heavy investments in new techniques and equipment. The boom in cotton textiles which was taking place, however, caused a surge in the demand for identical spinning and weaving machines.

CHAPTER TWO

A Networked Community Built by Cotton Textile Machinists

> The next step, from the American saw gin, is, to the system of machinery to make cotton yarn.
>
> *Tench Coxe, "Digest of Manufactures, 1810"*

Wood craftworkers and blacksmiths possessed the requisite skills to build the simple spinning wheels and looms that households and artisans used to make cotton textiles. Under the direction of immigrant mill managers such as William Pollard, a Philadelphia merchant who learned about worsted weaving in his home in Yorkshire, England, these craft workers and blacksmiths transferred their skills to building components of early Arkwright machinery. Around 1794 Pollard built a water-frame cotton-spinning machine for John Nicholson's speculative venture at the Falls of the Schuylkill River. A local whitesmith—that is, a worker of cold iron—made spindles and flyers, a local blacksmith turned out wrought-iron parts, and a New York iron founder supplied cast-iron weights. The switch to powered spinning and weaving, however, inexorably shifted the loci of construction to skilled machinists, machinery builders, and iron foundries. During the 1790s this transformation proceeded slowly, but the pace accelerated after 1800 as growing numbers of cotton mills needed machinery and the equipment became more sophisticated.[1]

SKILLED MACHINISTS REQUIRED

The Arkwright and Crompton spinning processes and power loom weaving operated on flow principles of production within each successive stage.[2] This placed a premium on increasing production speed and balancing the differential output of machines at each stage—carding, roving, spinning, warping, dressing, and weaving. Improvements in one stage could introduce bottlenecks in another, and this technological disequilibria sometimes set off compulsive sequences of innovations to improve machinery performance. Greater machine speed challenged machinists to establish static and dynamic relations among parts, including speeds, settings, and weights, and to cut, bend, and grind metal parts precisely to reduce friction and wear. The rapid speeds of Arkwright spinning machines, for example, required accurately ground spindles to reduce wobble. The growing complexity of machines, especially following the introduction of power looms, placed a premium on access to specialized knowledge about innovations in textile machinery, thus shifting innovation from individual machinists to networks of them and of their firms, because participants in these networks gained competitive advantages from greater access to knowledge about technical advances compared to those who remained isolated from the networks.

English immigrant machinists transferred textile machinery technology to the United States. Beginning in the late 1780s and early 1790s, Americans made numerous efforts to acquire cotton-spinning technology, taking trips to England to recruit workers, listing job advertisements in American newspapers that were distributed in England, and offering financial incentives to immigrant machinists and mill managers. Between 1800 and 1815 leading power loom innovators such as William Horrocks worked in England; thus, American cotton manufacturers were eager to seek assistance from English immigrant machinists such as John Murray, who built power looms near Bloomfield, New Jersey, in 1811. American merchants visited England to observe power looms and record information about their construction; Francis Lowell took such a trip during 1810–12. Around that time Americans began to challenge English dominance of power loom inventions. As early as 1810, Philo Curtis, from Oneida County, New York, and Thomas Robinson, from Newport, Rhode Island, experimented with building power looms. During the period from 1796 to 1815 at least forty-five different inventors took out that same number of power loom patents in the United States, and, with the exception of two patents, all of them were awarded after 1805. When Lowell and Paul Moody completed their loom in 1815, Americans and their English immigrant allies stood foursquare in the development process of power looms.[3]

As with steam engines, the social, political, and economic elite provided consequential support for the emergence of the textile machinery industry. These elite consisted of those in the major metropolises, especially the Boston merchants (such as Lowell), and those in subregional metropolises such as Providence, which had social and political leader Daniel Lyman and the Brown merchant family, among others. Likewise, financial backing also came from lesser elite in prosperous agricultural areas, including the rich farmers and the professionals (doctors and lawyers), large-scale retailers, owners of processing mills, and heads of teamster firms in the villages, towns, and small cities. At the same time, the mobility of textile machinery mechanics, the growth of small machine shops and foundries in areas of rising agricultural prosperity, the emergence of clusters of textile machinery shops, and knowledge networks linking machinists in communities of practice also contributed to the establishment of the textile machinery industry.

THE PROVIDENCE HEARTH
Machinist Networks within the Subregion

The Providence area—which went unchallenged as the core of the nation's cotton textile manufacturing from 1790 to 1820—had begun inauspiciously in 1790 when the merchant firm of Almy and Brown hired Samuel Slater, a mechanic who had trained in an English Arkwright spinning mill, to build machinery for a new cotton mill. In 1793 the mill became operational in Pawtucket, near Providence. Until the mid-1830s shortages of skilled mechanics, as well as of managers with technical expertise, impeded the growth of the cotton textile industry. Consequently, the mobility of textile machinists determined the industry's expansion, and their critical role gave them leverage to extract lucrative pay packages, which sometimes included ownership stakes (map 2.1). Slater participated in the partnership of Almy, Brown, and Slater, which operated the 1793 mill. As prominent East Coast merchants, Almy and Brown possessed the commercial contacts to sell cotton yarn in Providence's subregion of Rhode Island and adjacent parts of Massachusetts and Connecticut and in ports from Maine to New York. This access quickly set the mill apart as a successful competitor to British imports. The mill's machine shop, which Slater ran, became known as a place for ambitious mechanics to receive training.

Because Slater's production system of integrated machinery took time to perfect, the partners waited until 1799 to add new mills. Almy and Brown acquired a 50 percent ownership stake in the struggling Warwick Spinning Mill and promptly ordered the mechanic John Allen to train at the Almy, Brown, and Slater mill in Pawtucket. This tight bond between direct work on building textile machinery and the training of mechanics by skilled machinists—the artifact-activity couple—would govern much of the spread of textile machinists' skills. Also in 1799, Slater joined with the Wilkinson family machinists to found Samuel Slater and Company in Pawtucket. The machine shop in this mill and in the original 1793 mill, which operated under Slater's supervision, became premier training grounds for textile machinists.

Benjamin Walcott Sr., who worked with Slater on the 1793 mill, ranked among the most prominent associates of Slater. Subsequently, Walcott partnered with others

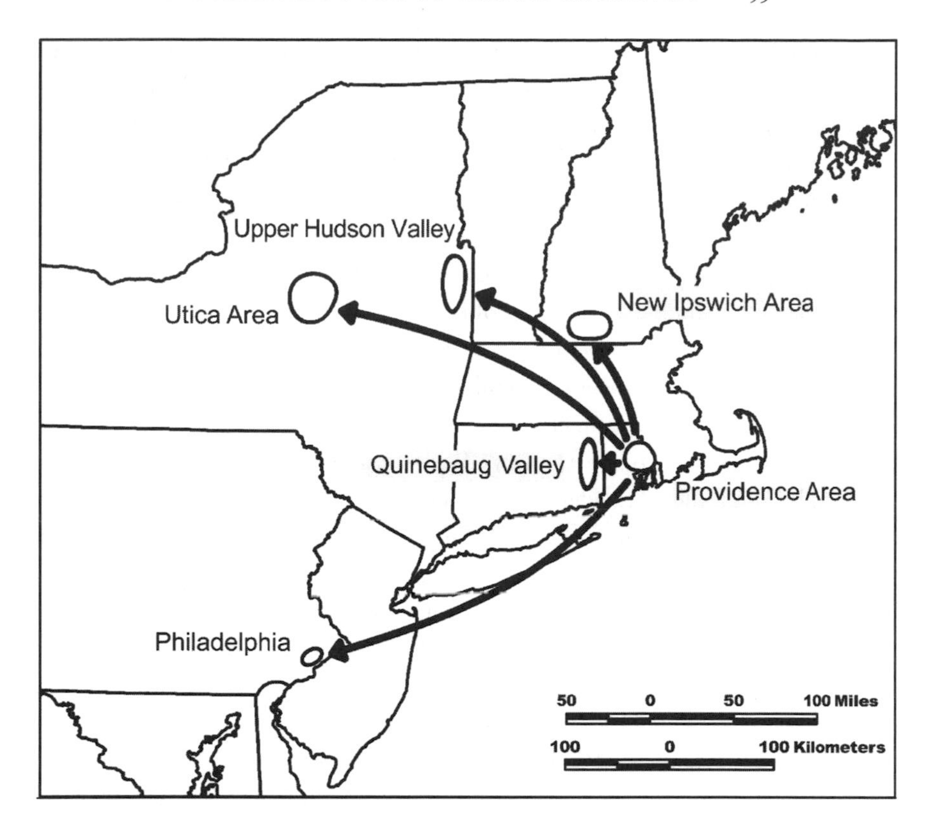

Map 2.1. Mobility of Slater-Trained Machinists from the Providence Area Hearth, 1800–1815

to found the Cumberland Mills (1802), Pawtucket Cotton and Oil Manufacturing Company (1805), and Smithfield Cotton Manufacturing Company (1805), at which he worked with Stephen Jenks Jr., a mechanic from the Jenks family of machinists. Benjamin Walcott Jr. worked with his father on all three mill projects, gaining valuable experience for his own imminent venture. In 1807 Slater expanded again when he joined the partnership of Almy, Brown, and Slaters, which also included Samuel's brother, John, a recent immigrant mechanic from England. Thus, from the beginning the Providence hearth participated in an international network of textile machinists engaged in know-how trading which gave its machinists competitive advantages.

Between 1805 and 1809 cotton mills proliferated in the Providence hearth, and the number of trained mechanics multiplied. Precise training origins and the identity of machine builders for some mills remain unknown, although several can be identified as English immigrant mechanics. Overlapping sets of social, political, and economic elite from Providence and vicinity, whose cohesive ties facilitated the organization of firms, created most of the mills. The Coventry Manufacturing Company, which commenced production in 1807, employed machinists Perez Peck and Samuel Ogden, who was English. The same year Peck joined with mechanics Peter Cushman, John White, and Joseph Hines in finishing machinery for the start of production at the Natick Manufacturing Company. Jonathan Adams, a partner in the merchant firm of Adams and Lathrop, which had invested in the Natick mill, joined with investors from his hometown of Northbridge, Massachusetts, to establish a mill there in 1809. This project drew on the services of William Howard, a mechanic from the Natick mill machine shop and a trainee of Peck and Cushman. Howard went to Northbridge to instruct Paul Whitin in machinery building, and after 1820 the Whitin family established one of the nation's premier textile machinery firms.

The artifact-activity couple was thus effective as a mechanism for conveying technical skills from leading Providence mechanics such as Peck and Cushman to Northbridge, through the Whitin family. Cushman, along with Ogden, who had worked on the machinery for the Coventry Manufacturing Company (1805–7), became the chief machinists for the Lyman Manufacturing Company, which commenced production in North Providence in 1810. As mechanics worked together in various shops, they tacitly shared component knowledge about individual machines and about the architectural system of integrated textile machines for yarn manufacturing. Their collaborative work and the subsequent departure of some machinists to other shops generated overlapping, though not highly redundant, social networks that served as conduits for spreading knowledge about job opportunities among mechanics.

The Wilkinson family of machinists extended the reach of the Slater group to the Quinebaug Valley of eastern Connecticut (see map 2.1). They allied with the Rhodes brothers of Warwick, Rhode Island, to found the Pomfret Manufacturing Company at a large waterpower site in Putnam. The fame of the Wilkinson family guaranteed that the Pomfret firm, which started production in 1807, would be the hub of machinist networks in the Quinebaug Valley. Within two years the Sterling Manufacturing Company was formed, reinforcing the imprint of Providence area mechanics. Initially, this project included Ogden, who had participated in the Coventry and in the Lyman Manufacturing companies. By 1812 at least seven other cotton mills organized by local and by Providence investors and machinists started production in the valley, accounting for half of the twenty-four mills built in Connecticut between 1809 and 1818 which survived until at least 1832. The proliferation of successful mills testified to the tacit exchange of expert machine-building knowledge among skilled machinists with network ties to the Providence area.

By the first decade after 1800 the subregion of the immediate Providence area and its hinterland extension to the Massachusetts border and to eastern Connecticut contained multiple machinist hubs. Although Slater's firms exerted enormous influence as training grounds, the movement of these trainees to other firms and the local emergence of talented machinists such as the Wilkinson family, Peck, Cushman, and Ogden created new hubs of technical networks, which inhibited any single machinist or group from exerting structural constraint on the growth of machinist networks. In this multi-hub network structure hub machinists forged bridges across networks, thus ensuring know-how trading among communities of practice. Slater and the Wilkinson family achieved such prominence as separate hubs—though linked to one another—that they forged the widest array of bridges. Individuals came to them for contacts and advice, and they maintained contacts to other networks of machinists.

These networks did not have extensive redundancies. Technical knowledge about innovations in textile machinery building circulated across networks, rather than a small number of ideas recirculating among mechanics with cohesive ties. The same lack of heavy redundancy characterized networks of the social, political, and economic elite who financed cotton mills; instead, they joined shifting coalitions of investors. Because they made decisions about which machinists to hire, the elite's network structure contributed to the multi-hub nature of machinist networks. The inability of individuals or groups to exert structural constraint over technical knowledge and the multi-hub, relatively nonredundant character of machinist networks propelled the expansion of textile machinery networks and supported the proliferation of successful cotton mills.[4]

Network Expansion Outside the Hearth

Merchant wholesaling channels and migration created long-distance bridges for the extension of textile machinist networks. New hubs formed, and their networks possessed bridges to Providence's multi-hub networks (see map 2.1). Samuel and Nathan Appleton, sons of a wealthy farmer in the southern New Hampshire town of New Ipswich, owned a retail store there as one component of their merchant wholesaling business, which they ran from their Boston headquarters. By 1800 the Appleton brothers had established close family and business contacts with leading Boston merchants through which they accessed information about the cotton yarn output of mills owned by Almy, Brown, and Slater. The Appletons probably sold the yarn to Massachusetts and New Hampshire retailers.

When Charles Barrett Sr., a wealthy New Ipswich resident, decided to start a cotton mill, he accessed the job mobility channel that connected New Ipswich (Charles Jr. had worked in the Appleton store in the late 1790s), the Appletons' headquarters in Boston, and Providence's cluster of Slater-trained mechanics. The Appletons thus served as the bridge across the structural holes separating these local networks. In 1804 Charles Barrett Sr. invited Charles Robbins, a Slater employee, to construct machinery and manage the New Ipswich Cotton Factory. As a skilled textile machinist, Robbins could tacitly communicate component knowledge of building textile machinery and architectural knowledge of the production system. He undoubtedly contributed to the emergence of a network of machinists—a knowledge cluster—in the New Ipswich area. Among the thirteen cotton mills established before 1820 in New Hampshire which lasted until at least 1832 (a sign of successful mills), six operated within twelve miles of New Ipswich.

Providence also had merchant wholesaling bonds with Albany, the subregional metropolis of the Upper Hudson and Lower Mohawk valleys. Sometimes these links were direct, but most of them passed through New York City merchants. By 1800 Albany merchants distributed cotton yarn from the Almy, Brown, and Slater mills; thus, information about skilled machinists must have percolated along with this commercial information and goods. Furthermore, migrants from New England had settled the valleys in New York by the mid-1780s, generating family and friendship channels of information with Rhode Island. In 1804 Job Whipple, owner of a flour mill in Greenwich, thirty miles north of Albany, probably used these channels when he journeyed to Rhode Island to find a mill manager and mechanic for his proposed cotton mill. He hired William Mowry, a young mechanic who had moved from Woodstock, Connecticut, in 1799 to work in Slater's mill. Mowry's lucrative financial package consisted of a half-interest in Whipple's waterpower privilege, and he provided all the capital to run the mill. Within a few years Mowry married Whipple's daughter, and several years later he purchased his father-in-law's share of the mill.

This culmination of events underscores the leverage skilled textile machinists possessed in acquiring benefits while their talents remained in short supply. Like southern New Hampshire and the Quinebaug Valley of Connecticut, the Upper Hudson Valley witnessed a proliferation of future successful mills following the arrival of a top-notch textile machinist, testimony to the creation of a subregional community of practice built around hub machinists who had trained in the Providence area. Between 1804 and 1815 investors started eleven cotton mills that achieved success (lasting until at least 1832), accounting for about one-quarter of such mills started in New York state before 1820.[5]

Migrants from the Providence subregion to the Utica area in central New York created information channels to transfer the premier textile machinist, Benjamin Walcott Sr., from Rhode Island to Utica in 1808. Dr. Seth Capron arrived in Utica two years earlier from Uxbridge, Massachusetts, just north of the Rhode Island border; he had previously lived in Cumberland, Rhode Island. Thus, Capron possessed family and friendship ties with people in the Providence hearth. His invitation to Walcott Sr., who had worked with Slater and had founded three textile mills of his own, suggests that Capron anticipated establishing a major textile project. Walcott became a partner in Walcott and Company, and a large mill and separate machine shop were finished by 1809. That same year Walcott returned to Rhode Island, and his son, Benjamin Walcott Jr., replaced him; the son had worked with his father on the three Walcott mills. Walcott Jr. became the mill manager and a partner, a rich financial package consistent with his skills, and by 1810 the firm became the Oneida Manufacturing Society. Within three years he had joined with his father-in-law, a prominent member of the local and state-wide elite, to found the Whitestown Cotton and Woolen Manufacturing Company. This firm, along with the Oneida company, became the basis for one of the nation's greatest textile manufacturing empires. Walcott Jr. must have become a hub in a network of talented machinists, many of whom probably trained in his machine shops. These machinists constituted a community of practice: during the 1809–16 period approximately twenty cotton mills, accounting for about half of the future successful mills founded in New York before 1820, commenced production in the Utica subregion.[6]

Between 1799 and about 1810, therefore, textile machinists with roots to Slater and his early associates and trainees had created a multi-hub structure consisting of Slater, the Wilkinsons, and several other machinists in the immediate Providence area; and various machinists in the Quinebaug Valley of Connecticut, southern New Hampshire and the New Ipswich area, the Upper Hudson Valley, and the Utica area (see map 2.1). Each hub had one or more talented machinists who served as leaders in textile machinist training—transferring knowledge through the mechanism of the artifact-activity couple—and each of them spawned a network of machinists in its subregion, thus creating a community of practice. Pivotal machinists served as bridges to other hubs, and, at least until 1820, Providence and its subregion remained the hearth from which bridges reached to the other networks.

The cumulative growth of competitive cotton mills (those lasting until at least 1832) founded from 1790 to 1820 reveals this proliferation of machinists and their network ties (fig. 2.1). In New England and New York each state's sharp rise in cumulative numbers commenced within a few years after the arrival of a leading mechanic who had trained in the Providence hearth. The lucrative financial packages, which often included major ownership stakes without any contributions of capital, reflected the continued shortage of talented machinists. Because machinists readily moved in order to take advantage of new opportunities, any efforts to control technological knowledge would have been futile. The multi-hub structure of textile machinist networks, with bridges linking them, facilitated rapid industrial expansion and fluid know-how trading of technical innovations. A diverse set of skilled metalworkers built textile machinery, and the lines separating textile machinists from other mechanics often had little consequence because, given the vicissitudes of episodic demand, specialization could not be sustained.

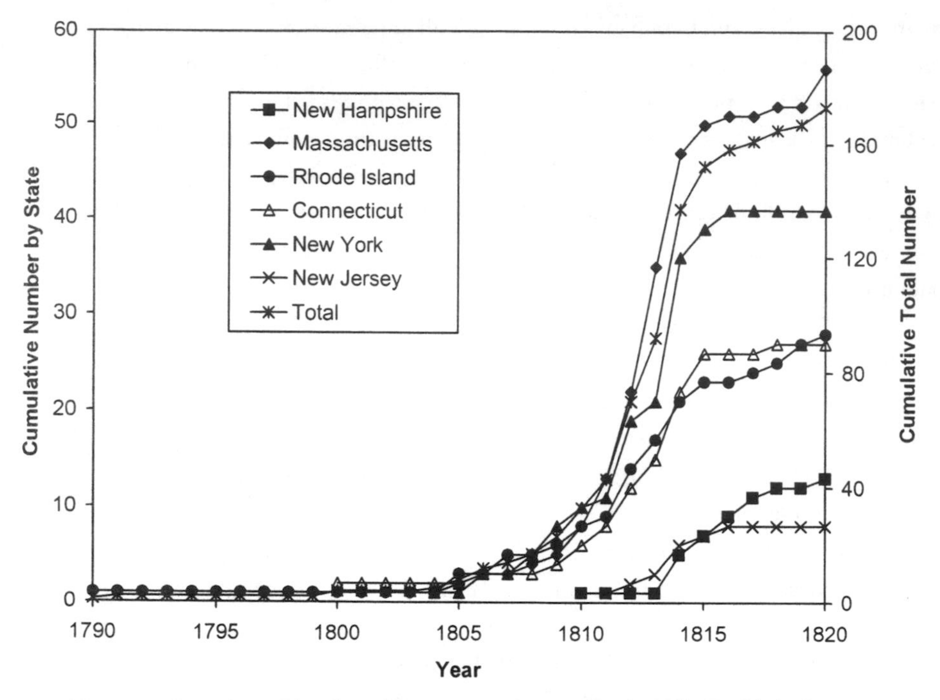

Figure 2.1. Cumulative Number of Competitive Cotton Textile Mills Established, 1790–1820. *Source:* Louis McLane, *Documents Relative to the Manufactures in the United States Collected and Transmitted to the House of Representatives, 1832, by the Secretary of the Treasury*, House Doc. No. 308, 22nd Cong. 1st sess. (Washington, D.C.: Duff Green, 1833).

THE SUPPORTIVE INFRASTRUCTURE OF MACHINISTS

Rising agricultural prosperity generated broad demands for metal durables, such as rods, screws, plates, and axles needed in agriculture, processing, transportation, and construction. The blacksmith met small-scale local demand for these goods, but growing demand opened opportunities for metalworkers to start small forges and foundries that instituted a simple division of labor to turn out larger volumes of similar goods. This spread the costs of labor and capital across many units, thus lowering the unit cost of metal goods. Firms had incentives to reduce costs by using better equipment, such as machines that cut, headed, and pointed nails. The number of patents for these machines rose from twenty-three in the 1790s to forty-seven in the 1810s. When textile machinists and mill managers arrived in an area of prosperous agriculture to erect a mill and build machinery, therefore, they encountered an existing infrastructure of metalworking firms.

Most textile machinists did not follow Slater's early approach to creating cohesive ties to machinists. When he arrived in Pawtucket in 1790 to build Arkwright spinning machinery for the mill of Almy and Brown, Slater boarded with Orziel Wilkinson, a prominent ironworker, and Slater promptly married Hannah, Orziel's daughter. The Wilkinson family of machinists produced metal components, such as spindles, rollers, and other iron parts, for the textile machines that Slater constructed during the 1790s. This work, along with supplying equipment for the few other cotton mills that also began operating that decade, however, did not provide much business. Instead, the increasing prosperity of Providence and of its agricultural subregion of Rhode Island, eastern Connecticut, and neighboring parts of Massachusetts fueled expansion of the Wilkinsons' machinist business. In 1791 Orziel built a furnace to make wrought-iron products, and three years later he established a rolling and slitting mill and furnished castings for the drawbridge in Cambridge, Massachusetts. Sometime between 1794 and 1796 his son, David, perfected a lathe to cut large industrial screws for oil mills, clothier's presses, and paper mills. From 1802 to 1804 the Wilkinson firm made castings for the lock gates of the Middlesex Canal, north of Boston. When growth in the number of new cotton mills accelerated after 1805, the Wilkinson family stood ready with a formidable capacity to build machinery (see fig. 2.1).

Although not as extensive as the infrastructure of machinists in the Providence area around 1800, other clusters of machinists in southern New England and the Hudson Valley also established close links to early textile machinery manufacturing. Taunton, Massachusetts, twenty miles east of Providence, contained a thriving metalworking industry built around a copper rolling mill and iron rolling and slitting mills that made nail rods. In 1806 investors in these firms enticed Silas Shepard, a talented mechanic from Wrentham, Massachusetts, about fifteen miles northwest of Taunton, to build cotton machinery for a new firm, the Taunton Cotton Manufacturing Company (the Green Mill). The expansion of textile machinery manufacturing in Taunton drew on extensive metalworking skills embedded in local mechanics and firms. Aside from building textile equipment, Shepard shared in a patent for a nail-heading machine in 1808.

In the wealthy agricultural and resource processing area between Boston and southern New Hampshire, cotton mills accessed skilled metalworking establishments. Jacob Perkins, one of the nation's outstanding industrial inventors (with fifteen patents by 1813), invented a semiautomatic nail machine by 1793. Within two years he and his partners established a large nail factory near an iron rolling mill in Amesbury, Massachusetts. Paul Moody, an equally brilliant machinist, trained under Perkins around 1800, including working in his Amesbury factory, and Perkins later recommended Moody to the elite who founded the Boston Manufacturing Company in 1813. Moody's career opportunity clearly benefited from Perkins's adherence to bourgeois virtue and from his prominent position in the social and business elite of the Boston area.

Builders of machinery for cotton mills in the vicinity of Hartford, Connecticut, tapped the machinist skills of the Pitkin family business. Elisha Pitkin—a Yale graduate (1753), member of the state's social and business elite, wealthy farmer, and Hartford merchant who engaged in the West Indies trade—partnered with his son, Samuel (Yale graduate, 1779), in a forge, iron foundry, and machine shop in nearby East Hartford. They produced anchors, cannon, nail rods, iron and brass castings, and large screws for paper, cider, and saw mills, and by the early 1790s this metalworking complex attained a large scale. It met demands for metal goods from farms and processing mills in the Connecticut Valley. When John Warburton, a skilled English machinist, arrived in East Hartford around 1794 to begin building Arkwright machinery for cotton mills, the Pitkins hired him.

Likewise, cotton mills in the productive farming areas of the Hudson Valley drew on local metalworking shops for equipment. Mowry, the Slater-trained machinist who started a cotton factory in Greenwich in 1804, developed ties to other mechanics in the valley. In 1815, when traveling to England to search for better machinery, he took along a machinist from Hudson, a town south of Albany. At Hempstead, immediately north of New York City, the Ramapo Works began as a nail and metalworking firm. It switched to cotton manufacturing about 1817 and quickly started building textile machinery, including power looms. In wealthy farming areas this juxtaposition of machinists skilled at building textile machinery and of forges and foundries energized communities of practice that crossed metalworking sectors, thus providing fertile exchanges of knowledge which transformed the machinery of textile manufacturing. The supply of inventors (and innovators), the affluent economies that underwrote the cost of risk taking on new ideas, and the concern with improving production technologies to turn out larger quantities of goods provided receptive settings for inventiveness. The Hudson Valley inventors, along with their peers in the rich agricultural areas of Massachusetts, Rhode Island, and Connecticut, contributed to the surging number of inventions during the first decade after 1800.[7]

NESTS OF TEXTILE MACHINERY BUILDERS

Rural households engaged in hand spinning and hand weaving constituted an enormous latent demand for cotton textiles, and meeting this demand would spur transformation of the textile machinery business. During the 1790s rising agricultural prosperity enlarged the number of rural and urban households that could afford to purchase factory cotton yarn. Inadequate machinery in American mills, however, made their yarn uncompetitive with British yarn. As the supply of skilled textile machinists increased, especially trainees of Slater and his disciples, mills reduced their production costs. Large declines in selling prices encouraged households to stop spinning yarn, and these price cuts gave a much greater boost to demand for yarn than slowly rising income. The upturn in the number of skilled machinists began about 1805, and their greater availability made it possible to establish large numbers of cotton mills (see fig. 2.1). These new mills spurred demand for textile machinery, and the expansion of existing ones and the shift to larger mills augmented it.

Creating Nests

The movement of textile machinists to areas of prosperous agriculture in southern New England and into New York state, as well as the emergence of secondary clusters in Philadelphia and near New York City, created many sites for making textile machinery. Most mills built their own equipment, and the infrastructure of metalworking shops, forges, and iron foundries supported machinery construction. When demand for textile machinery began to surge around 1805, its manufacture was being transformed even as most mills continued hiring their own mechanics to build equipment. Some machine shops in cotton mills started selling to other mills, and mechanics began forming firms to build machinery; nevertheless, few, if any, of these firms acquired sufficient business to make only textile machinery.

Sellers of textile machinery still faced competition because mills could hire their own mechanics to build equipment. To be successful, therefore, sellers had to make machinery that competed on the basis of a combination of price, improved technology, and quality. Economies of scale in the manufacture of machinery could be attained through the division of labor in a shop and the purchase of simple metalworking equipment such as a small forge. The minimum size of a small foundry with a cupola furnace, however, precluded its incorporation into a small machine shop. Equally, if not more critical, machinery builders had to offer equipment that incorporated better technology and quality than what was embodied in machinery built by a mill mechanic. This requirement placed a premium on participation in a community of practice of skilled machinery builders.

Technological change, such as invention and innovation, often represented a new combination of preexisting knowledge to satisfy demand for better machinery. "Breakthroughs" contributed to change, but small improvements arising from the workaday world of the textile machinist had greater significance. Typically, inventions and innovations occurred for one or both of two related reasons, recognition of a profitable opportunity or of a technological problem that required a resolution. The booming market for textile machinery lowered the risks of searching for and experimenting with technical improvements. Isolated textile inventors and innovators made few contributions to technological change. Instead, textile machinery mechanics and others, such as merchants who also devoted extensive time to machinery building, generated most of the improvements.

According to the mechanism of the artifact-activity couple, optimal learning occurred under the tutelage of a skilled teacher; thus, a shop with a talented textile machinist attracted other mechanics. In time they might leave to establish their own shops or be hired to run the shop of a cotton mill, and, if many cotton mills were being started nearby, some of the machinists stayed in the area. Network ties of these mechanics to their mentors created channels of communication, and the booming market for machinery reduced competitive risks of sharing innovations. Mentors served as network hubs, and the shared experiences of trainees who worked together at the mentor's shop created network bonds for future exchanges of know-how. This system supported the emergence of local and subregional communities of practice through which these machinists would perhaps engage in collective invention and innovation. Social network ties among machinists made possible job mobility among machine shops, and this movement allowed textile machinery builders to respond flexibly to variable demand for equipment. Skilled machinists in each nest of textile machinery builders also knew about the existence of other nests through the fluid mercantile information networks that tied metropolises to their hinterlands and linked different metropolitan regions. Any time a machinist moved to another nest, network ties between the nests strengthened, and these channels for exchanging knowledge created larger communities of practice.

John Clark, a machinist who learned about Arkwright spinning technology in Rhode Island (perhaps with Slater), contributed to creating communities of practice in and around Providence and in Paterson, New Jersey, and through his mobility between them he reinforced their joint mechanic community. From 1800 to 1807 Clark occupied the basement of the shuttered factory of the "Society for the Establishment of Useful Manufactures" (SUM) in Paterson. One of his first orders came from James Beaumont, an English manufacturer who constructed a cotton factory in Canton, Massachusetts, in 1801. Clark built three carding machines, a drawing and roving frame, and a mule spinning machine. About the same time, he may have finished six Arkwright spinning machines, with seventy-two spindles each, for a cotton mill erected in Beverly, Massachusetts. In 1808 Clark returned to the Providence area and built several Arkwright spinning machines for the Union Manufacturing Company of Maryland. The next year he returned to Paterson and established a small machine shop and a factory for carding wool. When John Clark Sr. retired in 1816, John Clark Jr. joined with Thomas Rogers, who had arrived in Paterson in 1812, to form the firm of Rogers and Clark. Despite the firm's accomplishments, the modest amount of textile equipment manufacturing in Paterson at this time meant the area could not qualify as a nest.[8]

The First Nests
Providence and Vicinity

Clark's return to Rhode Island in 1808 symbolized the stature of Providence and vicinity—situated in the heart of the earliest large center of cotton mill construction—as the first major nest of textile machinery builders. By that time Slater had directly trained or influenced a large cadre of local machinists, including Allen and Walcott Sr., and, the Wilkinson family of machinists had become major builders of textile machinery. Other talented machinists, such as Peck, Cushman, White, Hines, and Ogden, built machinery for Providence area mills. Many of these mechanics interacted with Slater and his trainees, and the larger collectivity often collaborated in changing groups of associates, thus setting the stage for the emergence of specialized shops around 1810. In that year Peck and his brother began manufacturing textile machinery in a shop that they rented from the Coventry Manufacturing Company, and by that time David Wilkinson's textile machinery production had become an important part of his diversified metalworking business. In 1813 Eleazer Jenks established a machine shop in Pawtucket to do heavy forging and to manufacture spinning machinery; Wilkinson used Jenks's shop to build textile machinery. That same year Larned Pitcher began producing cotton machinery in Pawtucket, and shortly afterward P. Hovey and Asa Arnold joined Pitcher.

In 1815 the Providence area attracted William Gilmour, a newly arrived English mechanic who carried knowledge about the path-breaking power weaving loom of William Horrocks. Gilmour built power looms at the Almy, Brown, and Slaters mill in Slatersville, but these looms did not work well. Daniel Lyman, owner of the Lyman Manufacturing Company, hired Gilmour, and, under Lyman's sponsorship and assisted by Wilkinson, talented head of a machine shop and a textile machinery builder in Pawtucket, Gilmour perfected Horrock's loom. For Wilkinson's efforts and the token sum of ten dollars, he acquired the right to build the Gilmour loom and started production by late 1817. Gilmour also built his own looms, which operated in the Lyman and Coventry mills and in mills in Taunton and Utica, New York, by 1818. He returned to Slatersville and completed looms for the Almy, Brown, and Slaters mill. Within about a year Pitcher and Ira Gay partnered to build cotton machinery, and in various subsequent incarnations their machine shop became one of the leading firms. At this time the factory village in North Providence housed six machine shops, many of which probably produced equipment for textile mills.

Taunton, a short distance east of Providence, functioned as an outlier of the Providence area's nest of machinery producers. Within a few years of equipping the Green Mill in 1806, Silas Shepard began selling his patented machines from the mill shop to the large number of cotton factories that started production in surrounding towns during the Embargo and War of 1812. He built his first power loom in 1811 and patented it a year later, and that same year John Thorp also received a patent for a power loom; the two men joined forces to build looms for sale. After Shepard returned from the War of 1812, the partners began building another loom, which they patented in 1816. They sold them to several Rhode Island mills and also licensed Wilkinson's textile machinery shop in Pawtucket to build and sell them. In addition, Shepard and Thorp patented a socket bobbin winder in 1816.[9]

The fluid mobility of machinists among firms, collaboration on projects, and shifting groups of partners in the nest of textile machinery builders in and around Providence thus created many strong ties for tacitly sharing the component knowledge of the machines and the architectural knowledge of the system of machines. These bonds created a powerful community of practice that enhanced collective invention and innovation.

Philadelphia

The nest of textile machinery builders in Philadelphia grew more slowly than in the Providence area. Before 1810 few Philadelphia mechanics built waterpowered textile machines for sale because local immigrant spinners and weavers worked on hand-powered machines in their homes and in small workshops. In 1794 James Davenport received one of the earliest patents issued for textile machines, and two years later he built machinery for the Globe Mill, which he probably owned. Someone named Eltonhead reportedly manufactured machinery in 1803, and four years later Daniel Large, the steam engine builder, turned out double-speed roving frames for sale. The arrival of Alfred Jenks in 1810, following a training period with Slater in Rhode Island, set a firm foundation for Philadelphia's textile machinery industry (see map 2.1). He possessed drawings for all types of textile equipment and started producing them, along with other machines, in the Holmesburg section of the city. His timing was shrewd or fortuitous; Philadelphia's textile firms had begun a shift to factory production. During the War of 1812 Jenks's business grew, and by the end of the decade he had moved to larger quarters in the Bridesburg section of the city. Since the 1790s mill villages had sprouted in the Brandywine Valley, about twenty-five miles southwest of Philadelphia. Around 1811 a small shop started producing textile machines, and after 1820 the valley's firms would expand their manufacturing of this equipment.[10]

Thus, the migration of textile machinists and their deep contacts with previous employers (or even partners) established network bridges between the communities of practice of textile machinery builders in Providence and vicinity and in Philadelphia. Because these bridges often had roots in previous peer or trainee relations within machine shops, followed by job mobility, they became effective channels to exchange component and architectural knowledge of textile machinery and to support the future mobility of skilled machinists.

Boston's Environs

The third nest of textile machinery builders to appear before 1820 consisted of Boston's environs, defined as its immediate satellites and other places within a radius of about fifty miles (excluding Providence and vicinity). Future machine shops of the great textile projects of the Boston Associates would constitute important parts of this nest. Before 1820, however, the mill shop of the Boston Manufacturing Company at Waltham was the sole representative of this type, and a web of ties bound it to other machinery and metalworking firms in eastern Massachusetts. These firms manufactured most of the machine tools used to make the Waltham textile equipment and built some of the machinery parts. The Easton foundry of Gen. Shepard Leach supplied a roller lathe, fluting lathe, and cutting engine, and small forges in nearby towns produced gearing and shafting. Paul Moody, the brilliant mechanic in charge of building the textile machinery, traveled to neighboring industrial centers to discuss ideas, making several trips to Amesbury to consult with his mentor, Jacob Perkins, who also installed the water wheel at Waltham. With this advice Perkins, who had trained Moody and recommended him for the position of senior machinist at Waltham, displayed his bourgeois virtue toward his protégé.

Initially, the Boston Manufacturing Company purchased most of its textile machinery from nearby firms. Luther Metcalf and Company, a machinery firm, built ten carding machines, five throstle spinning machines (with 60 spindles each), two mule spinners (with 192 spindles each), four reels, two winding blocks, and other accessories. In 1814 these items arrived at Waltham for installation, and other nearby firms supplied drawing frames, a stretcher, and additional spinning machinery. The mill's machine shop built power looms and dressing machines, and the company also purchased models from both E. Stowell and J. Stimpson.

After Moody perfected an operational power loom in 1814, he quickly turned his mechanical genius to the other machinery, first focusing on improving the dressing machine. Along with Lowell, Moody perfected the double-speeder roving frame for winding yarn on the bobbins, and he invented a filling frame and the "dead-spindle" system of spinning. These advances reached practical use by 1817, even though Moody did not receive the last of his major patents until two years later. During the decade following 1814 he was the country's foremost inventor and innovator of textile machinery. The Boston Manufacturing Company made its first sale of cotton machinery to the Poignand Plant and Company in Lancaster, Massachusetts, in 1817, and for the next three years sales of the machine shop averaged almost twelve thousand dollars annually. For the most part they consisted of power looms but also included the rights to build and use looms, warpers, double-speeders, dressers, and throstle spinning frames. Nonetheless, the shop's sales amounted to a tiny share—well below 10 percent in most years—of total sales (cotton goods and machinery); the significance of the big mill shops as sellers of textile machinery followed 1820.[11]

The establishment and early successes of this premier machine shop of the Boston Manufacturing Company, the path-breaking venture of the Associates, rested on a community of practice in Boston's environs which encompassed iron foundries, general machine shops, and shops that also built textile machinery. Hub members of this community consisted of some of the leading machinists in the East, including Perkins, Moody, Leach, and Metcalf. These individuals and their firms or shops maintained network bridges among themselves and to other machinists who had previously worked for them. Their community of practice structured the tacit exchanges of knowledge which Paul Moody (and probably several of his other senior mechanics) had in face-to-face meetings and subcontracts with other machine shops, which formed critical foundations of the first shop of the Associates' textile projects.

TEXTILE MACHINERY PATENTS

Knowledge about unsolved production problems and about technological advances in equipment circulated within and among the nests, and this knowledge reached both textile mechanics outside the nests and other machinists in areas of agricultural prosperity. If inventors formalized their solutions as patents, codified knowledge about them moved through the machinist networks. Because inventors possessed property rights in the patents that allowed them to sell, assign, or license them, they often encouraged exchanges of that knowledge. When patent disputes ended up in the courts, lawyers, judges, and expert witnesses acquired insights into the patent, and they in turn spread that knowledge through their networks.

The distribution of spinning machine and power loom patents from 1791 to 1820 confirms that these forms of knowledge exchange were operating at that time (tables 2.1 and 2.2). Before 1805 most cotton mills could not compete effectively against British imports; therefore, there was little incentive for domestic inventors, and, as a result, few patents were obtained. Slater spent much of the 1790s perfecting his production methods, and he trained few mechanics. During that decade Philadelphia, a budding textile center, and Mount Pleasant, New Jersey, along the Delaware River north of Philadelphia, reported a few spinning patents; and, Massachusetts, home of numerous small cotton mills, reported several power loom patents. The surge of machinery patents for both spinning and weaving followed 1805, consonant with the proliferation of successful cotton mills (see fig. 2.1).[12]

Across the East's prosperous agricultural areas numerous individuals participated

TABLE 2.1

Spinning Machine Patents, 1791–1820

Boston and Its Hinterland			
Metropolis	Patentee and Year	Satellites	Patentee and Year
Boston	S. Baldwin, 1815	Fitchburg, Mass.	N. Giles, 1812
	P. Moody, 1819	Providence, R.I.	J. Brown, 1813
			J. Brown, 1814
			W. Bald Jr., 1820
Inner Hinterland	Patentee and Year	Outer Hinterland	Patentee and Year
Coventry, Conn.	P. Arnold, 1818	Vermont	S. Baldwin and E. Town, 1812
		Barre, Vt.	G. Brewster, G. Trumbull, and J. Mathes, 1812
		Hartford, Vt.	S. Clements, 1812
		Durnham, Vt.	W. Ayres and J. Cochran, 1815
		Lyman, N.H.	P. Paddleford, 1816
		Richmond, N.H.	Z. Wheeler, 1816
New York City and Its Hinterland			
Metropolis	Patentee and Year		
New York City	W. Griffin, 1812		
	W. Shotwell and A. Kimball, 1813		
	J. Chappins, 1814		
Inner Hinterland	Patentee and Year	Outer Hinterland	Patentee and Year
Albany, N.Y.	E. Hearrick, 1810	Brookfield, N.Y.	D. Read, 1811
Greene County, N.Y.	A. Webster, 1810	Madison County, N.Y.	N. Jones, 1812
Humphreysville, Conn.	W. Humphreys, 1811	Hartwick, N.Y.	L. Bissell, J. Hinman, L. Hinman and B. Gains, 1813
Woodbury, Conn.	C. Crafts, 1812	Paris, N.Y.	E. Smith, 1814
Greene County, N.Y.	A. Webster, 1812	Otsego, N.Y.	L. Bissell, L. Hinman, and S. Willson, 1815
		Scipio, N.Y.	J. Richardson, 1816
Philadelphia and Its Hinterland			
Metropolis	Patentee and Year	Satellites	Patentee and Year
Philadelphia, Pa.	G. Parkinson, 1791	Mount Pleasant, N.J.	O. Herbert, 1792
	W. Pollard, 1791	Wilmington, Del.	J. Alricks, 1809
	J. Baxter, 1811	Burlington, N.J	B. Allison, 1813
	B. Allison, 1812		B. Allison, 1814
Inner Hinterland	Patentee and Year		
Harrisburg, Pa.	S. Comstock and M. Pike, 1813		

TABLE 2.1
continued

South	Patentee and Year	Midwest	Patentee and Year
Wythe, Va.	J. Sprinkel, 1816	Flemingsbury, Ky.	N. Foster, 1809 N. Foster, 1813
		Shelbyville, Ky.	C. Shillideay and G. McCaslon, 1816

SOURCE: Martin Van Buren, "Patents Granted by the United States," Secretary of State, Communicated to the House of Representatives, January 13, 1831, 21st Cong. 2nd sess., Doc. No. 50, *New American State Papers, Science and Technology*, vol. 4: *Patents* (Wilmington, Del.: Scholarly Resources, 1972), 208–9.

TABLE 2.2
Power Loom Patents, 1791–1820

Boston and Its Hinterland			
Metropolis	Patentee and Year	Satellites	Patentee and Year
Boston	N. Perry, 1811 R. Sugden, 1813 F. Lowell and P. Jackson, 1815 F. Hall, 1817	Providence, R.I. Franklin, Mass. Canton, Mass. Dedham, Mass. Brookline, N.H. Worcester, Mass. Taunton, Mass. Townsend, Mass. Smithfield, R.I.	J. Thorp, 1812 A. Ware Jr., 1813 J. Brasier Jr., 1814 C. Whitney, 1814 W. Wright, 1814 S. Blydenburgh and H. Healy, 1815 S. Shepard and J. Thorp, 1816 B. Cummings, 1817 W. Gilmour, 1820
Inner Hinterland	Patentee and Year	Outer Hinterland	Patentee and Year
Massachusetts	A. Whittemore, 1796 D. Grieve, 1797	Walpole, N.H.	E. Cutter, 1811
Ashford, Conn.	W. Janes, 1810		
Maine	R. Rogers, 1811 R. Rogers, 1812		
Ashford, Conn.	W. Janes, 1813		
Augusta, Me.	E. Robinson, 1813		
Newport, R.I.	T. Williams, 1813		
New London, Conn.	T. Mussey, 1817		
Canterbury, Conn.	W. Levally, 1819		
New York City and Its Hinterland			
Metropolis	Patentee and Year	Satellites	Patentee and Year
New York City	W. Squire, 1815 E. Warren, 1818	Newark, N.J.	R. Crosbie, 1812

TABLE 2.2
continued

New York City and Its Hinterland

Inner Hinterland	Patentee and Year	Outer Hinterland	Patentee and Year
Canaan, Conn.	S. Briggs, 1813	Rochester, N.Y.	P. Bennet, 1806
Stockbridge, Mass.	E. Herrick, 1813	Paris, N.Y.	P. Curtis, 1810
Canaan, Conn.	D. Briggs, 1814	Manlius, N.Y.	J. Phelps, 1812
Kent, Conn.	S. Bronson, 1814	Walton, N.Y.	C. Hathaway, 1814
Ridgefield, Conn.	L. Forrester, 1819	Butternuts, N.Y.	J. Jelleffs, 1815
		Otsego, N.Y.	S. Chapman, 1816
		Paris, N.Y.	P. Curtis, 1816
		Attleboro, Pa.	R. Jackson, 1816
		Queensbury, N.Y.	A. Porter, 1816

Philadelphia and Its Hinterland

Metropolis	Patentee and Year	Satellites	Patentee and Year
Philadelphia, Pa.	R. Lloyd, 1810	Hempfield, Pa.	J. Hess, 1813
	T. Mussey, 1811	Bristol, Pa.	T. Siddall, 1814
	C. Shepard, 1812		
	S. Craig, 1814		

Inner Hinterland	Patentee and Year	Outer Hinterland	Patentee and Year
Lebanon, Pa.	C. Cooper, 1808	Kingston, Pa.	A. Campbell, 1812
	C. Cooper and G. Shaake, 1812		
	C. Cooper and G. Shalk, 1815		

Baltimore and Its Hinterland

Metropolis	Patentee and Year
Baltimore, Md.	J. Guiramond, 1814

Inner Hinterland	Patentee and Year	Outer Hinterland	Patentee and Year
		Berkeley County, W. Va.	J. Coulter and S. Gano 1814

South	Patentee and Year
Montgomery County, Va.	J. Heavin, 1812
Albemarle, Va.	B. Brown, 1814
Montgomery County, Va.	J. Heavin, 1814
Wythe, Va.	J. Spinkle, 1814

SOURCE: Martin Van Buren, "Patents Granted by the United States," Secretary of State, Communicated to the House of Representatives, January 13, 1831, 21st Cong. 2nd sess., Doc. No. 50, *New American State Papers, Science and Technology*, vol. 4: *Patents* (Wilmington, Del.: Scholarly Resources, 1972), 185–88.

in patenting, although few acquired more than one (see tables 2.1 and 2.2). In these areas many people engaged in spinning and/or weaving, so inventors had easy access to knowledge about technical processes. They also could work with the blacksmith shops, forges, and small foundries, which possessed metalworking skills. Occasionally, inventors in subregions outside the areas of major textile manufacturing generated several patents within a short time, suggesting that technical knowledge circulated among inventors. In Virginia's Montgomery County, John Heavin acquired power loom patents in 1812 and 1814, and three other Virginia inventors also took out patents in 1814. Across much of the East networks of merchants provided channels for technical knowledge because they invested in cotton mills. Nevertheless, most of the spinning machine and power loom patenting directly or indirectly related to textile mechanic networks, even if the actual inventors did not work in mills.

The concentration of steam-related patenting in New York City and its satellites did not carry over to textile machinery work (see tables 2.1 and 2.2). The reason is simple—before 1820 these places turned out small amounts of cotton textiles. Similarly, the limited textile output in the South and the Midwest created few inducements for inventors. Nonetheless, spinning machine and power loom patenting was much more concentrated than the many sites of inventors suggests. The centers of cotton textile mills and of nests of textile machinists housed the most active patentees of textile machinery. Although Boston produced few textiles, it housed numerous investors in the industry, some of them technically proficient, and they probably retained network ties to local machinists. Patentees in Boston thus contributed to the accumulation of textile machinery inventions.

Boston's satellites and inner hinterland, extending out as much as one hundred miles, constituted the single greatest arena of textile machinery patenting in the East, especially in power looms (see tables 2.1 and 2.2). This territory also included the textile manufacturing of Providence and vicinity, which operated in a coequal way with the Boston-inspired cotton mill expansion. These cotton mills led the switch to integrated spinning and weaving built around machinery. Thus, the Associates' efforts to establish integrated production at the Boston Manufacturing Company in Waltham replicated a larger move in eastern Massachusetts to improve spinning and power loom weaving. Paul Moody acquired a patent, as did the merchants Francis Lowell and Patrick Jackson, and, other inventors participated, including those in Canton and Dedham, near Boston.

The Providence nest of textile machinists and inventors associated with them, however, dominated patenting in Boston's satellites and inner hinterland. The large number of patentees in Providence and in nearby towns, such as Taunton and Townsend in Massachusetts and Smithfield and Newport in Rhode Island, capture the ties between Slater and his trainees. Beginning around 1807, the migration of Slater-trained mechanics and of Wilkinson family members and their trainees to the Quinebaug Valley and other towns of eastern Connecticut to build machinery for cotton mills inspired an inventive force in Ashford, New London, Canterbury, and Coventry. The simultaneous proliferation of inventions encompassing Boston, its satellites, and its inner hinterland suggests that weak ties connected an overarching community of practitioners. In the more narrowly defined communities of practice of Boston and its immediate environs and of Providence and vicinity, subsets of individuals may have engaged in collective invention.

Inventors in Philadelphia, its satellites, and its inner hinterland contributed a smaller number of textile machinery patents, and the patentees' efforts divided about equally between cotton-spinning machines and power looms, whereas the patentees in the Boston region focused more on power looms (see tables 2.1 and 2.2). That differentiation reflected the greater relative importance of cotton-spinning mills and of hand-loom weaving in the Philadelphia region. This community of practitioners might have engaged in collective invention because these machinists retained many ties with one another. Furthermore, numerous long-standing business relations existed between the metropolis and its region through which technical knowledge could be channeled. The clustering of these patents within a few years (1809–14) suggests that this effort was not happenstance, yet this community of practice did not act autonomously from those in New England. Following a training and employment stint with Slater, Jenks arrived in Philadelphia in 1810, quickly establishing himself as a leading textile machinery manufacturer and probably playing a role in know-how trading of ideas about equipment innovations.[13]

Like with steam-related patenting, rich farming areas distant from the metropolises housed textile machinery inventors (see tables 2.1 and 2.2). The most prominent of these areas consisted of New York City's inner hinterland of the Hudson Valley and bordering areas and its outer hinterland in central New York state, places that maintained ties to Providence and vicinity through the migration of Slater-trained mechanics. In the Hudson Valley inventors in Greene County and in Albany took out spinning machine patents, and inventors in the hill towns of Canaan, Kent, and Ridgefield in Connecticut and of Stockbridge in Massachusetts contributed patents in power loom weaving. The numerous patents in cotton spinning and in power loom machinery from central New York, however, dwarfed the efforts of inventors in the Hudson Valley area. With one exception—P. Bennet in Rochester—this patenting followed the arrival in Utica of Benjamin Walcott Sr. in 1808 and of his son, Benjamin Jr., who arrived the next year and stayed to become one of the nation's leading textile industrialists. The son's machine shops must have been magnets for young apprentices and experienced machinists from throughout central New York. Besides leaving to establish cotton mills, these machinists and their acquaintances certainly spurred the surge in textile machinery patenting. This effort coincided with the rise in steam-related patenting in the same subregion, suggesting that this rich agricultural area housed a large community of practice of machinists (see table 1.1). Its sectoral subcommunities—steam engines and textile machines—engaged in extensive know-how trading and, possibly, collective invention, thus epitomizing the gains from cluster knowledge.[14]

TECHNICAL SKILLS EMBEDDED IN NETWORKS

Because the supply of top textile machinists fell significantly below the demand for their expertise, they had many opportunities to job-hop among firms locally and over long distances. Typically, they received lucrative financial packages (salary and sometimes ownership stakes) to encourage them to switch employers. Nonetheless, hub machinists such as Slater or the Wilkinson family members, as well as firms that employed other leading machinists, had difficulty exerting structural constraint over other mechanics or firms; thus, many companies had access to the best machinists. The high mobility of mechanics and the multi-hub nature of their networks facilitated the spread of technical skills among firms and subregions. Training experiences and job mobility networks contributed to the formation of powerful communities of practitioners which were organized by subregions and which interlinked across them to form larger communities. Consequently, textile machinists' networks became conduits for tacitly sharing component knowledge of machines and even architectural knowledge of systems of machines.

Textile machinists joined forces with their peers in the burgeoning metalworking shops (blacksmith shops, forges, and small foundries) in the same areas and shared solutions to machining problems. This interchange deepened, particularly, in the early nests of textile machinery building in Providence and vicinity, Philadelphia, and Boston's environs. Thus, textile machinery firms served as a second tendril of the pivotal producer durables. Because some textile machines required the construction of many identical metal components, this equipment posed technical challenges to produce large numbers of parts and machines cheaply. In contrast, steam engine manufacturing involved many fewer identical components. Consequently, textile machinery served as a stronger root of the machine tool industry than steam engines did.[15] Federal armories and private firearms firms constituted the third tendril of the pivotal producer durables and materially influenced the machine tool industry.

CHAPTER THREE

The Federal Armories and Private Firearms Firms Operate in Open Networks

> Mills for the making of gun barrels have been erected; gun locks, and every other article in a gun, have been made in the best manner, and of the most substantial kind. The workmen, the execution, the machinery, and the demand, have all progressed apace with each other.
>
> *Subscribers, Gun Manufacturers, Borough of Lancaster, Pennsylvania, "Memorial to the House of Representatives of the United States"*

The executive and congressional branches of the new national government retained fresh memories of the difficulties the Revolutionary Army had experienced acquiring adequate supplies of pistols and muskets from the states' public armories and gunsmith shops. After the war the government continued purchasing firearms from gunsmiths, but the supply remained dangerously low. Consequently, in 1794 Congress passed an act establishing national armories, and that same year President George Washington settled on sites for armories at Springfield, Massachusetts, and Harpers Ferry, Virginia. The Springfield Armory began firearms production in 1795, and construction of the Harpers Ferry Armory started four years later.

ACHIEVING A "SYSTEM OF UNIFORMITY"

During the formative years of the federal armories and private firearms firms, the United States Army embraced a revolutionary philosophy of arms manufacture which originated in the French military. In 1765 General Gribeauval began to advocate the "system of uniformity"—a comprehensive program to reform the organization and production of armaments. It aimed to standardize equipment and parts and to develop new methods of manufacturing based on model arms, improved techniques, and rigid inspection procedures. Gribeauval promoted the efforts of Honoré Blanc, a talented armorer who designed labor-intensive procedures to manufacture muskets with uniform parts using the innovative techniques of die forging, jig filing, and hollow milling. As ambassador to France, Thomas Jefferson witnessed Blanc's methods in 1785 and promoted the procedures in correspondence home over the next several years.

Another channel for the system of uniformity to enter the United States followed Major Tousard, a French artillerist and military engineer who was steeped in the Gribeauval-Blanc approach. In 1793 he moved to the United States and, two years later, joined the new Corps of Artillerists and Engineers. Drawing on research originally commissioned by President Washington, he published a work in 1809 which stressed the importance of developing a system of uniformity and regularity which became a standard text for American officers. This system thus entered the American political and military arena under the patronage of influential leaders. The goal of achieving interchangeable parts, especially those so perfectly interchangeable that soldiers could repair guns under battlefield conditions, never formed the core of the program; these repairs remained unrealistic. The manufacture of parts that perfectly interchanged among a large number of firearms either required a labor-intensive, prohibitively costly effort that could never supply a sufficient number of firearms for an army or required precision machine tools that would not appear until the late nineteenth century. Instead, the army followed a practical, though still challenging, approach, establishing a system of uniformity—a broad program aimed at improving the organization and production of armaments.[1]

The federal government and state militia needed many more firearms than skilled gunsmiths could manufacture using craft techniques, and these arms cost too much; instead, federal and private armories supplied the arms. They reorganized production, such as employing a division of labor and paying wages based on the quantity finished; mechanized manufacturing, such as adding waterpowered machinery and devising new equipment; and developed measuring devices (gauges) and tools (fixed jigs) to produce uniform, but not perfectly interchangeable, components. Between the late 1790s and 1820 the federal armories and private firms mostly completed the transformation of firearms manufacturing from craftwork into an industrial system. The achievement of substantial mechanization and uniformity followed 1820. During the time span from the 1794 act to establish national armories until the 1815 act to reorganize the Ordnance Department under the War Department, federal armories and private firms for the most part followed separate development paths, even though both expanded under the stimulus of federal appropriations. Paradoxically, after 1815 the Springfield Armory became integrated into an innovative network of mechanics and firms which served as the seedbed of machinery and machine tool firms, whereas Harpers Ferry Armory, even under the later leadership of a brilliant mechanic, remained peripheral to these networks.

THE FEDERAL ARMORIES

The Springfield Armory

David Ames, the first superintendent of the Springfield Armory, and Robert Orr, the first master armorer, had trained as gunsmiths. Consequently, they hired gunsmiths and organized a traditional apprenticeship program, but this approach did not raise productivity and merely substituted one large gunsmith shop for several small ones. In 1799 the armory reached full-scale production and completed about forty-six hundred muskets; Ames failed to raise production from that level and left three years later (fig. 3.1). During Henry Morgan's three-year tenure as superintendent, he introduced a fourfold division of labor consisting of barrel makers and forge men, filers, woodstockers and assemblers, and grinders and polishers; changed management procedures; and started reducing the apprenticeship program. These efforts caused a sharp drop in the unit cost of manufacturing muskets (1802-05), compared to traditional craft organization (1798–1801), but they failed to raise production volume even though the number employed stayed about the same (figs. 3.1 and 3.2). Productivity, measured by the number of muskets produced per worker, fell, and this poor performance led to Morgan's dismissal (fig. 3.3). The increased division of labor failed to raise productivity, and falling unit costs probably resulted from a relative shift in the labor force to lower-wage, lesser-skilled workers who handled the simpler tasks created by the division of labor.

The next superintendent, Benjamin Prescott, served from 1805 until 1813, when the War Department fired him. He eliminated most remaining vestiges of the craft tradition and finished the armory's transformation into a disciplined factory organization. The piece rate system of paying wages, which he introduced, placed a premium on routinized production of parts. By 1810 about one-third of the workers completed tasks that way, a level maintained for the next five years. The real wages of armory workers rose steadily from 1805 to 1811, whereas wages of other industrial workers such as in the Brandywine region of Delaware (mostly employees of DuPont, the gunpowder manufacturer) stagnated and the nation's unskilled nonagricultural laborers experienced volatile wages (fig. 3.4). To regulate costs and productivity, Prescott started a comprehensive record-keeping system organized by workshop. The early stages of a subdivision of work by operation (i.e., grinding, boring, and polishing) rather than by

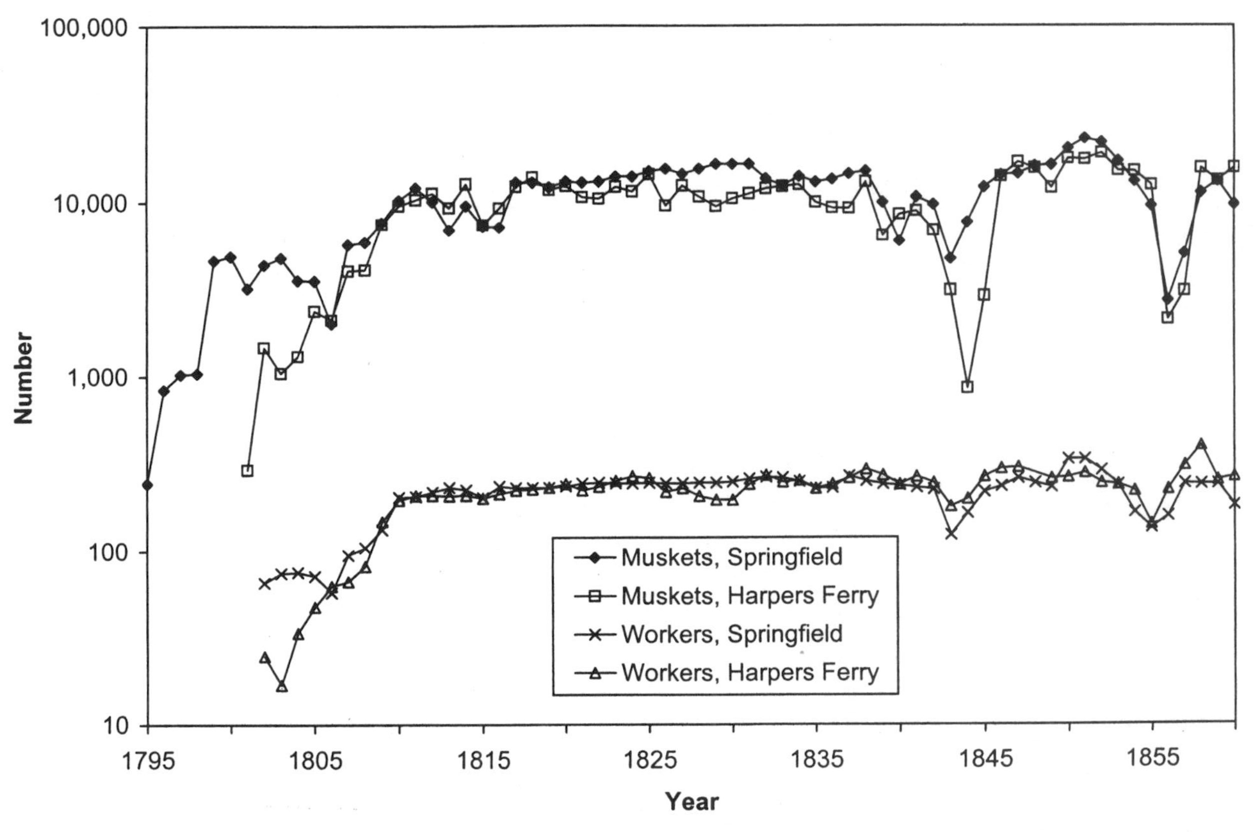

Figure 3.1. Number of Muskets Finished and Number of Production Workers at the Springfield and Harpers Ferry Armories, 1795–1860. *Sources:* Felicia J. Deyrup, "Arms Makers of the Connecticut Valley: A Regional Study of the Economic Development of the Small Arms Industry, 1798–1870," *Smith College Studies in History* 33 (1948): 233, app. B, table 2; 245, app. D, table 4; Merritt R. Smith, *Harpers Ferry Armory and the New Technology: The Challenge of Change* (Ithaca, N.Y.: Cornell University Press, 1977), 342–45, table 1.

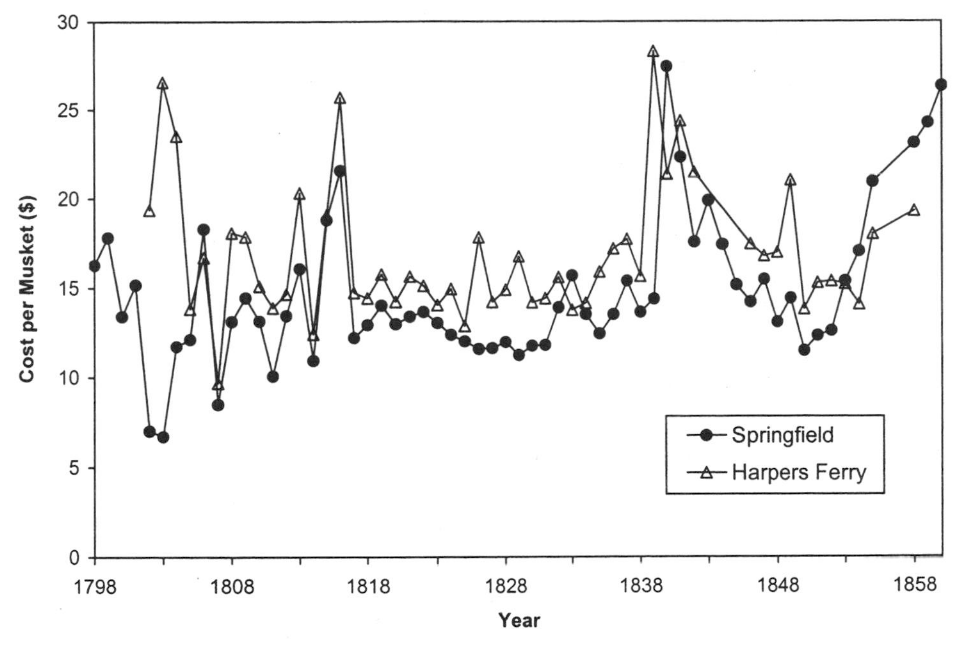

Figure 3.2. Cost per Musket at the Springfield and Harpers Ferry Armories, 1798–1860. *Sources:* Felicia J. Deyrup, "Arms Makers of the Connecticut Valley: A Regional Study of the Economic Development of the Small Arms Industry, 1798–1870," *Smith College Studies in History* 33 (1948): 229–32, app. B, table 1; Merritt R. Smith, *Harpers Ferry Armory and the New Technology: The Challenge of Change* (Ithaca, N.Y.: Cornell University Press, 1977), 342–45, table 1.

"limb," or the whole part (i.e., barrel, lock [firing mechanism], and wooden stock), appeared by 1806 and was implemented as eleven occupational specialties; by 1815 the number had risen to thirty-four. The shift to process manufacturing further undermined the craft tradition, preparing the way for greater machinery operations.

The introduction of machinery proceeded slowly, albeit punctuated by flurries of activity. Handwork, especially filing, dominated labor time in the armory, as it would throughout the antebellum. Before 1800 machinery chiefly consisted of trip hammers, several lathes, and, possibly, power-driven grindstones and polishing disks. An 1810 survey of techniques for manufacturing a musket conveyed the level of mechanization achieved following the demise of the gunsmith craft tradition. The 204 workers, who were distributed across twenty specialties with a master armorer supervising them, completed 10,300 muskets. Furnaces and forges in the Salisbury district of northwestern Connecticut supplied wrought-iron bars, rods, and flat barrel skelps. Workers used a trip hammer to mold barrel skelps around a mandrel into tubes; barrel welders then closed the tubes using hand-powered hammers. They shaped the inside

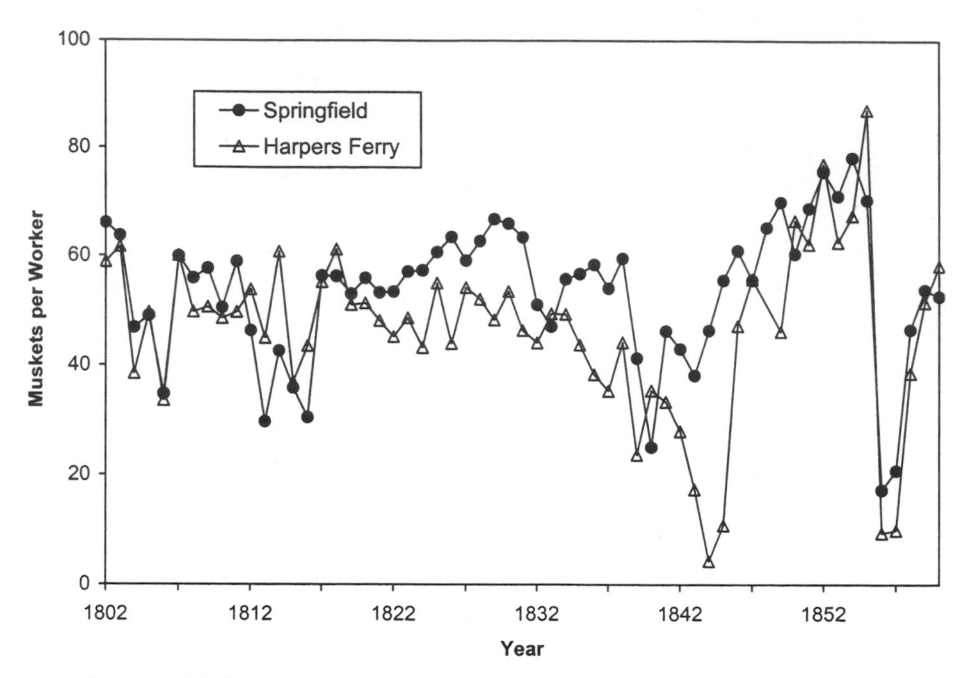

Figure 3.3. Muskets per Worker at the Springfield and Harpers Ferry Armories, 1802–1860. *Sources:* Felicia J. Deyrup, "Arms Makers of the Connecticut Valley: A Regional Study of the Economic Development of the Small Arms Industry, 1798–1870," *Smith College Studies in History* 33 (1948): 233, app. B, table 2; 245, app. D, table 4; Merritt R. Smith, *Harpers Ferry Armory and the New Technology: The Challenge of Change* (Ithaca, N.Y.: Cornell University Press, 1977), 342–45, table 1.

of barrels with a powered boring machine and used powered grinders and polishers to finish the outsides of barrels. Smiths wielded sledges to hot-forge bayonets and the twenty-nine iron parts for locks and mountings. Filers smoothed and shaped parts and hand-punched holes for screws, and they worked on all of the lock parts so that they fit into a working lock mechanism. Stockers shaped walnut gunstock blanks by hand, using chisels, gouges, and sandpaper. The hollow miller that formed the arbor, pivot, and sides on the tumbler represented the only power-driven machinery used for the lock mechanism. The Springfield Armory probably was one of the earliest adopters of it, along with Eli Whitney's armory and several other private ones. Workers made little or no use of gauges to measure parts and of filing jigs to hold them; visual inspection and hand testing sufficed for checking the muskets' quality.

When Prescott, who held the superintendent's position temporarily again in 1815, left Springfield for the final time, he could claim substantial accomplishments. He had introduced a piece rate system, initiated a comprehensive record-keeping system

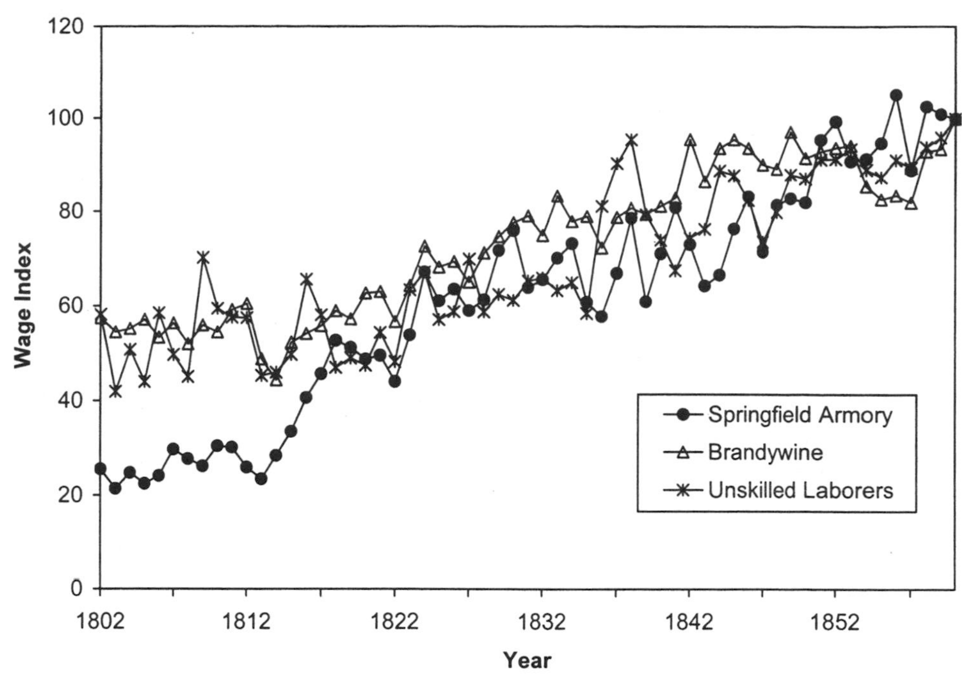

Figure 3.4. Real Wages of Springfield Armory Workers, Manufacturing Workers in the Brandywine Region of Delaware, and Unskilled Nonagricultural Laborers in the United States, 1802–1860. *Sources:* Donald R. Adams Jr., "The Standard of Living during American Industrialization: Evidence from the Brandywine Region, 1800–1860," *Journal of Economic History* 42 (December 1982): 904–5, table 1; Paul A. David and Peter Solar, "A Bicentenary Contribution to the History of the Cost of Living in America," *Research in Economic History* 2 (1977): 16–17, 59–60, tables 1, B.1; Felicia J. Deyrup, "Arms Makers of the Connecticut Valley: A Regional Study of the Economic Development of the Small Arms Industry, 1798–1870," *Smith College Studies in History* 33 (1948): 241–43, app. D, table 3.

to regulate costs and productivity, expanded occupational specialization, started subdividing work by operation rather than part, and modestly increased mechanization. The number of muskets finished annually soared from just over three thousand in 1805 to over ten thousand by 1815 (see fig. 3.1). Nevertheless, the unit cost of muskets exhibited an unmistakable uptrend, and productivity, measured by the number of muskets per worker, stabilized and then fell (see figs. 3.2 and 3.3). Still, the Springfield Armory had become an industrial enterprise. It made large numbers of firearms without skilled gunsmiths, compared to the small output, though higher quality, of the gunsmith shop.[2] In contrast, its sister federal armory at Harpers Ferry made less progress toward the industrial form and still retained important vestiges of the gunsmith tradition.

The Harpers Ferry Armory

Powerful patrons backed Harpers Ferry Armory. As president, Washington had chosen the site at the confluence of the Potomac and Shenandoah rivers because it lay near the new Federal City yet was far enough inland to be secure against foreign invaders. Tobias Lear and George Gilpin, merchants based, respectively, at Georgetown and Alexandria on the Potomac River, capitalized on their close friendship with Washington to ensure that Harpers Ferry came to fruition. They expected that the armory would stimulate development of the Potomac basin, and their mercantile houses would profit from importing supplies and exporting the basin's products. During 1794–97 they participated in purchasing land for the armory, giving them the opportunity to speculate on property nearby.

The isolated setting of Harpers Ferry appeared to pose future problems. Frederick and Hagerstown, the nearest towns of moderate size, did not have large mercantile houses to provide supplies. Armorers' tools and other specialized goods needed to be purchased in distant eastern seaboard cities, and bar and pig iron for guns had to be transported about one hundred miles from furnaces and forges in central Pennsylvania; often these supplies came overland by wagon. The rural culture of the armory's environs seemed to undermine its development as a disciplined factory. The Virginia slave plantation gentry dominated the economy and the politics, and the large number of small freeholders with forty acres to one hundred acres of land, and perhaps a few slaves, occupied the lower socioeconomic ranks. The rural calendar ruled social life, and the population presumably abhorred factory discipline.

This interpretation of the impact of isolation and rural culture on the future success of the Harpers Ferry Armory, however, may overstate the case somewhat. The distance to eastern seaboard cities did not present insurmountable challenges. The mercantile houses of Lear in Georgetown or Gilpin in Alexandria, only forty-five miles away, could handle orders for supplies, and this distance did not greatly exceed the thirty miles separating the Springfield Armory from Hartford, Connecticut, wholesale houses. Most goods that Harpers Ferry purchased, including bar and pig iron, had high value relative to their weight, which made wagon transportation feasible. Straight-line overland distances of the armory from Philadelphia (150 miles) and Baltimore (65 miles) compared favorably to the distances of the Springfield Armory from New York City (110 miles) and Boston (90 miles). Cheap water transportation made the cost of circuitous routes from coastal cities to each armory similar for high-value goods. The hundred miles separating central Pennsylvania furnaces and forges and Harpers Ferry did not drastically exceed the forty-five miles from the furnaces and forges around Salisbury, Connecticut, to the Springfield Armory. They supplied

Springfield until 1816, when central Pennsylvania furnaces and forges, about three hundred miles from Springfield, began competing with Salisbury firms as suppliers of bar and pig iron.

The rural culture may not have severely retarded the shift of Harpers Ferry to disciplined factory production. Between 1790 and 1810 the household-based economy in the Connecticut Valley north of the Springfield Armory presumably operated with an ethic of reciprocal local exchange built on obligations to cooperate. This area did not have a marketable agricultural staple for export and thus could not be dominated by a landed elite. Nevertheless, this rural culture bore remarkable similarity to the area around Harpers Ferry at the same time.

The economic structures of the armories' rural environs diverged, impacting allocations of capital investment. Within a fifty-mile radius of Springfield increasing agricultural wealth and urban development generated demands for rural and urban craft products and for small workshop manufactures. Capital investment flowed from prosperous farmers, professionals (e.g., physicians and lawyers), and business people (e.g., retailers and millers) to these manufactures. In contrast, a rural economy built on slave plantation agriculture, as existed in parts of Virginia and Maryland near Harpers Ferry, posed different investment incentives. Plantation owners controlled the slaves' demands for manufactures, and their farms either internalized production or imported cheap goods designed for slave markets, thus undercutting local manufacturers. Prosperous farmers, professionals, and other business people invested their capital in slaves rather than in fixed infrastructure (such as towns and transportation) or in manufactures. Consequently, Harpers Ferry Armory could not embed itself in an extensive network of mechanics, craft shops, and small factory workshops which facilitated development of the industrial technology of arms manufacturing, whereas the Springfield Armory sat foursquare in such a network.

The establishment of Harpers Ferry Armory proceeded slowly. From 1794 to 1801 it had weak managers and experienced delays and problems covering all facets of the project, including land acquisition and construction of buildings, machinery, and a canal to carry water to power machinery. Before 1805 the armory finished far fewer muskets than Springfield, but they had similar rates of production per worker (see figs. 3.1 and 3.3). The elegant "short" rifle (Model 1803) epitomized the Pennsylvania gunsmiths' traditions that influenced production at Harpers Ferry until 1816. Most of the early gunsmiths hailed from Philadelphia, and subsequent workers came from other areas in southeastern Pennsylvania and nearby parts of Maryland. Superintendent Joseph Perkins instituted a division of labor which followed traditional European craft methods. Individual artisans worked on a limb of the gun, such as the barrel, lock, or wooden stock, and the master armorer coordinated output so appropriate numbers of each part were ready at the right time. This craft system rested on the gunsmiths' manual dexterity with tools rather than on the use of machines for forging, grinding, or boring.

In 1807 James Stubblefield became superintendent, and the following year the War Department responded to the war fears of the Embargo and ordered him to increase production to fifteen thousand muskets annually, a level it was unable to achieve before 1822. He traveled to the Springfield Armory and Whitney's factory in New Haven to examine machinery and the organization of production. As a result of this trip, Stubblefield commenced a reorganization of Harpers Ferry in 1809. Borrowing from Prescott's innovations at Springfield, he introduced a piece rate accounting system and expanded occupational divisions from the six instituted in 1807 to about twenty specialties, similar to Springfield's, in 1810. Within six years he had created fifty-five occupational titles, compared to the thirty-four at Springfield. An individual artisan making the wooden stock, however, still completed all processes from cutting to boring to oiling. A dramatic rise in appropriations fueled the early stages of Harpers Ferry's transformation. From 1807 to 1810 annual expenditures more than tripled to $145,000, the number of buildings doubled to twelve, and the number of armorers increased from sixty-seven to almost two hundred.

Nevertheless, the introduction of piece rate accounting and greater occupational specialization did not eliminate all vestiges of the craft tradition as they had at the Springfield Armory. A noncontractual training program gradually replaced the formal apprenticeship system, yet armorers continued to hire their sons and train them. This cohesive employment network isolated Harpers Ferry armorers from know-how trading with other communities of practice of firearms machinists. Enhanced occupational specialization simplified production procedures, thus compensating for shortages of skilled labor; but between 1811 and 1816 Stubblefield added few mechanical innovations to shift the production emphasis from limbs (whole parts) to processes. Instead, his changes represented a graft onto the trunk of the gunsmith's craft tradition; he had no enthusiasm for challenging it. From 1809 to 1820 this hybrid system allowed Harpers Ferry to keep up with Springfield in total volume of production using approximately the same number of workers, which meant productivity for both armories remained similar (see figs. 3.1 and 3.3). Compared to Springfield, however, Harpers Ferry consistently had higher costs per musket and lower quality during the 1802–20 period (see fig. 3.2). Still, the quality of its firearms compared favorably with those from private armories.[3]

The Springfield Armory demonstrated that an industrial system of musket production surpassed a large-scale craft approach. Neither armory produced muskets equal in quality to those made by a skilled gunsmith, but that worker turned out few firearms at a high cost. By 1815 federal armories had not achieved interchangeable manufacture. They did not possess tools and equipment to mechanize production of interchangeable parts, and laborious hand filing to achieve interchangeability remained too costly. Private armories faced similar struggles as federal armories building industrial enterprises from a base of craft traditions in gunsmithing, yet they could not rely on the government to appropriate funds to cover cost overruns. Their decisions about organizational structure, division of labor, and manufacturing technology had to meet the final test of profitability; therefore, they followed a different industrial path.

THE FEDERAL CONTRACT SYSTEM

Politicians and the military recognized that federal armories possessed insufficient capacity to meet the need for arms. Although the War Department sporadically awarded contracts to gunsmith shops, they could not make enough firearms. In 1798, under the threat of war with France, Congress appropriated eight hundred thousand dollars to purchase cannons, firearms, and ammunition. That same year the War Department formalized a contract system, and, for several years, awarded contracts. This approach atrophied until the renewal of war threats during the Embargo, and in 1808 Congress passed a law appropriating two hundred thousand dollars annually to arm and equip the state militia. The War Department awarded five-year contracts to eighteen firms and individuals to deliver a total of eighty-two thousand muskets.

The federal contract system initiated in 1798 contained the seeds of a major policy decision; to ensure a supply of firearms, private firms would become quasi-public institutions. Because they could not readily shift to other manufacturing, they required subsidies to allow them to remain in business. The War Department would provide this support in the form of advances on contracts which permitted contractors to establish factories; sometimes ongoing payments were made. The Connecticut contractors Eli Whitney of New Haven and Simeon North of Middletown received the first advances in 1798 and 1799, but other contractors do not seem to have received them. Before 1820 the amount of capital needed to set up a private armory probably ranged between five thousand and fifteen thousand dollars.

Most contractors came from the ranks of gunsmiths and, secondarily, blacksmiths because they possessed the requisite skills, but the capital needed exceeded their financial resources. The shrewdest investors—merchant capitalists in New York, Boston, Philadelphia, and Baltimore and in subregional metropolises such as Providence, Hartford, and Albany—shunned arms manufacturing. They made a prescient decision; only one private armory that started before 1830 in New England, the leading center of arms making, survived the Civil War. Typically, government-mandated prices of contract firearms fell below actual production costs at federal armories. In this artificial public economy contractors confronted difficulties estimating and bargaining for adequate reimbursement to cover depreciation charges for their plants and equipment. The contract system required factories to follow a model arm, and government inspectors had to certify that firearms matched it and met standards established at federal armories. Firearms failing inspection were rejected, thus raising the costs for private firms.[4]

NEW ENGLAND FIRMS

Private armories would confront these difficulties; nevertheless, problems remained in the background during the early flush of enthusiasm to obtain contracts under the congressional appropriations of 1798 and of annual ones commencing in 1808. Before 1815 Pennsylvania gunsmiths located in Lancaster County, near Philadelphia, ranked among the leaders outside New England in acquiring contracts, but they held steadfastly to craft traditions. In most contracts the number of firearms ranged from twenty-five to two hundred; however, six musket makers received contracts that ranged between about one thousand and four thousand muskets.

New England firms led the shift from gunsmith shop to firearms factory, and they accounted for ten of the eighteen firms to win five-year contracts under the 1808 appropriation bill. Before 1820 prominent New England factory owners included, in Rhode Island, Stephen Jenks (operated 1770–1814) in North Providence and Pawtucket; in Massachusetts, Lemuel Pomeroy (1790–1845) in Northampton, then in Pittsfield, and Asa Waters and Company (1798–1841) in Millbury; and, in Connecticut, Nathan Starr Jr. and Sr. (1798–1845) and Simeon North (1799–1852) in Middletown and the Whitney Armory (1798–1888) in New Haven. With the exception of Jenks, these contractors, along with the Johnson family (1820–54) in Middletown, remained among the handful of contractors on which the government relied in the early 1820s. Most of the New England firms disappeared by mid-century, yet a community of practice of entrepreneurs and mechanics emerged, establishing firms that, according to every industrial measure—employment, capital, and the value of the product—dominated firearms manufacturing into the twentieth century. New England firms accounted for 33 percent of the industry in 1850, over 50 percent in 1860, and 60 percent or more from 1870 to 1940.

The concentration of successful armories in or near the Connecticut Valley (Pomeroy, North, Starr, and Whitney) and in the Blackstone Valley (Jenks and Waters) was not accidental. Their close proximity to the federal armory at Springfield—all firms were within a sixty-five-mile radius—suggests that contact with it helped guarantee success. With the exception of the Whitney armory, however, these contacts did not flower until after 1815, and every private firm had begun by 1799, too early for the federal armory to influence them directly.

From the late 1790s to 1820 the private firearms firms shared conditions that supported their growth. They originated in rich agricultural valleys with farmers, professionals, and business people who indirectly provided capital to their armories by endorsing notes for loans and posting bonds for federal arms contracts. Whitney, the New Haven manufacturer, for example, tapped ten local members of the social and political elite, mostly Yale graduates, to post bonds, and, similarly, North relied on wealthy residents of Middletown. The same farming areas also housed numerous metalworking shops—nail-cutting firms, forges, and small foundries—which served farms and mills. Private armories, therefore, operated in communities of practice consisting of firms with extensive metalworking experience whose family, friendship, and business ties bound firms in the same sector and across them. Owners and top mechanics used their networks to share tacit knowledge about building metal parts and equipment and to share architectural knowledge about organizing equipment in a production process. They experimented with a division of labor in shops and small factories and with a mechanization of production through hand-powered equipment or waterpowered machinery.

The private armories also could learn from the small firms in the Connecticut Valley which increased production of metal consumer goods, such as tinware and hardware, for sale beyond their shops' immediate vicinity. Similarly, investors and mechanics in this valley, as well as in the Blackstone Valley, started manufacturing cotton yarn for nonlocal sale, and their mills housed machine shops. These firms challenged the craft ethos by focusing on increasing total production for sale in larger market areas.[5]

New England armories drew on these skills and ideas to surmount limitations of the craft traditions and to meet the War Department's increasingly stringent demands for a system of uniformity in manufacturing firearms. Producers of arms for private consumers faced one avenue to success: compete with other firms to make guns that consumers would choose to purchase. When the War Department contracted to buy firearms, however, the manufacturer confronted alternative avenues to success: try to win the contract through social and political influence, the route Whitney chose, or through merit, which North chose.

Eli Whitney

As the son of a moderately prosperous farmer in Westborough, Massachusetts, Whitney worked in his father's shop and learned basic mechanical skills, such as wood turning on a lathe and nail forging. He prepared for college by studying with two Yale alumni, one of whom was a close friend of Ezra Stiles, the school's president. Before starting at Yale, Whitney spent a month studying mathematics with Elizur Goodrich, a professor at the school. During his college years in New Haven, from 1789 to

1792, Whitney inserted himself into Yale's network of social elite and graduated at the age of twenty-seven with a cadre of faculty, friends, and alumni contacts who aided his career.

Nonetheless, he started raggedly. A six-month visit to a Georgia plantation managed by Yale graduate Phineas Miller provided the setting for inventing the cotton gin, and during 1793–94 Whitney corresponded with Secretary of State Thomas Jefferson to acquire a patent. That effort succeeded after Professor Goodrich provided a letter of introduction for him to meet with Comptroller of the Treasury Oliver Wolcott of Connecticut, who was a Yale alumnus. Secretary of State Edmund Randolph granted Whitney the patent for the cotton gin in 1794. Over the next four years Whitney and Miller expended large sums of money, but they failed to build suitable cotton gins, market them, or protect the patent.

In 1798, facing potential bankruptcy, Whitney penned a letter to his new friend, Secretary of the Treasury Wolcott, offering to undertake a government contract to use waterpowered machinery to produce between ten thousand and fifteen thousand stands of arms. Each stand included a musket, bayonet, ramrod, and flint, plus selected parts to repair it. Whitney and Miller had failed to manufacture cotton gins successfully, a simple machine that paled next to the complex musket; thus, this firearms proposal stretched credulity. The contract Whitney received in 1798 called for delivery of ten thousand muskets two years later, and a crucial paragraph differentiated the contract from others awarded that year. It stipulated that he would receive an advance of five thousand dollars, plus an equivalent amount as soon as he demonstrated minimal effort in preparing for production. This easy ten thousand dollars would keep him from ruin and might have lured him into arms manufacturing.

Subsequent events suggest that Whitney possessed vague ideas about manufacturing firearms. He made efforts to contract with other firms to make parts of muskets, intending to assemble firearms in his factory, but that approach revealed his misunderstanding of the principles developed by Blanc to achieve a system of uniformity. Blanc's principles required that the manufacturer closely monitor all production stages, whereas subcontracting made such supervision difficult. After signing the contract, Whitney traveled to the Springfield Armory to meet with the superintendent and to observe production methods, and, in a letter to Wolcott in 1799, he admitted trying to bribe federal armorers to work in his New Haven factory. After eighteen months he still had not finished building equipment for the factory. Shortly before the contract's expiration in 1800, he had not delivered a single musket, even after receiving thirty thousand dollars in advances from compliant federal officials.

Whitney's legendary demonstration of interchangeable musket locks before Washington officials in 1801 therefore has questionable features. The factory had recently commenced production, and it completed only five hundred muskets that year. The lock, the most difficult mechanism to manufacture, could not have been assembled through random selection of lock components during the demonstration, unless Whitney assigned workers to hand-file over twenty parts of each lock, laborious work. Machine manufacture of interchangeable lock parts, with minimal filing, did not succeed until the late 1840s. Alternatively, if he assembled completed lock mechanisms in the muskets while implying that lock components also interchanged, he had carried out a clever ruse.

By 1803 Whitney finally reduced his efforts to sell cotton gins and devoted serious attention to arms manufacturing, yet he seldom produced more than fifteen hundred muskets annually, a number that paled next to the output of federal armories (see fig. 3.1). He continued making firearms for state militia and received a federal contract for fifteen thousand stands of arms in 1812, completing delivery eight years later. Although he contributed design improvements to existing machines, jigs, and fixtures, he did not design any new machines. His factory's equipment resembled the Springfield Armory's, which had few machines before 1815. He proselytized for the principle of interchangeability, but his factory never achieved it. Musket parts, especially those in the lock, required extensive filing to make them fit properly, and they could not be interchanged with parts in other guns.

Nonetheless, Whitney contributed to the transformation of firearms production from craft shops to factories. His architectural knowledge of firearms manufacturing as a production system covered improved machinery design, the proper layout of equipment in the factory, and turning out the correct volume of each part to maintain the pace of production. Federal armories sought his advice on improving their organization of production. His access to the social and political elite allowed him to proselytize for the principle of interchangeability, which helped keep government officials focused on it. In this way he helped the army make progress toward its more realistic goal of having a system of uniformity, which required sweeping improvements in the organization and production of armaments.[6]

Simeon North

By the late 1790s towns within thirty miles of Berlin were poised to become a hotbed of manufacturing metal goods (such as tinware, hardware, and castings) in forge shops and small foundries, which would lead to a vigorous community of practice of metalworkers. In 1795 the thirty-year-old North purchased a farm with a mill and waterpower rights in Berlin, Connecticut; besides sawmilling, he started manufacturing scythes and other farm implements. Precisely when and how North acquired metalworking skills remains obscure, though he might have learned the rudiments of the gunsmith craft from Elias Beckley, a neighbor in that trade. Gunsmiths such as Oliver Bidwell (Hartford) and Elisha Buell (Marlborough) operated in the vicinity, and some gunsmiths, including Abijah Peck (Hartford), Nathan Starr (Hartford and Middletown), and Amos Stillman (Farmington), joined the rush to obtain arms contracts under the appropriation bill of 1798.

Federal Contracts

North's firearms career started in 1799 with a federal contract to manufacture five hundred horse pistols, and he quickly demonstrated the talent to produce high-quality guns in timely fashion, in contrast to Whitney's struggles to make acceptable muskets and meet contract deadlines. North probably completed the contract by 1800, and that year he received a second one for fifteen hundred pistols to be delivered within two years. The War Department certified his stature by advancing him two thousand dollars, possibly making him the second contractor, along with Whitney, to receive that funding. Over the next few years North probably obtained additional contracts, and his revenue supported construction of a two-story factory, with a basement housing a forge shop, sometime before 1808.

The North contract of 1808 to produce two thousand ship-boarding pistols contained an advance of four thousand dollars. This contract signified that a private firearms factory participated in the effort to improve the efficiency of arms manufacturing. After four months of production North articulated key points in a letter to Robert Smith, secretary of the navy: "I find that by confining a workman to one particular limb of the pistol untill [*sic*] he has made two thousand, I save at least one quarter of his labour . . . . I have seventeen thousand screws & other parts of the pistols now forged & many parts nearly finished." This application of the division of labor and batch production to firearms manufacturing united that industry with others in Connecticut. Knowledge exchanges through subregional social networks must have been a prime mechanism for the transfer of technical skills. When North wrote his letter to Secretary Smith, Eli Terry's innovations to expand clock production using similar methods had been under way for several years in nearby Plymouth. North's ties to talented clockmakers must have been conduits for making him aware of Terry's approaches. Sometime before 1791 Terry had apprenticed with Timothy Cheney, a leading clock maker in nearby Manchester. His nephew, Elisha Cheney, a prominent clock maker in Berlin, where North had his factory, was his brother-in-law.

Because skilled mechanics moved up and down the Connecticut Valley, North also knew about the piece rate system of paying wages and about the enlargement of occupational specialization which Prescott had instituted at the Springfield Armory. North's batch production approach, however, exceeded the limits of Springfield's piece rate system. That system maintained a balance in the volume of parts made by each worker, whereas North's workers probably continued to produce parts until they had finished the contract's entire volume. In 1810 the War Department rewarded North with an extension of his contract for another one thousand pistols.

The War Department's 1813 contract with North to produce twenty thousand pistols to be delivered over a period of five years became the focal point for serious consideration of interchangeability under the rubric of the system of uniformity. North agreed to the contract stipulation that "component parts of pistols, are to correspond so exactly that any limb or part of one Pistol, may be fitted to any other Pistol of the Twenty thousand." The contract wording failed to specify whether or not additional filing might be necessary to permit the exchange of parts. No filing implied perfect interchangeability, but this extreme approach during wartime (War of 1812), without any proven technical foundation, struck army officers, federal armory superintendents, and contractors as quixotic. Even additional filing to exchange parts presented a difficult challenge because no amount of filing could make parts fit if they deviated too much from a pattern. Normally, North did not make foolish claims. His acceptance of the contract's terms must have reflected a reasoned judgment that technology existed to raise the standards of uniformity, even if he could not achieve perfect interchangeability.

North's twenty thousand dollar advance on the 1813 contract, guaranteed upon posting a bond, provided the wherewithal to plunge into the project. He purchased fifty acres of land in Middletown, near Berlin, to obtain more waterpower, and he built a three-story factory that housed over fifty workers. Eventually, the fully equipped factory represented an investment of one hundred thousand dollars, equal to the largest textile factories in New England at that time. Reuben, Simeon's son, managed the old Berlin factory as the forge shop for the firm. But the work did not proceed smoothly. New and improved machinery and methods may have caused delivery delays, and the army became dissatisfied with the model pistol it had given North to duplicate. In 1816 it requested that he submit a proposal for the new model. The revised agreement stipulated that locks interchange, and the absence of reference to its components implies that North and the War Department had lowered their sights. An additional advance of twenty-five thousand dollars helped finance the changes, and employment rose to seventy workers prior to the contract's completion in 1819. The degree of interchangeability which North achieved under the 1813 contract remains uncertain. The lock mechanism posed daunting challenges, and, in the absence of sophisticated metal-cutting and grinding machines, it required laborious hand filing. Nevertheless, by 1816 his contributions to firearms manufacturing and to machine tool technology included the introduction of gauging and of special-purpose machinery to make components.[7]

Gauging

Before approving the revised pistol contract, Col. Decius Wadsworth—head of the Ordnance Department, which had authority to inspect, supply, and repair firearms—

sent the federal armory superintendents, James Stubblefield (Harpers Ferry) and Roswell Lee (Springfield), to inspect North's two factories. What they saw impressed them: "he has made an improvement in the lock by fitting every part to the same lock which insures a more rigid uniformity than they have heretofore known." North may have acquired ideas for gauging from Terry and other clock makers who had developed that process for wooden clock manufacture. They used gauges to produce batches of similar parts, but they did not require exact precision because wooden clock parts fit together easier than metal parts. North's ties to the Cheney family, who knew Terry and other clock makers, probably provided this information.

North may have been the first to make a conceptual leap from simpler principles of gauging used in wooden clock production to the complexity of gauging in firearms manufacturing. The worker used a gauge with dimensions appropriate to the part in the model firearm to make a part such that it could replace the part in the model. Following this procedure, parts would interchange with those in the model and with one another, or at least do so with only modest filing. Previously, armories had used crude measuring devices to check the dimensions of parts to supplement the trained eyes of workers. They would make parts similar to those in the model firearm, such as parts of the lock, and then file parts to fit together, producing a unique lock. Lee, impressed by North's innovation of gauging, introduced it at the Springfield Armory in 1817, and the elaboration of the gauging principle became fundamental to the later achievement of interchangeable manufacture.

Special-Purpose Machinery

Before 1813 the federal armories and most private firms employed traditional forging, boring, grinding, and filing equipment to make parts. These differed little from those used in gunsmith shops, except that factories made greater use of waterpowered machinery. North added three types of special-purpose machinery: stocking machines, milling machines, and a barrel-turning lathe. On the inspection trip to North's firm in 1816, Stubblefield and Lee noted: "Also his mode of stocking they consider an improvement, inasmuch as it gives more accuracy and uniformity, and by the help of water machinery facilitates the work and lessens the expense, and may advantageously be adopted and brought into uniformity with arms manufactured at the public armories." The federal armories still did all the work on gunstocks by hand; they did not add machinery until the 1820s.

Wooden clock makers gave North insights into the advantages of employing waterpowered machinery for woodworking. His brother-in-law, Cheney, used such machinery, and the news of Terry's success in lowering production costs had reverberated throughout nearby towns. On August 22, 1814, just two years before the inspection trip by Lee and Stubblefield, six individuals from the Bristol-Plymouth area (less than twenty miles from Middletown) received, in affiliation with Terry, six patents for special-purpose machinery to make wooden clocks. The mechanic networks would have informed North about the dramatic events transpiring at Terry's clock factory in Plymouth.

The milling machine may not have been ready for use in March 1816, when Stubblefield and Lee inspected North's works. They made no comment about it, but most evidence suggests that his factory made innovations to broaden its use that year. A table, which fed into a rotary multiple-toothed cutter, held the metal workpiece. The milling machine increased the speed and accuracy of removing excess metal from a part to bring it closer to final dimensions, thus reducing the amount of filing. This effort posed difficult challenges, including creating cutting teeth that did not break, maintaining sharp teeth, removing pieces of cut metal, and controlling the feed of the workpiece. Solving these problems occupied machinists for many decades.

The exact sources of North's milling innovation remain obscure; nevertheless, by the time he introduced it, metalworking technology in Massachusetts, Connecticut, and Rhode Island had reached new heights. Since the mid-1790s inventive activity had surged, and the supply of machinists in forge shops and small foundries had expanded. Communities of practitioners engaged in know-how trading through networks of workshop and factory owners, and peripatetic mechanics carried technical advances to their new jobs. Early productivity gains from milling machines, however, remained too small to offer advantages over filing. Similar to wood-stocking machinery, federal armories lagged in introducing milling machines; Harpers Ferry added them during the 1820s, and Springfield did so the next decade.

In 1816 North added a third special-purpose machine, the barrel-turning lathe, to his factory. Its screw-driven carriage held the musket barrel and turned while the cutting tool removed metal from the barrel's surface. This replaced the dirty, dangerous practice of grinding the barrel's surface. North's barrel-turning lathe was an adaptation of the lathe with a screw-driven carriage which David Wilkinson, the Pawtucket, Rhode Island, machinist, had built from 1794 to 1796. The fame of the Wilkinson machine shop and of the machinists it trained ensured that knowledge about the lathe circulated widely through machinist networks.[8]

ASSESSING PROGRESS ON THE SYSTEM OF UNIFORMITY

From the mid-1790s to 1815 private firearms factories such as those of North and Whitney and the federal armories at Springfield and Harpers Ferry made progress on achieving a system of uniformity; production volume rose, and quality improved somewhat. Any gains came mainly from the increased division of labor, while maintaining many hand processes, but this approach offered diminished opportunities for more advances in productivity and quality. Hand motion speed had finite limits, and greater exchange of parts among workers raised the time spent handling components, apart from working on them. Creating a cadre of skilled gunsmiths might generate additional improvements in quality, yet that approach was discredited because not enough gunsmiths could be trained to keep pace with the demand for firearms at reasonable prices. The War of 1812 highlighted the defects of the existing approach to firearms manufacturing. The federal and private armories could not supply enough guns to meet sharply rising demand, and they turned out too many faulty weapons that failed under wartime conditions.

The achievement of the system of uniformity remained a distant dream in 1815. Even if federal and private armories improved uniformity and quality, maintaining their separate development paths would hinder progress. Although the War Department supplied model patterns, the methods and technology had so little precision that firearms differed among producers. The occasional exchange of visits among federal armory superintendents and factory owners and the War Department's supervisory role over federal armories and its monitoring of private firms proved inadequate to standardize production and raise quality.

NETWORKS OF FIREARMS PRODUCERS

A rich resource existed to transform firearms manufacturing: long-standing social and business networks bound friends and acquaintances among the War Department, federal armories, and private firms. Left to their own devices, these individuals passed along knowledge leading to gradual improvements in firearms production, but the pace delayed solutions to many of the daunting technological challenges in woodworking and metalworking. In order to generate major advances in machinery, exchanges of technical knowledge needed to accelerate and attention be focused on important problems. An organizational catalyst that encouraged know-how trading within and among communities of practitioners stepped into this void.

The Organizational Catalyst

In response to the inadequacies of military armaments during the War of 1812, Congress passed "An Act for the Better Regulation of the Ordnance Department" in 1815. This legislation gave the department clearer, expanded authority to manage arms contracts, run the national armories, and devise regulations to achieve uniformity in armaments. Committed army officers, principally Col. Decius Wadsworth and Lt. Col. George Bomford, relentlessly followed the act's aims to completion. Wadsworth brought formidable background to his position as head of the Ordnance Department.

As a Yale graduate, he tapped that college's influential alumni, including Whitney, with whom Wadsworth had developed a long association as a professional and friend. This dated from at least 1799, when Wadsworth began service as a military inspector to prove barrels and inspect muskets. He often interceded for Whitney with the War Department. Wadsworth served in the Corps of Artillerists and Engineers and worked closely with Tousard, one of the key transmitters to the army of the French Gribeauval-Blanc philosophy and methodology for achieving a system of uniformity.

After passage of the Ordnance Act, Wadsworth quickly appointed Col. Roswell Lee to replace Benjamin Prescott as superintendent of the Springfield Armory. Lee possessed long-standing ties to leading figures who developed the system of uniformity. In 1803 Whitney hired Lee to serve as a mechanic, and, following Whitney's recommendation, Wadsworth brought Lee into the Ordnance Department. In June 1815 Wadsworth organized a meeting in New Haven to set an agenda to advance the system of uniformity. Those attending constituted the inner social network of leaders dedicated to this effort: Lee and Stubblefield, respectively, superintendents of the Springfield and of the Harpers Ferry armories; Whitney, host of the meeting and owner of a private armory; Prescott, former superintendent at Springfield; and Wadsworth. Whitney and Stubblefield had developed cordial relations after Whitney was offered the superintendent's position at Harpers Ferry in 1807 and declined it, paving the way for Stubblefield; he visited Whitney's New Haven factory the following year. These leaders maintained their personal and professional ties through visits and wide-ranging correspondence covering "business, shop talk, office rumors, and personal matters." Because each of them constituted a hub of their own social network and bridged to other networks of mechanics, they reached many machinists along the East Coast. This extensive network possessed extraordinary capacity to exchange knowledge about organizational and technological innovations, thus supporting swift advances in the system of uniformity.

Participants in the New Haven meeting agreed on a phased plan, first, to make muskets uniform and then to extend the principle to sidearms. Wadsworth selected Harpers Ferry to make the first model muskets, and, after testing, these muskets, along with pattern rifles, would become products of the federal armories. If the effort achieved success, private contractors would join the program. Wadsworth chose Bomford to supervise the program from the Ordnance Department. He possessed impressive credentials—superb engineer, West Point graduate, and member of the upper crust of Washington society. That social position came courtesy of his wife, who was from a leading political family active in Pennsylvania and at the federal level. After Wadsworth's death in 1821, Bomford became chief of the Ordnance Department and held that position until 1842. The combination of his skills and long tenure at the pivot of power made him a prominent figure in the project of achieving the system of uniformity.

Ordnance Department officers exerted bureaucratic control over operations within each armory and coordinated relations between them. Starting in 1816, each armory had to submit quarterly returns detailing: receipts and expenditures for buildings, raw materials, and equipment; work performed; inventories; and wage accounts by each worker and type of job. The department focused on production quotas for a guaranteed market and appropriations determined revenue. It had no incentive to insist on comprehensive measures of production costs such as accurate charges for interest, insurance, and depreciation but simply coordinated the allocation of production levels and type of weapons between the armories. Bomford expected Lee to provide leadership and to take the initiative in sharing knowledge with Stubblefield. The exchanges would help upgrade Harpers Ferry, because it lagged Springfield in musket quality and efficiency. Lee grasped the chance to improve communications with Stubblefield; the two men became friends, exchanged visits, and discussed their full range of operations. At Lee's instigation these exchanges included mechanics, machinery, and raw materials.

Proper inspection of the quality and uniformity of completed arms went to the heart of the system of uniformity, because this implied that production processes contained inspection methods to check the precision of parts. The evolution of these methods, especially the gauging system, proceeded slowly because they required concurrent improvements in machinery in order to make precise parts. The variation inherent in hand processes raised difficult problems of matching parts to gauge measurements. Machines with built-in consistency for making parts reduced the inaccuracies of handwork, and handwork, especially filing, focused on the final stage of making parts fit the gauge. These advances in gauging and machinery innovation took time. The broader aim of achieving uniformity, especially improvements in quality, stayed at the forefront, whereas the goal of achieving perfect interchangeability became an ideal to realize later.

The Ordnance Department implemented astounding policies that spurred rapid technological change in firearms production, machinery, and machine tools. It ordered federal armories to allow private, as well as government, visitors to observe manufacturing freely, make drawings of machinery and the layout of shops, borrow patterns for machinery, and acquire any other knowledge they wished. Federal establishments were encouraged to exchange mechanics not only with one another but also with private armories. The department capitalized on its power to award contracts to private firms, and it created an implicit understanding that contractors share innovations with federal armories on a royalty-free basis. These policies established a free-flowing network of knowledge exchange among federal and private armories within a community of practice, thus speeding up the adoption of technological innovations. The extensive networks around each of the hub participants in the New

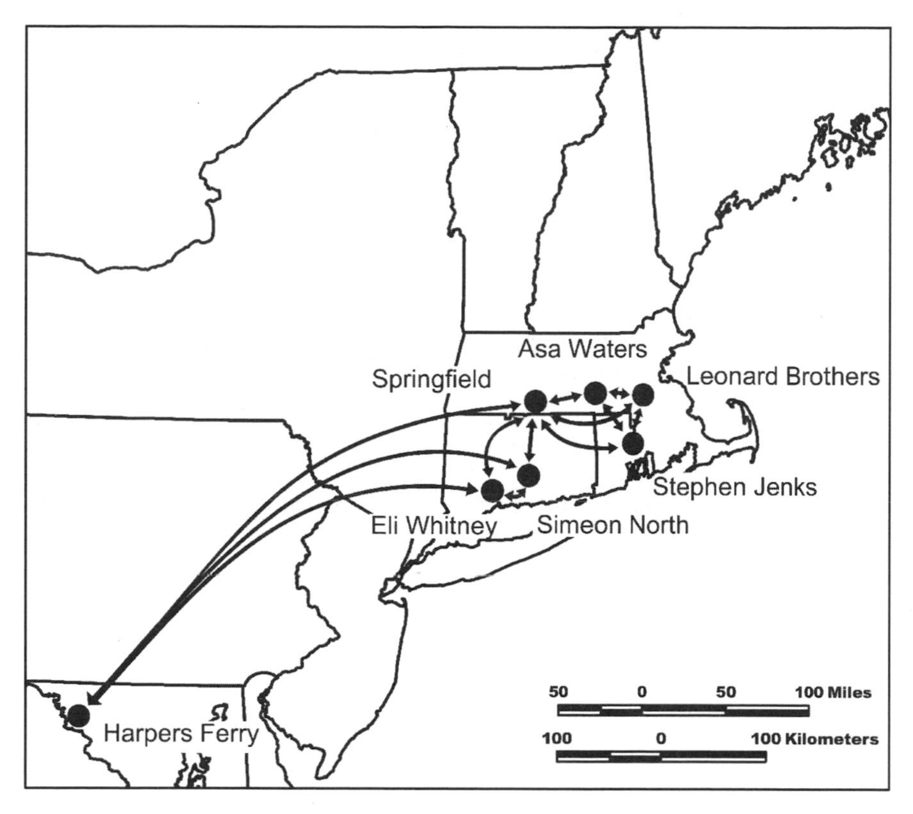

Map 3.1. Networks of Firearms Armories, 1810–1820

Haven meeting of 1815 prevented anyone from exerting structural constraint on the exchanges of knowledge and on the job mobility of machinists (map 3.1).[9]

The Springfield Armory Clearinghouse

Lee grasped the opportunity to make the Springfield Armory a leader in achieving a system of uniformity. Between 1815 and 1820 he established the armory as a clearinghouse to collect and distribute the technical knowledge and innovations which, since the 1790s, had been emanating from the vigorous metalworking and machinery shops and the private armories in New England. The armories and their machinists had motivation to exchange knowledge. They dealt with a complex, difficult-to-machine lock; trouble welding barrels; and labor-intensive, costly methods to shape gunstocks. Lee transformed the armory from being a passive observer to being an active participant in the networks of knowledge exchange. It originated few innovations in the critical technologies of gauging, forging, and machining; its efforts remained mostly derivative, elaborating on borrowings.

Springfield's efforts from 1815 to 1822 demonstrate that progress in interchangeable manufacture fell lower on the Ordnance Department's agenda of advancing the system of uniformity. By 1815 innovations in filing jigs, hollow milling of screws, and drilling machines had improved the uniformity of lock mechanisms, the most complex limbs of firearms. Springfield, along with other firearms factories, added more power-driven tools for working on lock mechanisms, such as a machine to strike metal components on a die, heavier hollow milling machines, drilling machines, and a clamp machine to mill cock pins. Nevertheless, these tools, mostly in place by 1821, did not represent major technological advances. Instead, Lee directed Springfield's efforts to solving problems of making barrels, the musket's most expensive limb. Workers wasted large amounts of raw materials while making them, and many were defective. He also focused on developing inspection gauges to ensure that firearms matched model muskets more closely.

The Community of Practice of Barrel Making

Following 1806 the productivity of barrel welders, measured by the number of barrels welded annually, had stagnated then fell during the War of 1812 (fig. 3.5). Improvements in barrel welding and turning which Lee introduced at the Springfield Armory between 1815 and 1819 had originated in private armories and machine shops, and these advances generated substantial gains in productivity and quality and significantly reduced costs. Waters, the well-known firearms contractor in Millbury, Massachusetts, innovated the trip-hammer welding technique in 1808. Within a year the firearms contractors Jenks (in Pawtucket, Rhode Island, thirty miles southeast) and the Leonard brothers (in Canton, Massachusetts, thirty miles east) adopted it. The Springfield Armory, under Superintendent Prescott, did not adopt this machinery at the time. Within three months of becoming superintendent, however, Lee installed a waterpowered trip hammer to weld musket barrels. The trial proved so successful that he added three more the following year; two years after that he had ten in operation. Compared to hand welding, the new method produced a stronger seam, cut labor time in half, and eliminated one of the two assistants, thus significantly reducing costs. Quick, dramatic gains from Lee's adoption of trip-hammer welding demonstrated that Prescott's accomplishments had left Springfield in a laggard state compared to private armories, which had adopted that machinery much earlier.

The adoption process of trip-hammer barrel welding operated through the networks that bound the hub machinists in the private firms (see map 3.1). Waters, the technique's innovator, had his gun factory in Millbury. As early as 1793, this thriving industrial center of small workshops and factories contained over forty firms that

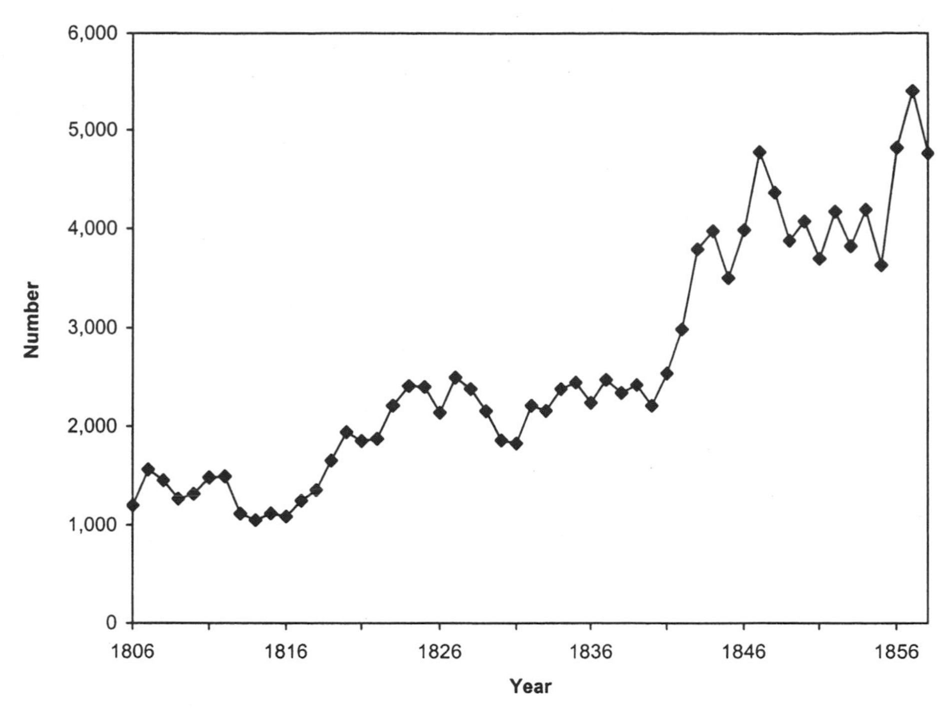

Figure 3.5. Average Number of Barrels Welded Annually per Welder at the Springfield Armory, 1806–1858. *Source:* Felicia J. Deyrup, "Arms Makers of the Connecticut Valley: A Regional Study of the Economic Development of the Small Arms Industry, 1798–1870," *Smith College Studies in History* 33 (1948): 247, app. D, table 5.

used waterpowered machinery. One diffusion path from Millbury followed mechanic networks in and near the Blackstone Valley to the firearms factory of Jenks in Pawtucket. By 1809 the Blackstone Manufacturing Company and Almy, Brown, and Slaters, among the nation's greatest textile factories and which contained major machine shops and mechanics, operated in this valley. Jenks's Pawtucket operations sat at the hearth of textile machine shops and cotton mills, and, along with his firearms factory and ironworking business, he had invested in the nearby Smithfield Cotton Manufacturing Company, established in 1805. He had convenient access, therefore, to metalworking innovations through local contacts with mechanics in cotton mills and machine shops.

In Canton the Leonard brothers had several possible sources of knowledge about Waters's trip-hammer welding machine. As firearms contractors, they participated in the network of similar firms receiving knowledge via federal inspectors who traveled among contractors. Their knowledge about Waters's machine also could have come from other industrial firms clustered in Canton. The Revere Copper Company, based in Boston, opened a rolling mill in town around 1801, and this local mill had iron foundry connections and merchant ties to Boston, which kept it informed about business developments in the hinterland town of Millbury. By 1809 Canton also housed two cotton mills, whose owners included Richard Wheatley, a Boston merchant, and James Beaumont, a well-to-do English immigrant with textile experience. These factories accessed knowledge about metalworking innovations through Wheatley's mercantile networks. Furthermore, Beaumont had ordered machinery from Paterson, New Jersey's Clark, and he in turn had ties to Providence-area textile mechanics; Clark had spent time there before 1808 as well as in that year. Thus, the Leonard brothers also could have learned about Waters's trip-hammer welding machine from mechanics in Canton's cotton mills, who possessed ties to textile machinists in the Providence area; they in turn participated in networks with Jenks. Consequently, Lee's quick adoption of the trip-hammer innovation at the Springfield Armory, following his appointment, reflected acceptance of machinery already well-known among leading metalworkers, and he had intimate acquaintance with Waters's gun factory through the visits of federal musket inspectors.

The adoption process of the barrel-turning lathe also coursed through networks that linked hub machinists, but in this case Lee acted as a hub. The barrel-turning lathe that North introduced in 1816 at his Middletown pistol factory rested on the principle of the screw-driven carriage that David Wilkinson had developed in his family's Pawtucket machine shop around 1794–96. Between 1816 and 1819 mechanics transformed the concept of the barrel-turning lathe into four other versions, each with its own improvement. Lee knew about North's lathe, either from his inspection with Stubblefield of North's factory in 1816 or from regular contacts between North and Lee. Sylvester Nash, a mechanic at the Springfield Armory, developed a barrel-turning lathe shortly after North introduced one, and Bomford (at the Ordnance Department) told Lee to send Nash to Harpers Ferry to build a lathe. Between his arrival there in 1816 and his return to Springfield in 1818, Nash made an improved lathe and patented it. His work at Harpers Ferry underscores the significance of the artifact-activity couple as a mechanism for the transfer of technical skills to build a machine, in this case interregionally.

In 1817 the Ordnance Department instructed Lee to try out the barrel-turning lathe designed by Daniel Dana and Anthony Olney in Canton, where the Leonard brothers had their armory. Knowledge about the lathes of North (Middletown) and Nash (the Springfield version) probably reached Canton through the Wilkinson firm in Pawtucket because the Dana-Olney lathe used a "roller engine" designed by Wilkinson. Other sources in Canton—the Revere copper works, mechanics in cotton mills, or the firearms factory of the Leonard brothers—could have supplied knowledge about the lathe. Lee ordered the same roller engine from Wilkinson that Dana and Olney used, and Springfield mechanics, under the supervision of master armorer Adonijah Foot, perfected an improved version in 1818 which combined ideas from Dana-Olney, Nash, and Wilkinson. Not to be outdone, Millbury's Waters built his version of the barrel-turning lathe by the spring of 1818 and patented it by December. He may have heard about avenues to pursue from Lee, with whom he corresponded regularly, or from the armory inspectors, who visited him; from Wilkinson through machinist ties in the Blackstone Valley; and from Dana and Olney and their neighbors, the Leonard brothers in Canton, who adopted the trip-hammer barrel welder of Waters in 1809.

Waters's lathe, as well as the other versions, however, could not cut the complete barrel; the mechanic had to finish the flat and the oval end at the breach by hand. In 1818 Waters hired Thomas Blanchard, a local mechanic in Millbury, to solve this problem, and he adapted an English innovation of the cam motion principle for turning irregular surfaces. News about Blanchard's innovation quickly reached Lee, probably through visits by federal arms inspectors. He invited Blanchard to Springfield to build his barrel-turning lathe as well as a machine for draw-grinding barrels. In July 1818 he finished the work, and that same month, with support of the Ordnance Department, Lee sent Blanchard to Harpers Ferry to build a lathe and a draw-grinding machine. Within three months he built an improved lathe that turned the entire barrel, but neither federal armory permanently adopted it. Springfield used the improved Dana-Olney lathe and the original Blanchard lathe that turned only the breach end of the barrel.

The adoption processes of the trip-hammer welding machine and the barrel-turning lathe suggest that a macro community of practice linked machinists and firms from southwest of Boston to Rhode Island and west to the Connecticut Valley in Massachusetts (Springfield area) and in Connecticut. Within this macro community smaller communities of practice were aligned by industrial sectors, including firearms, textile machinery, and iron foundry and machine shops. Networks of machinists and firms provided the communication media for technical skills and innovations. Multiple hub individuals of these networks acted as nodes of smaller networks, bridging to other hubs, thus linking the communities of the different industrial sectors. No hub machinist monopolized knowledge, and network ties among mechanics and firms were not heavily redundant. This situation maximized exchanges of diverse component knowledge about trip-hammer welding and barrel-turning lathes.

Sometime in 1817 or 1818 Lee turned the attention of the Springfield Armory to improving inspection procedures for muskets, and he probably acquired the idea for gauges on his inspection tour of North's factory in 1816. Lee's master armorer, Foot, along with several other workers, built on North's contribution and developed gauges for use during production and afterward for final inspection. By 1819 they had devised gauges for as many as twenty-seven parts, most of them probably used for final inspection rather than during manufacturing. Although the Springfield Armory may have led gauging efforts among firearms producers by 1819, the key advances came later.

Private armories such as those owned by North and Waters seem to have surpassed the federal armories in metalworking technology by 1815. Over the next five years Lee borrowed existing technologies from private firms and relied on them for innovative ones. To carry this out, he drew on wide-ranging networks of machinists and firms to stay abreast of developments. By hiring skilled machinists such as Foot as well as by inviting talented mechanics such as Blanchard to come to the armory to build machinery, he made considerable progress incorporating this technology. His approach underscores the critical role of machinists' mobility among shops as a mechanism for transferring technical skills and component knowledge of machines.

The use of greater numbers of special-purpose machines to improve quality and productivity impacted work organization. The jump in the number of occupational specialties from thirty-four to eighty-six between 1815 and 1820 constituted the greatest absolute and relative five-year increase in specialization over the time span from 1806 to 1860. The emphasis on incorporating new barrel-making technologies, such as trip-hammer barrel welding and barrel-turning machines, resulted in sixteen barrel-making tasks by 1820. The greater division of labor allowed Lee to expand the piece rate system of paying wages. By 1815 the share of workers paid this way was 31 percent, and it surged to 82 percent five years later, staying above 70 percent for the next forty years. Thus, within five years of becoming superintendent Lee had expunged any remaining semblance of the gunsmith craft system. The real wages of the workers soared at this time, suggesting that they benefited handsomely from his changes (see fig. 3.4).

The results of Lee's early efforts to advance the system of uniformity demonstrate that the goals of improving firearms quality and of achieving broad dimensional uniformity of arms ranked higher than goals such as achieving greater productivity and lower costs, which private firearms firms considered more important. The introduction of trip-hammer barrel welding in 1815, for example, produced an increase in the average number of barrels welded per worker from about eleven hundred to almost fourteen hundred during the 1816–18 period. Not until a year later, when the average reached almost seventeen hundred barrels, did welders who used machines exceed the highest number done with hand methods. Achieving sustained productivity improvement outside of experimental conditions, while maintaining or improving quality, took time. Springfield's overall measures of productivity, the cost per musket and the number of muskets produced per worker, revealed no significant gains during Lee's early years (see figs. 3.2 and 3.3). By 1820, however, he could boast that the muskets had higher quality and greater uniformity. The addition of waterpowered machinery permitted Lee to allocate more effort to hand filing in order to achieve these improvements. Nevertheless, the share of workers engaged in machine work stayed around 10 percent and remained at that level until 1850, when the armory reached 20 percent.

The Harpers Ferry Armory Lags

Between 1815 and 1820 the small number of machinists and metalworking firms within about fifty miles of Harpers Ferry constituted a meager community of practice, hindering the armory's progress in achieving the goals of the system of uniformity. The wide contacts of Marine Wickham, master armorer at Harpers Ferry from 1808 to 1810, remained atypical. In 1811 he relocated 150 miles northeast to Philadelphia, where he served as inspector of contract arms and troubleshooter for the federal armories. Within five years he commenced a career, lasting until 1834, as an arms manufacturer and wholesaler of arms supplies, during which he maintained contact with Harpers Ferry and Springfield. The two ironworks upriver on the Potomac and the Foxall-Mason foundry downriver in Georgetown constituted the only firms of any consequence near Harpers Ferry which used machinery.

The Ordnance Department implicitly recognized the problem of Harpers Ferry's isolation from machinist networks when Bomford assigned Lee the task of encouraging Superintendent Stubblefield to make improvements. Over the years Lee sent as many as twenty-five workers to Harpers Ferry, but few stayed permanently; in contrast, the reverse flow to Springfield totaled no more than three workers from 1816 to 1829. Lee also temporarily loaned machinists, millwrights, and pattern makers to Stubblefield. On the surface he favored cooperation, albeit a mostly one-way flow of technology from Lee; nevertheless, Stubblefield lacked the fortitude to implement improvements effectively.

The laggard status of Harpers Ferry appeared in many guises. After Wickham's departure for Philadelphia, Stubblefield did not have a senior machinist. James Greer, the best of the remaining machinists, failed to keep abreast of mechanical developments, and Stubblefield possessed little grasp of machining. On a visit to the Springfield Armory in 1816 the superiority of trip-hammer barrel welding impressed him, and he boldly announced plans to install four machines at Harpers Ferry, the same number as at Springfield. The following year Stubblefield obtained the help of David Parsons, a Springfield mechanic, to build trip hammers for barrel welding. Yet none were finished until at least 1834, and hand methods did not cease until 1840; this adoption pace lagged Springfield by almost twenty-five years. The adoption of the barrel-turning lathe at Harpers Ferry proceeded better. Stubblefield acquired the lathe that Nash built on his visit between 1817 and 1818, and by 1821 he added the early version of the Blanchard lathe for cutting the breach end of the barrel. Nonetheless, he did not replicate the mechanical synthesis at Springfield, which incorporated ideas from

Dana-Olney, Wilkinson, and Nash. That would have required participation in the know-how trading among hub machinists in different communities of practice. Harpers Ferry machinists possessed few network ties other than those with Springfield.

The retarded specialization of Harpers Ferry reflected an inability to incorporate mechanization effectively into the production process. In 1816 its occupational specialties numbered fifty-five, whereas Springfield had thirty-four the previous year, but by 1820 its total soared to eighty-six, while Harpers Ferry had increased by only five. It fell behind in using gauging to advance uniformity and did not adopt the technique until 1823, over five years after Springfield had started using gauges. Harpers Ferry had higher costs per musket and a lower, as well as declining, number of muskets finished per worker (see figs. 3.2 and 3.3).

During the five years following Lee's appointment, the Springfield Armory rapidly caught up with private armories in mechanization, and he made the armory a clearinghouse for technological innovations in New England. In contrast, Stubblefield made a halfhearted challenge to craft-based gunsmith traditions by adopting some of the machinery that Lee informed him about. The quality and uniformity of Harpers Ferry's firearms remained poor, and it fell farther behind Springfield in advancing the system of uniformity. Stubblefield struck a deal with his workers which persisted from 1815 through much of the next decade. Employees gained lenient work practices, maintained craft traditions, and received high wages. In return, Stubblefield, along with several families who constituted the inner circle of privilege—called the "Junto"—retained control over the armory and over politics, social life, land, and commerce in the village of Harpers Ferry. At the Ordnance Department Wadsworth and Bomford recognized Harpers Ferry's problems, but they lacked on-site influence, short of removing Stubblefield and installing a new superintendent to challenge the established order. In 1825 Bomford commenced a determined effort to improve the armory, but it took four more years to acquire political support in Washington to force Stubblefield's resignation.

Stubblefield and the armory's workers resisted change so long because their views coincided with the local craft ethos in their environs. In Massachusetts, Connecticut, and Rhode Island, however, numerous firms aimed to sell their products outside the local economy. They broke with craft traditions and formed communities of practice which developed new ways to manufacture goods in workshops and factories. Gunsmiths at the Springfield Armory witnessed these challenges to craft traditions throughout the Connecticut Valley and elsewhere in southern New England; their resistance crumbled shortly after 1800. Consequently, the armory became an integral network participant in transforming firearms production from craft workshops to industrial enterprises.[10]

THE NEXT STEP

Between 1790 and 1820 the pivotal producer durables sector emerged out of broad demands of growing agricultural and industrial economies. The social and political elite, entrepreneurs, and machinists created wide-ranging social networks that served as channels of know-how trading within and among key communities of practice, especially iron foundries, textile machinery firms, and federal and private armories. They used similar processes in metalworking (turning, cutting, grinding, boring, milling, and polishing) and faced equivalent problems of power transmission (gearing and belting), control and feed mechanisms, and metallurgy (strength and heat). This technological convergence continued after 1820. The heavy capital equipment sectors (iron foundries, steam engines, and locomotives), textile machinery, and firearms accelerated the pace of technological change, and solutions reverberated throughout these industrial sectors. The machine tool industry emerged as one institutional response to these changes. The vibrant know-how trading within and among communities of practice and the machinists' networks which reached throughout the prosperous farming areas and urban-industrial centers in the East spurred these technological advances.[11]

PART TWO

The Elaboration of the Networks, 1820–1860

CHAPTER FOUR

Iron Foundries Rule the Heavy Capital Equipment Industry

> The [Novelty Works] forms, in fact, a regularly organized community, having, like any state or kingdom, its gradations of rank, its established usages, its written laws, its police, its finance, its records, its rewards, and its penalties. The operation of the principles of system, and of the requirements of law, leads, in such a community as this, to many very curious and striking results.
>
> *Jacob Abbott, "The Novelty Works," Harper's New Monthly Magazine*

The national economy expanded vigorously after 1820. Real gross domestic product grew at a compound annual rate of about 4 percent, accelerating from just below that figure in the 1820s to 4 percent during the next two decades to 5 percent in the 1850s. The average person's well-being increased as real gross domestic product per capita rose a little over 1 percent annually. Real value added in manufacturing gained somewhat over 9 percent annually during 1839–49 and almost 6 percent annually during the following ten-year period. This growth spurred the heavy capital equipment sector because it served broad demands in the economy. The sharp upward trajectory of Pennsylvania anthracite coal production and of shipments on the leading coal canals after 1820 implies that the heavy capital equipment industry, which increasingly relied on anthracite for fuel, readily met its enlarged fuel requirements (fig. 4.1). By 1840 many eastern households and firms had access to anthracite. The subsequent retardation in the growth rate of anthracite production and of shipments on the leading coal canals, even as economic growth accelerated and manufacturing production

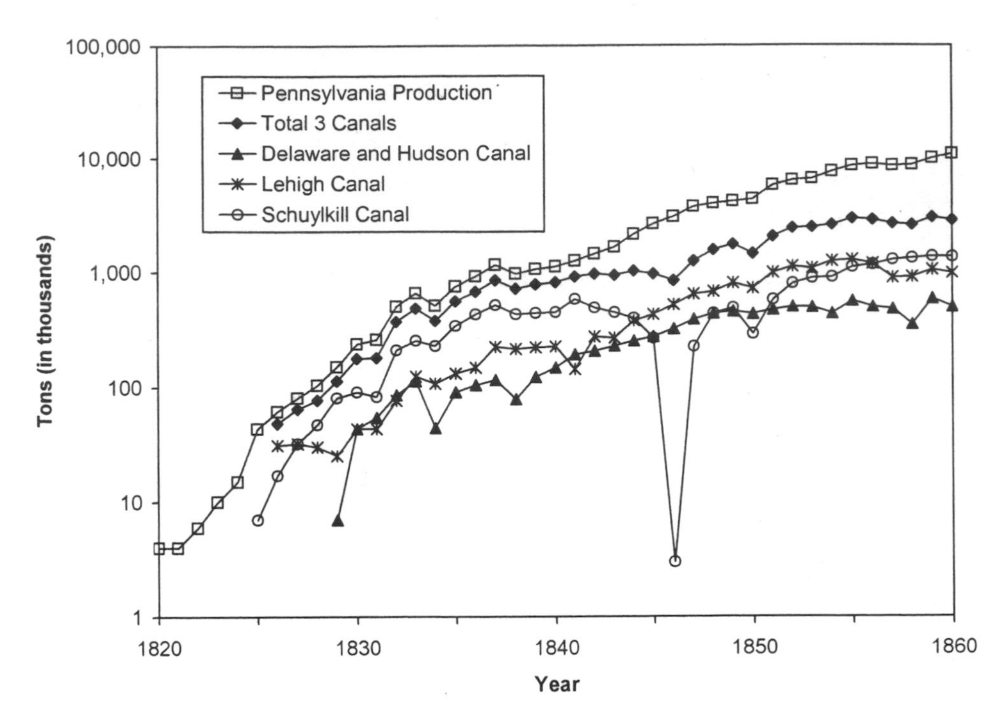

Figure 4.1. Pennsylvania Anthracite Coal Production and Shipments on the Leading Coal Canals, 1820–1860. *Sources:* Chester L. Jones, *The Economic History of the Anthracite-Tidewater Canals,* University of Pennsylvania Series on Political Economy and Public Law, no. 22 (Philadelphia: John C. Winston, 1908), 28–29, 35–36, 86, 155–56; U.S. Bureau of the Census, *Historical Statistics of the United States, Colonial Times to 1970, Bicentennial Edition,* 2 pts. (Washington, D.C.: Government Printing Office, 1975), ser. M123.

surged, indirectly suggests that the continued expansion of the capital equipment industry was not hindered by supply constraints on the availability of anthracite.

Heavy capital equipment firms, principally iron foundries, produced steam engines. The number of stationary engines used in the nation's manufacturing soared from forty-three in 1820 to twenty-six thousand in 1860 (fig. 4.2). By the 1850s the nation and most regions began to see a slowdown in the rate of growth in the number of engines. The United States produced most of its engines, and the regional data indicate where the engines were used, not their manufacturing sites. From a market perspective the Middle Atlantic and the South dominated in using steam engines before 1840. New England caught up with the South around that date, and they tracked each other's growth over the next twenty years, then the South passed New England as a market. The Middle Atlantic market continued to exceed that of New England after 1840. The number of engines in the Midwest soared after 1840, matching the

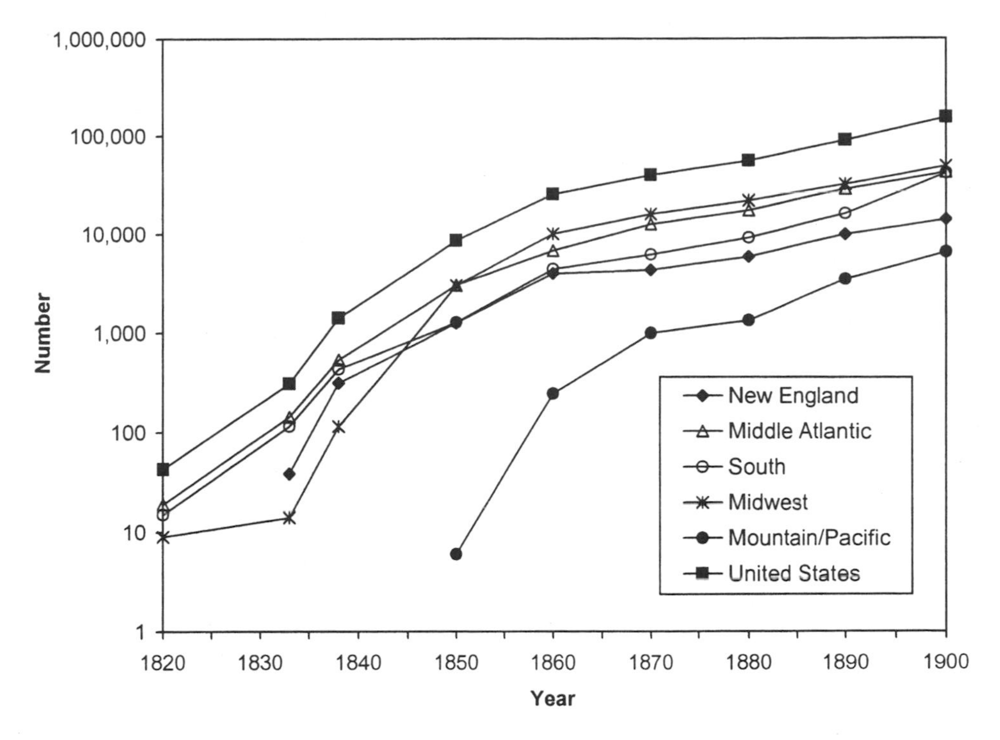

Figure 4.2. Stationary Industrial Steam Engines, 1820–1900. *Source:* Jeremy Atack, Fred Bateman, and Thomas Weiss, "The Regional Diffusion and Adoption of the Steam Engine in American Manufacturing," *Journal of Economic History* 40 (June 1980): 285, table 1.

Middle Atlantic within ten years; then the Midwest surged past to become the largest user of steam engines for the remainder of the century.[1]

IRON FOUNDRIES PROLIFERATE
Serving Diverse Markets

By 1820 foundries supplied various castings for manufactures, and some of them either made machinery or, occasionally, made castings for the small number of specialized machinery builders. Several large foundries operated in New York City and Philadelphia or in their satellites, but their small numbers testified to limited demand for their products. Coastal cities and urban places in broad river valleys located away from waterpower sites needed some industrial steam engines but provided limited outlets for foundries. In 1820 factories that used steam power constituted only 1 percent of all factories that used steam or waterpower in the nation, and this share rose to only 6 percent by 1838. Then the share jumped to 18 percent in 1850 and to 35 percent a decade later. Nonetheless, neither water nor steam power achieved a clear competitive advantage in New England and the Middle Atlantic during the antebellum. Transportation (chiefly steamboats) employed about two-thirds of the nation's steam power by 1838; therefore, buyers along coastal waterways and navigable rivers acquired most of the engines.

In small local and subregional markets blacksmith shops that engaged in simple metal forging and iron furnaces that produced castings met most metalworking needs. In contrast, an iron foundry required a sizable local and subregional market to generate sufficient business to cover its fixed costs (land, buildings, and equipment) and its variable costs of materials (pig iron, wrought iron, and fuel) and of labor. Consequently, after 1820 foundries expanded in the East Coast metropolises and their immediate environs, areas where they had concentrated previously. They also emerged in prosperous agricultural and industrializing areas: southern New England and outliers in coastal Maine and interior New Hampshire and Vermont, the Hudson Valley, the Erie Canal Corridor of New York and borderlands of Lake Ontario, and the valleys in Pennsylvania, including the western part, which looked to midwestern markets.

Foundries reduced their average costs if they produced many identical castings from patterns by spreading the fixed costs of making patterns over a large volume. Competitor foundries, as well as iron furnaces, however, could readily duplicate simple patterns such as those for kettles, hollow ware, and sash weights. For these widely demanded products, average costs ceased declining well before large volumes had been turned out. Much of the cost of making castings consisted of the molten pig iron—buying pig iron and reheating it in a cupola furnace—and of the labor cost of finishing castings. Transportation costs for heavy castings kept distribution within local and subregional markets. Foundries possessed competitive advantages producing custom castings for local and subregional manufacturers, transportation companies, and construction firms because they needed close contact with customers in order to design patterns and to meet special requirements. This custom work also extended to the production of various machinery (including steam engines) and equipment for manufacturers. Before 1840, therefore, most foundries distant from the environs of the metropolises had difficulty breaking out of local and subregional markets.[2]

Local and Subregional Markets

During the economic upturn of the 1820s iron foundries proliferated in areas of rising agricultural and industrial prosperity. By the early 1830s Boston's hinterland housed many foundries, some of them within fifty miles of the city. About twenty to thirty miles south of Boston as many as five foundries, including three large ones, operated near one another. Machinists in this cluster operated in networks—owners shared ideas, and workers moved among the firms carrying technical skills, thus building a local community of practice. Shepherd Leach's venerable foundry in Easton had twenty employees, and its annual sales of machinery to Massachusetts and Rhode Island amounted to about twenty-five thousand dollars. Nearby, an even larger foundry employed one hundred workers and sold sixty thousand dollars of various castings in Massachusetts. The Bridgewater firm of Lazell, Perkins, and Company (founded in 1813) produced almost one hundred thousand dollars of castings for local markets and for customers in Pennsylvania, Maryland, and Virginia. The small foundries in this cluster typified those outside major cities; they had fewer than two dozen employees and sold less than twenty thousand dollars of goods. General Learhe's Iron Foundry (started in 1824), about thirty-five miles northwest of Boston, operated in Chelmsford, near the rapidly growing textile city of Lowell; its one hundred workers mainly produced machinery castings.

Worcester and vicinity, about fifty miles west of Boston, housed foundries that served the nearby farmers, small factories, and cotton and woolen machinery firms. During the 1820s several foundries commenced operations in Worcester, producing various castings such as plows, kettles, and machinery components. By 1832 Smith's Industrial Foundry had four employees, who turned out four thousand dollars of machinery castings for the Massachusetts market. A short distance to the west, the Worcester and Brookfield Iron Foundry employed twenty-five men and produced fifteen thousand dollars of machinery castings, stoves, plows, and hollow ware for the subregion and other parts of New England. In the southern New Hampshire counties of Rockingham and Merrimack, between 50 and 75 miles from Boston, three small iron foundries produced various castings for local markets, each with several employees. The Bath Iron Foundry (started in 1828), an equally small firm farther north along the Maine coast, produced castings mostly for the local market, but it also sent a quarter of its output by coastal vessels to Boston, 150 miles away.

The Connecticut Valley of Massachusetts, about 90 miles west of Boston, housed productive agriculture and small-scale manufacturing firms, and the valley reached north along the border between New Hampshire and Vermont. Over a half-dozen foundries operated in the valley and adjacent hills, most of them producing various castings for local and subregional markets of consumers (hollow ware), farmers (plows), and factories (machinery). The American Hydraulic Company, owned by Francis Phelps, however, diverged. This Windsor County, Vermont, foundry, started in 1827, had twenty employees turning out as much as seventy thousand dollars annually of specialized machinery (fire engines and pumps) for sale across the East. Several smaller foundries and machine shops also operated in the county, foreshadowing this machinery center's future and suggesting that a community of practice of machinists had emerged there by the early 1830s.

After 1820 productive agriculture continued in Rhode Island and Connecticut, but manufacturing led their economic growth—textiles and machinery in Rhode Island and textiles and metal consumer and producer durables in Connecticut. Foundries thus faced many industrial demands. By 1832 the ten iron foundries in Rhode Island served the state's many factories, and the Fall River Furnace Company (founded in 1825), fifteen miles southeast of Providence, had sixteen employees producing fifteen thousand dollars of castings annually. It sold them mostly to factories in Fall River, a rapidly growing Massachusetts industrial city, and their machinists maintained network ties to textile mill shops and machine shops in Rhode Island. Connecticut had foundries in Norwich (one), Middletown (two), and Hartford (two).

Similar synergy operated between foundries and the prosperous agriculture and industry beyond the immediate environs of the large metropolises. Outside New York City "upstate" New York contained one hundred foundries, and about a quarter of them were large (20 to 50 employees); five huge foundries each employed between 80 and 360 workers (table 4.1). Almost two-thirds of these foundries operated in the rich agricultural and industrial subregions, either in the Hudson Valley or in the Erie Canal Corridor stretching between Albany and Lake Erie, which accounted for 80 percent of the upstate foundry employment. The Albany-Schenectady-Rensselaer County conurbation (including Albany, Troy, and other small cities), at the intersection of the Hudson and Mohawk rivers and the Erie Canal, housed an immense foundry complex. Its eleven foundries employed 14 percent of upstate area's foundry workers, almost two-thirds as many as in New York City. Large foundries also operated in the Erie Canal counties of Herkimer (cities of Little Falls and Herkimer), Oneida (Utica), Cayuga (Auburn), Monroe (Rochester), and Niagara (Buffalo), confirming that these budding industrial centers possessed a strong producer durables sector by 1830, only five years after completion of the canal.

The agricultural, industrial, and iron and coal mining areas of central and western Pennsylvania contained numerous small foundries with sales of one thousand to ten thousand dollars, and their employees numbered between three and fifteen. They sold castings for households (stoves and hollow ware), farmers (plows), and factories and mines (machinery and mill gearing). In Pittsburgh and surrounding Allegheny County about twenty foundries operated, almost half of them specializing in steam engines for local mills and midwestern steamboats and factories. The largest firm, McClurg, Pratt, and Wade, had eighty employees and produced almost one hundred thousand dollars of castings and machinery; steam engines accounted for more than half of its sales. These foundries beyond the immediate environs of Boston, New York, and Philadelphia provided critical components, such as castings, and machinery for their subregional economies, and some of them became important firms. Nonetheless, the metropolises and their satellites constituted the greatest foundry centers.[3]

TABLE 4.1
New York Counties with More than 20 Employees in Iron Foundries by Subregion, 1830

			Employment in Three Largest Foundries		
County	No. Foundries	Total Employed	1	2	3
Hudson Valley	27	944			
Putnam	1	360	360		
Dutchess	3	122	110	6	6
Columbia	3	55	28	15	12
Albany	5	159	68	30	27
Rensselaer	5	98	36	30	16
Schenectady	1	40	40		
Saratoga	5	75	20	20	16
Washington	4	35	15	10	8
Erie Canal Corridor	32	715			
Montgomery	3	20	15	3	2
Herkimer	9	138	44	32	16
Oneida	7	143	100	16	9
Madison	2	26	16	10	
Onondaga	2	48	24	24	
Cayuga	4	84	40	20	16
Ontario	1	24	24		
Monroe	2	124	100	24	
Niagara	2	108	80	28	
Subtotal of Hudson Valley and Erie Canal Corridor	59	1,659			
New York County (City)	9	465	200	90	50
Other counties	40	393			
Total New York State	108	2,517			

SOURCE: Louis McLane, *Documents Relative to the Manufactures in the United States Collected and Transmitted to the House of Representatives, 1832, by the Secretary of the Treasury*, House Doc. No. 308, 22nd. Cong., 1st sess. (Washington, D.C.: Duff Green, 1833), doc. 10, no. 48.

PREMIER IRON FOUNDRY CENTERS

During the 1820s and 1830s large foundries in New York and Philadelphia and in their satellites strengthened their positions as heavy capital equipment manufacturers, and others in Boston and its satellites joined them. These firms continued to maintain diversified product portfolios—including a wide range of types of castings, large anchors, heavy anvils, structural castings, steam engines, and mill machinery. Limited demand for each type of heavy equipment and machinery hindered specialization; before 1840, for example, only Pittsburgh and a few other places housed foundries that specialized in steam engines. With the exception of several other products such as stoves and locomotives, most large foundries did not specialize until after 1860. This absence of specialization, however, did not reflect primitive operations. These firms—they came to be called "heavy engineering works"—developed a cadre of sophisticated foundry managers, talented engineers/machinists, and skilled mechanics, and they created communities of practice that acquired capacities to make larger, better-quality castings, to build more complex equipment and machinery, and to improve metalworking techniques and machines, such as huge planers and boring machines, large metal cutters, and heavy grinders.

New York

The leading foundries in the New York region were hubs in networks of heavy capital equipment firms, and, arguably, the city's Allaire Foundry possessed the greatest influence into the 1840s. James Allaire had leveraged his steam engine work with Fulton to make his foundry the second largest in the city by 1820, after McQueen's. During the subsequent decade Allaire's foundry more than doubled in size, innovating the use of planing machines and the design of steam engines. It attracted as apprentices several of the nation's most talented engineers, or machinists, including Louis De Coudres, who stayed to run Allaire's brass foundry; John Roach, who became a leading builder of iron steamships; and Charles Haswell, who became the first engineer-in-chief of the U.S. Navy. During De Coudres's fifty-year eminence in local machinist networks and through his contact bridges to engineers leaving for other positions, the Allaire Works stayed attuned to innovations and opportunities to build equipment, making it a hub of know-how trading of foundry practice. By 1830 Allaire had two hundred employees, more than twice as many as the second-rank foundry, Sabbaton, with ninety workers, whereas McQueen's foundry had only seventeen employees (see table 4.1).

Other New York City foundries grew rapidly during the 1820s. New York Iron Foundry, which started in 1816, employed fifty workers by 1830, and at least five other foundries employed from 10 to 40 workers. At that time New York's nine foundries, together, employed 465 workers, and these firms and their workers constituted a powerful community of practice for advancing the iron foundry industry. During the following decade Archimedes Iron Works began operations (1833), and two years later Peter Hogg and Cornelius Delamater took over the Phoenix Foundry; by the next decade both had become huge foundries.

The industrial satellites of New York contained some of the nation's largest foundries. At Cold Spring, in Putnam County, on the Hudson River, the West Point Foundry diversified from cannon into general castings, steam engines, sugar mill machinery, and pumping engines. By 1830 it employed 360 workers and consumed two thousand tons of anthracite annually, about four times as much as the Allaire Works, New York City's largest foundry (see table 4.1). The Matteawan Company (founded in 1814) at Fishkill in Dutchess County started as a cotton textile firm that also produced textile machinery for sale. During the 1820s it added a foundry and diversified into steam engines, sugar mill machinery, and other products. At the end of the decade its foundry employed 110, and two years later it built a separate machine shop. Hoboken's Stevens Works, an innovative engineering firm prominent since shortly after 1800, built steam engines and steamboats; it added locomotives in the early 1830s. In nearby Paterson, Paul and Beggs began foundry operations in 1826, and this producer of mill gearing and machinery employed 40 workers within six years. At that time Rogers, Ketchum and Grosvenor started in Paterson with a capitalization of fifty thousand dollars and 120 workers, making various castings for railroads, in addition to textile machinery, and within a year the foundry decided to enter locomotive manufacturing.[4]

Philadelphia

The large metropolitan industrial complex centered on Philadelphia also had a highly networked community of machinists. The leading current and future foundry owners, including Matthias Baldwin, Marine Wickham, James Rush, and Samuel Merrick, cooperated, along with other business, social, political, and educational elite, in founding the Franklin Institute in 1824. Many of these leaders participated in it for another three decades. Within two years of its founding the institute established the *Journal of the Franklin Institute,* which became the nation's most eminent technical journal. The institute aimed to disseminate knowledge of mechanical science and quickly began initiatives, including classes for mechanics, a lecture series, regular exhibitions for manufacturers to display products, and a program of awards. It served as a clearinghouse for know-how trading of foundry knowledge, and its boards and committees became vehicles for communication among hub individuals, complementing their multiplex network contacts through social, political, and religious organizations. Network bridges connected these hub individuals and firms, and each of them also had links to less central mechanics and smaller firms. These network relations speeded foundry innovations among firms and opened up business opportunities.

Philadelphia's foundries established a resilient community of practice that built skills across various sectors, setting the base for long-term competitive advantages in manufacturing heavy capital equipment. Baldwin worked in jewelry manufacturing, gaining skills in metalworking and mechanics (motion and gearing), and in 1824 he formed a partnership with David Mason to make engravers' and book-makers' tools. They soon added hydraulic presses, high-precision lathe slide rests, and engraved copper cylinders for printing calicos, and within three years Baldwin constructed a small stationary steam engine. Beginning in 1829, Baldwin became the firm's sole owner, and he continued to manufacture various tools and steam engines. Within three years he completed his first locomotive, and by 1836 he had moved into a new factory, the Baldwin Locomotive Works, valued at about two hundred thousand dollars. The workers built locomotives and other capital equipment. Following a similar trajectory, Merrick manufactured hand-operated fire engines in the 1820s and diversified into other machine work. In 1836 he partnered with John Towne and established the Southwark Foundry to make steam engines, machine gearing, and fire engines. During that decade Merrick, Baldwin, and Rush worked on steam boiler experiments under the auspices of the Franklin Institute, and they published reports about their projects.

Alfred Jenks, the Slater-trained mechanic who had moved from Rhode Island to Philadelphia in 1810, established the Bridesburg Machine Works around 1820. The firm, although known for textile machinery, actually operated as a diversified machinery firm; its employees numbered 110 by 1832. Levi Morris and Company commenced machinery operations in 1828, and, like other firms, it became a large capital equipment firm around the 1840s. During the 1820s Wilmington, Delaware, and its vicinity had expanded as a heavy capital equipment satellite of Philadelphia, and by 1832 it housed at least two foundries. The Mahlon Betts Iron Foundry began as a small machine shop in 1812 and added a foundry in 1829; within three years its employment totaled 18. Although Betts primarily sold castings nearby, Jonathan Bonney and Company (founded in 1831), with employment similar to Betts, also sold castings in Philadelphia.[5]

Boston

Many firms in the Boston metropolitan industrial complex achieved fame as textile machinery manufacturers, yet it also housed heavy capital equipment firms in the metropolis and in its satellites, especially within twenty miles of Boston. Cyrus Alger's father, a prominent owner of three foundries in Boston's satellites around 1800, provided training for his son at the Easton factory of the family business, thus helping Alger enter Boston-area foundry networks seamlessly. In 1809, at the invitation of a prominent individual who sought a partner, Alger moved to Boston to establish a foundry. After gaining substantial profits making cannonballs during the War of 1812 and becoming sole owner of the foundry, he diversified the business. Alger became a hub of Boston foundry networks through his firm's success and his prominence as an inventor and metallurgist. In 1827 he took on additional partners to form the South Boston Iron Company, a much larger foundry than the one he had operated previously. By 1839 he possessed at least five patents relating to cast-iron plows, cast-iron cannons, and casting rollers for casting iron, and he contributed design improvements to many metallurgical and metalworking processes.

Holmes Hinkley started work in a machine shop in 1823 and shortly thereafter established his own business. Within three years he built his first steam engine, and several years later it became a major product line of the firm, though he also made other machinery. By 1832 Boston housed at least nine foundries, their total employment of 436 approaching that of New York City. The $1.1 million worth of products included various types of machinery, engines, stoves, and miscellaneous castings for sale mostly in New England. At least three foundries operated in satellites within a few miles of Boston, including one that employed 80 workers who turned out various machinery, hydraulic presses, and engines for markets in New England and elsewhere.[6]

Over the two decades following 1820, machinist networks in the heavy capital equipment industry operated effectively as mechanisms for know-how trading of technical skills and innovations, spurring the establishment of iron foundries and expansion of existing ones throughout the areas of rising agricultural and industrial prosperity. Hand methods and brawn, supplemented by mechanical or simple powered lifting devices, dominated the practice of making heavy equipment. Advances in boring machines and large lathes for cutting metal paled next to improvements in casting and filing techniques and in learning processes dealing with machinery construction.

These technological approaches suggest that firms anywhere could participate in the heavy capital equipment sector, but that was not the case. By the late 1830s most foundries produced castings and simple machines for local and perhaps subregional markets, whereas large foundries with the capacity to construct big castings, mill machinery, and steam engines concentrated in the metropolises of New York, Philadelphia, and Boston and in their satellites. Other major foundry clusters existed in the Hudson Valley, the Erie Canal Corridor between Albany and Buffalo, and Pittsburgh. These centers constituted the hubs of machinist networks in heavy capital equipment, and they possessed extensive internal networks supporting their growth. Steam engine manufacturing became a leading edge of the heavy capital equipment industry, and the 1838 steam engine report prepared by Levi Woodbury benchmarks this industry.[7]

STEAM ENGINES IN 1838
Regional Sales

The transportation sector (mostly steamboats) used about two-thirds of the nation's steam power, whereas industry employed around one-fifth of it. As a share of total industrial steam power, early-stage processing—sawmills (19 percent), iron (13 percent), and sugar, rice, and flour/grist milling (17 percent)—accounted for almost half. This comports with the many stationary engines used in the resource-processing areas of the Middle Atlantic and the South and the imminent surge in numbers in the Mid-

TABLE 4.2
Number of Engines Produced in Region of Stationary Steam Engine Builder by Region of Steam Engine User, 1838

	Region of Engine User				
Region of Engine Builder	New England	Middle Atlantic	Midwest	South	Total
New England	260	2	1	2	265
Middle Atlantic	35	288	2	69	394
Midwest	0	11	343	38	392
South	0	2	0	55	57
Other	9	39	30	42	120
Total	304	342	376	206	1,228

SOURCE: Peter Temin, "Steam and Waterpower in the Early Nineteenth Century," *Journal of Economic History* 26 (June 1966): 191, table 1.
NOTE: Regions: New England (Maine, New Hampshire, Vermont, Massachusetts, Rhode Island, Connecticut); Middle Atlantic (eastern New York, eastern Pennsylvania, New Jersey, Maryland, Delaware, District of Columbia); Midwest (western New York, western Pennsylvania, Ohio, Michigan, West Virginia); South (Virginia, North Carolina, South Carolina, Georgia, Florida, Louisiana); Other (England, Unknown).

west (see fig. 4.2). In contrast, foundries and machinery firms used about 9 percent of total industrial power. With the exception of interregional sales of stationary engines to the South, users purchased virtually all of their steam engines from their region's manufacturers (table 4.2). A small number of firms in New York City and the Hudson Valley sold engines to nearby New England buyers, whereas most of that region's steam engine manufacturers operated in eastern Massachusetts and Rhode Island, far away from Middle Atlantic buyers; therefore, they sold few engines to them. Likewise, southern producers sold most of their engines within their region, but manufacturers in the Middle Atlantic and the Midwest captured significant market shares in the South.

This sale of engines to the South suggests that transportation cost did not constrain market areas of firms. The bulky heavy equipment could be disassembled for shipment, and it had high value relative to its weight. By 1838, however, demand for stationary steam engines remained too limited for firms to specialize in them or to produce large numbers each year. Instead, foundries manufactured engines as one line among a diverse portfolio of products. The failure of southern foundries to specialize in steam engines cannot explain their inability to meet intraregional demand; such specialization also remained rare outside the South. Its firms did not fully meet regional demand because few foundries possessed the capacity to build engines. Farm capital was invested in slaves rather than in land and machinery, and low market densities (slave demand remained constrained) limited consumer and urban infrastructure manufactures. This situation restricted markets for foundries, the chief producers of steam engines; thus, the South had fewer, more dispersed foundries capable of building engines.[8]

TABLE 4.3
Distribution of Stationary Steam Engine Builders by Number Built for Engines in Operation, 1838

No. of Engines Built	% of Builders	% of Engines Built
1	46.3	9.1
2–5	29.4	17.2
6–10	11.2	18.2
11–15	5.1	12.6
16–20	2.8	10.3
21–30	2.3	10.6
31–37	2.3	15.1
74	0.5	6.8
Total %	99.9	99.9
Total number	214	1,085

SOURCE: Louis C. Hunter, *A History of Industrial Power in the United States, 1780–1930*, 2 vols. (Charlottesville: University Press of Virginia, 1985), 2:230, table 24.

The Concentration of Builders

Machinist networks conveyed technological skills about building stationary engines to a wide array of firms and mechanics in the nation, and at least 214 builders had their steam engines operating in 1838. This dispersal of capabilities did not signify that all of these builders possessed the capital and industrial infrastructure, including equipment and machinists, to manufacture engines as market opportunities beckoned. Almost half of the builders made only one engine, probably for their own firm's use or for a one-time order from another firm, and their engines accounted for just 9 percent of the number turned out (table 4.3). This pattern of manufacture succeeded when a firm's machinists built its equipment or did metalworking as part of production processes—such as manufacturers of textiles, hardware, iron bars and plates (rolling mills and forges), and nails, and these machinists could order castings for the engine from a foundry. A small share of the builders turned out most of the steam engines: 52 of them (24 percent of the total) made more than five, accounting for three-fourths of the engines; and the 17 builders (8 percent) making more than fifteen each produced 43 percent of the engines. An elite group of 7 manufacturers based in the large metropolises (Boston, New York, Philadelphia, and Baltimore) and in Providence accounted for 3 percent of the builders, but they produced 22 percent (239 of 1,085) of the nation's stationary engines (table 4.4). This concentrated production comports with the agglomeration of the East's major foundries in these same centers.

TABLE 4.4
Market Areas of Selected Leading Steam Engine Builders as of 1838

Builder / Place of Use	Number of Engines		
	Stationary	Steamboat	Total
Holmes Hinkley, Boston, Mass.	75	1	76
Boston, Mass.	42	0	42
Portsmouth, N.H.	3	0	3
Portland, Maine	1	0	1
Bath/Brunswick, Maine	18	1	19
Bangor/Belfast, Maine	2	0	2
Nantucket, Mass.	1	0	1
New Bedford, Mass.	4	0	4
Providence, R.I.	2	0	2
New London, Conn.	1	0	1
Florida	1	0	1
Providence Steam Engine Co., R.I.	21	3	24
Providence, R.I.	12	0	12
Fall River, Mass.	7	0	7
Boston, Mass.	2	2	4
Hartford/Middletown, Conn.	0	1	1
James P. Allaire, New York, N.Y.	8	34	42
Perth Amboy, N.J.	0	3	3
Philadelphia, Pa.	0	3	3
New Haven, Conn.	0	3	3
Hartford/Middletown, Conn.	0	1	1
Providence, R.I.	2	1	3
Portland, Maine	0	1	1
Buffalo, N.Y.	0	2	2
Watertown, N.Y.	0	2	2
Erie, Pa.	0	1	1
Norfolk, Va.	0	1	1
Richmond, Va.	0	1	1
Wilmington, N.C.	0	2	2
Charleston, S.C.	1	7	8
Savannah/Brunswick, Ga.	2	4	6
Florida	3	0	3
Sandusky, Ohio	0	1	1
Michigan	0	1	1
West Point Foundry, N.Y.	31	13	44
Perth Amboy, N.J.	0	1	1
Philadelphia, Pa.	1	0	1
Fairfield, Conn.	0	1	1
Hartford/Middletown, Conn.	0	1	1
Fall River, Mass.	0	1	1
Nantucket, Mass.	0	1	1
Bangor/Belfast, Maine	1	0	1
Buffalo, N.Y.	0	2	2
Wilmington, N.C.	4	1	5
Charleston, S.C.	2	0	2
Savannah/Brunswick, Ga.	7	3	10
Florida	0	1	1
Louisiana	16	0	16
Cleveland, Ohio	0	1	1

TABLE 4.4
continued

Builder / Place of Use	Number of Engines		
	Stationary	Steamboat	Total
Levi Morris and Co., Philadelphia, Pa.	34	0	34
Philadelphia, Pa.	26	0	26
Trenton, N.J.	1	0	1
Harrisburg, Pa.	1	0	1
Pottsville, Pa.	3	0	3
Wilmington, Del.	1	0	1
Charleston, S.C.	2	0	2
Rush and Muhlenberg, Philadelphia, Pa.	37	8	45
Philadelphia, Pa.	28	5	33
Lancaster, Pa.	1	0	1
Harrisburg, Pa.	1	0	1
Perth Amboy, N.J.	0	1	1
Washington, D.C.	2	0	2
Portsmouth, N.H.	1	0	1
New Bern, N.C.	1	0	1
Wilmington, N.C.	0	1	1
Charleston, S.C.	2	1	3
Louisiana	1	0	1
Watchman and Bratt, Baltimore, Md.	33	36	69
Baltimore, Md.	14	9	23
Annapolis, Md.	1	1	2
Wilmington, Del.	0	1	1
Philadelphia, Pa.	0	1	1
Washington, D.C.	2	5	7
Salisbury, Md.	2	0	2
Norfolk, Va.	1	1	2
Petersburg, Va.	0	2	2
New Bern / Washington, N.C.	4	0	4
Wilmington, N.C.	2	3	5
Charleston, S.C.	4	7	11
Savannah/Brunswick, Ga.	1	5	6
Florida	2	0	2
Louisiana	0	1	1

SOURCE: Levi Woodbury, *Steam Engines*, Transmitted to the House of Representatives, 1838, by the Secretary of the Treasury, House Doc. No. 21, 25th Cong., 3rd sess., *New American State Papers, Science and Technology*, vol. 7: *Steam Engines* (Wilmington, Del.: Scholarly Resources, 1973), 11–482.

Market Areas of Leading Firms

The top manufacturers of stationary and of steamboat engines typically sold at least two-thirds of them in their local metropolitan and inner-hinterland market areas within about one hundred miles of the metropolis (see table 4.4).[9] Within their own metropolitan areas the top firms possessed dominant market positions based on their capitalization, foundry infrastructure, skilled machinists, and network relations with local buyers; competitors from elsewhere rarely overcame these advantages. Allaire and the West Point Foundry made large sales in New York City and in nearby parts of New York state, but the 1838 *Steam Engine* report gave no data for these areas. Boston's Holmes Hinkley rarely sold engines away from coastal New England, and the Providence Steam Engine Company's market area was concentrated within fifty miles of home. Baltimore's Watchman and Bratt, however, stood as an exception. The share of total sales made to its metropolitan area and inner hinterland markets fell below the shares of total sales that engine builders in other centers made to their proximate markets. The reason for the low sales was that Baltimore and nearby Maryland, portions of which had slave plantation agriculture, generated limited industrial demand.

Although Allaire focused on steamboat engines and the West Point Foundry specialized in stationary ones, they possessed remarkably similar market areas. With the exception of West Point's large sales of stationary engines in Louisiana—Allaire had no sales there—both sold engines all along the East Coast from Maine to Florida and several to the Midwest. West Point, however, did not dominate Louisiana markets; firms from the Ohio and Mississippi valleys effectively competed for that business. W. Tift's foundry in Cincinnati sold twenty-three steam engines to Louisiana sugar mills, dwarfing West Point's sale of eleven. Allaire and West Point far surpassed other leading steam engine builders in sales (adjusting for the missing New York data) and in market area reach, but they failed to sell more than a trivial number of engines in other metropolitan markets with major builders.

The Philadelphia firms of Levi Morris and Company and of Rush and Muhlenberg, both specialists in stationary engines, made over three-fourths of their sales within the metropolis and its immediate environs out to a radius of about fifty miles (see table 4.4). Along with New York City, the Philadelphia steam engine market ranked among the largest in the nation; these big foundries could therefore function in the same market. Morris rarely sold outside Philadelphia's inner hinterland and, then, only to the Charleston market, whereas Rush and Muhlenberg reached a larger territory in the South. In contrast, Baltimore's Watchman and Bratt, the most diversified engine builder (with about equal numbers of stationary and steamboat engine sales), compensated for its lower share of sales in the Baltimore area with a deep penetration of markets around Washington, D.C., and of southern markets from Virginia to Florida as well as one sale in Louisiana. With the exception of the latter, Watchman and Bratt outsold each of the New York giants, Allaire and West Point, in the South.

Competition in Local Markets

The leading steam engine firms possessed large capitalization, a big workforce of skilled machinists, and access to sophisticated technical networks, but they could not capture all of the local steam engine markets distant from their headquarters. Engine builders that participated in a rich, local community of practice could thwart the sales incursions of the leading firms. The subregion of Bath and Brunswick, Maine, which extended from these cities to the north along the Kennebec Valley to Augusta, specialized in lumber milling, wood products (furniture), and agriculture. The valley housed at least five foundries and two machine shops. A foundry purchased an engine from a local builder, and a furniture firm bought an engine from a local foundry which did not make its own stationary engine. James Pond, probably from Boston, built it, and he also made an engine for another local foundry. Freeman and McClinch, a local machine shop, produced three engines (one for itself); they constituted the largest local manufacturer. The local engine manufacturers, however, lacked the capacity to compete effectively with Holmes Hinkley, whose Boston foundry dominated the Kennebec Valley market. It produced three-fourths of the twenty-four engines in the valley and supplied lumber mills, furniture firms, grain mills, and foundries.

The limited success of Bath/Brunswick area firms in competing in steam engine markets contrasts markedly with the achievement of firms in the Connecticut Valley between Hartford and Middletown and in nearby towns. Since the 1790s this subregion had acquired a thriving metalworking complex of foundries and firms which produced machinery, firearms, tinware, and hardware. Numerous skilled machinists worked in these firms, participating in networks that connected them to other metalworking firms in Connecticut and to the north in the Connecticut Valley of Massachusetts. The Hartford/Middletown subregion housed three steam engine builders accounting for fourteen of the twenty-three stationary engines in operation in 1838. They built them for various manufacturers, including machinery firms, foundries, and metal consumer goods firms, among others (table 4.5). Guild and Douglass built half of the engines supplied by local firms, and Daniel Copeland turned out five. He traveled to New York City to hire people skilled in making engines, a normal use of machinist networks to acquire talent; the subregion had plenty of mechanics to implement steam engine designs. Ample water power in the area had discouraged engine building prior to the late 1820s, and therefore machinists had no need to acquire experience constructing them. The quick emergence of successful local steam engine firms after the transfer of New York City talent to the subregion testified to the skills of local machinists and to the networks of mechanics that facilitated know-how trading. These Connecticut machinists rapidly established a community of practice organized around building steam engines.

Aside from Copeland's local sales of stationary engines, he sold two engines for steamboats operating on the Connecticut River between Hartford and Springfield and six engines for steamboats in the Savannah/Brunswick, Georgia, subregion. These distant sales had roots in the long-standing network ties of Hartford/Middletown area

TABLE 4.5
Builders and Buyers of Stationary Steam Engines Operating in Hartford/Middletown, Connecticut, 1838

Builder (No. of Engines)	Buyer
Local (14)	
Daniel Copeland (5)	machinery factory
	2 iron foundries
	tin and pewter ware factory
	steam planing mill
Guild & Douglass (7)	machinery factory (own firm)
	firearms factory
	2 wood-turning factories
	braid factory
	pen factory
	tannery
Asa Richardson (2)	rule factory (own firm)
	wallpaper factory
External (9)	
John Gore, Brattleboro, Vt. (3)	2 cotton mills
	brassware factory
William T. James, New York (4)	printing firm
	state prison
	tin factory
	tannery
Novelty Works, New York (1)	tin factory
Paul A. Sabbaton, New York (1)	state prison

SOURCE: Levi Woodbury, *Steam Engines*, Transmitted to the House of Representatives, 1838, by the Secretary of the Treasury, House Doc. No. 21, 25th Cong., 3rd sess., *New American State Papers, Science and Technology*, vol. 7: *Steam Engines* (Wilmington, Del.: Scholarly Resources, 1973), D13.

businesses to Georgia and the Carolinas. Since the late-eighteenth century, merchant wholesalers in Hartford, Middletown, and New Haven had sent agricultural produce to these southern markets. Furthermore, from 1815 to 1830 many Connecticut industrial entrepreneurs in tinware, brass, and clocks had worked for a time as retailers or peddlers in Georgia and the Carolinas, and the factory owners continued to target these markets after 1830. Copeland probably drew on these business networks to access the southern markets, for which he built six engines in 1834 and 1836, after finishing three for the Hartford/Middletown area.

Daniel Copeland trained his son, Charles, who left for Columbia College in New York City for advanced education. With top-notch technical training as a machinist-apprentice under his father and access to elite social and educational networks through his college education, Charles was a prototypical engineer on a fast-track career. In 1836, at the age of twenty-one, he was appointed superintendent of the West Point Foundry. It had not penetrated the Hartford/Middletown subregional market, and, collectively, the dominant New York City firms—William James, the Novelty Works, and Paul Sabbaton—built only six engines (see table 4.5). John Gore, a Brattleboro, Vermont, machinist, accounted for three of the subregion's engines, and his market dominance extended throughout the Connecticut Valley between Brattleboro and Middletown. He built three engines for the Keene area of New Hampshire, just east of Brattleboro, and eight of the nine engines in operation in the Massachusetts portion of the valley. Like the builders in the Hartford/Middletown area, Gore succeeded against external competitors because he drew on the textile machinery firms, armories, and general machine shops, among others, in the valley. These firms and their machinists had created a vibrant community of practice in metalworking within and across industrial sectors.

At this time the network of machinists in New Haven was weaker than in Hartford and Middletown; local foundries therefore did not capture much steam engine business. New York City firms dominated, constructing nine of the eleven engines: Paul Sabbaton led with four, William James and the Novelty Works each supplied two, and Vicar and Snedeker sold one. Augustus Parker of Utica, New York, and Guild and Douglass, the leading engine builder in Hartford/Middletown, each built one engine. None of the three largest New York City builders that sold engines in New Haven achieved the total national sales and market reach of Allaire or West Point Foundry (see table 4.4). Sabbaton's twenty-one engines (excluding missing New York City data) made him the largest, and he focused on stationary engines. Outside the New York City area, New Haven constituted his major northern market, and he sold a few engines along coastal New England. The South (especially South Carolina), with almost 40 percent of his sales, represented his next largest market. James sold only stationary engines (a total of seven) in a limited area near New York City and in Connecticut. The Novelty Works sold about the same total number, divided equally between stationary and steamboat engines, and it had a sprinkling of sales in Boston, Virginia, Georgia, and Michigan. The success of these foundries in acquiring markets for their steam engines, even when faced with competition from the Allaire and West Point foundries, suggests that astute firms could exploit market niches. They probably leveraged business and machinist networks that gave them access to customers.

The stationary steam engine market in Newark (and the small market in Elizabethtown), across the Hudson River from New York City, existed in the shadow of that metropolis's foundries (table 4.6). Its firms captured half of the market; James and Sabbaton replicated the market power they exhibited in New Haven. Outside of Newark A. Birbeck sold three engines for steamboats in Buffalo. Elsewhere in New York state two firms each built an engine for Newark's market, and Baldwin, the prominent Philadelphia engine builder, also made one. The many talented machinists in Newark's firms accounted for just under one-third of the engines. Seth Boyden, owner of an engine and machinery firm, built his own and one for a wood-planing mill. Each of the other three engines was built by the firm using it, and these firms—a

TABLE 4.6
Builders and Buyers of Stationary Steam Engines Operating in Newark, New Jersey, 1838

Builder (No. of Engines)	Buyer
Local (5)	
Seth Boyden (2)	steam engine/machinery factory (own firm)
	wood planing mill
A. Connison (1)	iron foundry (own firm)
Joseph Dalrimple (1)	wood and iron turning factory (own firm)
H. Davis (1)	wood and iron turning factory (own firm)
External (11)	
Matthias Baldwin, Philadelphia (1)	brass foundry and axle factory
A. Birbeck, New York (2)	lumber mill
	cordage and rope factory
William T. James, New York (3)	wood and iron turning factory
	sash and blind factory
	machine shop
Paul A. Sabbaton, New York (3)	2 carriage factories
	machinery and axle factory
(Unknown), Auburn, N.Y. (1)	malleable iron factory
(Unknown), Van Buren, N.Y. (1)	malleable iron factory

SOURCE: Levi Woodbury, *Steam Engines*, Transmitted to the House of Representatives, 1838, by the Secretary of the Treasury, House Doc. No. 21, 25th Cong., 3rd sess., *New American State Papers, Science and Technology*, vol. 7: *Steam Engines* (Wilmington, Del.: Scholarly Resources, 1973), G12.

foundry and two wood- and iron-turning factories—employed their own mechanics to make the engine. Both factories probably obtained cast-iron components from nearby foundries.

Philadelphia's prominent engine builders penetrated local markets in Schuylkill County, a metal manufacturing and anthracite coal mining area about eighty miles northwest of the city. Levi Morris and Company constructed three engines, two for coal mines and one for an iron foundry, and Baldwin built one for a flour mill. An unknown firm from New York supplied an engine for an iron furnace, and J. May Jones and Company, a steam engine and machinery firm in nearby Reading, in Berks County, built two engines. Haywood and Snyder, however, a steam engine and machinery firm in Schuylkill County, dominated as the single, largest supplier. Its engines—one for itself, one for a flour mill, and five for coal mines—accounted for half of the county's total. The expansion of coal mining, iron-rolling mills, and nail works created opportunities for local foundries and machinery firms to supply equipment. Machinists in Schuylkill County, and nearby Berks County, must have accessed networks of engine builders to gain the requisite technical skills, and then, through know-how trading among themselves, they created a community of practice. Local mechanic networks also supplied knowledge about the special needs of the growing sectors such as coal mining. These assets probably contributed to the competitive success of the local engine producers vis-à-vis more prominent Philadelphia firms.

TABLE 4.7
Builders of Stationary Steam Engines Operating in Philadelphia, 1838

	Number of Engines	
Builder	Total	Own Use
Local	168	23
Matthias Baldwin	12	0
James Brooke	6	1
Stacy Costill	10	0
Garrett and Eastwick	3	1
M. W. and T. Grear	7	0
Jacob Green	2	0
Thomas Halloway	7	1
Hyde and Flint	15	0
Daniel and C. Large	8	2
Alexander McClausland	2	0
Prosper Martin	4	0
Merrick and Agnew	3	3
Levi Morris and Co.	24	1
Parrish and Johnson	9	1
Rush and Muhlenberg	26	2
C. Sellers and Son	2	1
Sutton and Blumner	11	0
Single engine builders for own use	10	10
Single engine builders for others' use	7	0
External	4	
Oliver Evans, Pittsburgh, Pa.	2	
West Point Foundry, New York	1	
(Unknown), Pittsburgh, Pa.	1	

SOURCE: Levi Woodbury, *Steam Engines*, Transmitted to the House of Representatives, 1838, by the Secretary of the Treasury, House Doc. No. 21, 25th Cong., 3rd sess., *New American State Papers, Science and Technology*, vol. 7: *Steam Engines* (Wilmington, Del.: Scholarly Resources, 1973), H5, H7.

Nonetheless, the metropolis's firms formed a mighty steam engine complex, and the huge local market consumed about 16 percent of the nation's stationary engines. Philadelphia County housed 172 steam engines whose builder or location can be identified; local firms made 98 percent of them (table 4.7). The Oliver Evans Works in Pittsburgh, another firm from that city, and the West Point Foundry were the only companies based outside the county which built engines for the local market. Philadelphia contained eleven firms that made more than five engines each for the local market, and they accounted for 21 percent of the nation's builders turning out that many. Of the seventeen builders who made one engine, seven constructed them for firms that they did not own. The buyers typically had few skilled machinists; these businesses included a blacksmith shop, surgical instrument firm, cotton mill, and shipyard. The ten engine builders with one engine to their credit which supplied their own operations mostly consisted of metalworking firms—producers of various machinery, fire engines, and steam engines, a brass foundry, and screw and auger factories—whose skilled machinists could make engines. Experience with engine building constituted an excellent way for mechanics to enhance their metalworking skills and to learn to deal with moving parts, and this knowledge could be transferred to all types of heavy capital equipment. This synergy, along with the enormous talent pool of steam engine machinists, enhanced the sophisticated community of practice in the city.[10]

Explaining a Conundrum

High-value steam engines could be disassembled and shipped to other places without unduly raising the engine's final cost, thus allowing builders to penetrate regional and, occasionally, interregional markets. Firms in Boston, New York, Philadelphia, and Baltimore competed effectively in other markets because their engines had greater sophistication, better quality, and lower prices, and these differentials overwhelmed the advantages of local builders, such as stronger network ties with nearby buyers and minuscule cost to ship the engine. Thus, in Bath/Brunswick, New Haven, and southern coastal ports, steam engine firms from metropolitan centers captured significant market shares. Nonetheless, this leaves a conundrum. The metropolitan manufacturers failed to penetrate one another's market beyond token sales, and they did not dominate in some small industrial centers that had engine builders.

The leading centers of steam engine production possessed ample capital to fund a foundry infrastructure and had a large supply of skilled machinists in their machinery sectors; nevertheless, their equivalence in the capacity to build engines had deeper roots. The community of practice of engine building which encompassed these centers had emerged before 1820, and the machinist networks that linked them maintained this community. Participants in these networks engaged in know-how trading, and the job mobility of machinists among firms and production centers transferred technical skills and innovations. As a result, steam engines produced in each center possessed similar levels of sophistication, quality, and price, and firms had difficulty competing in one another's local market. Ties to local buyers and the ability to meet special demands from them constituted the key differentials; external competitors could not break into the market. The small order flow and primitive metalworking equipment prevented firms from achieving significant economies of scale, thus preventing a few large firms from dominating the market.

The same rationale helps explain the capacity of steam engine manufacturers in places such as Hartford/Middletown and Schuylkill County to compete effectively in their local market against top firms from Boston, New York, and Philadelphia. These places distant from the metropolises contained major agglomerations of metalworking firms and machinists which operated in highly integrated networks of know-how trading within a strong local community of practice. This made their engines competitive based on sophistication, quality, and price with the engines built by prominent firms in the metropolises. The local firms' ties to nearby buyers provided the extra advantage over the metropolitan firms, but these local businesses possessed insufficient capital to support larger-scale production and sales efforts over wider territories in competition with the major steam engine builders.

THE MACHINERY INDUSTRY IN 1840

Like the subsector of steam engine production, machinery manufacturing as a whole concentrated in a small number of leading industrial centers in 1840 (table 4.8). Measured by either value of machinery production or employment, the metropolitan complexes of Boston, New York, Philadelphia, and Baltimore housed just over 40 percent of the nation's and 55 percent of the East's machinery manufacturing. Each of these complexes constituted a large community of practice, and their numerous mechanics gave them exceptional access to machinist networks, thus strengthening their competitive advantages. When these complexes are combined with the other major machinery counties in the East, this larger group accounted for about two-thirds of the nation's machinery industry and about 85 percent of the East's. This extraordinary concentration refutes the argument that a broad swath of the labor force—as reflected in the ubiquitous blacksmiths—possessed machinist skills. Although blacksmiths appeared in most local economies, their simple skills did not extend to making most machinery, other than one-time, crude prototypes. To produce machinery (including complex equipment) in large quantities required capital, skilled machinists integrated into a community of practice, and access to larger market areas than the local economy.

Outside the immediate environs of the metropolitan complexes, most of the other leading machinery centers were in the East's areas of agricultural prosperity: eastern Massachusetts and Rhode Island, the Connecticut Valley of Massachusetts (Hampshire and Hampden counties), Connecticut, the Hudson Valley, the Erie Canal Corridor, and southeastern Pennsylvania. The small variability in the value of machinery per employee across most of the production centers resulted because the cost of the inputs used in machinery manufacturing—capital, labor, and land—did not differ much throughout the East. It also probably hindered specialization. Pittsburgh, one of the largest machinery centers, had a high value of machinery per employee because its firms turned out many steam engines.

By this 1840 benchmark, therefore, most foundries turned out a diversified set of products, and they continued that practice for the next two decades. Yet around this time some foundries specialized in stoves, providing a glimmer of the specialization that gathered force in the late nineteenth century.

Table 4.8
Leading Centers of Machinery by Value and Number of Employees, 1840

Place/County	Value of Machinery ($)	Number of Employees	Value per Employee ($)	Percentage of Nation: Value (%)	Percentage of Nation: Employees (%)
Boston hinterland					
Boston complex	467,300	462	1,011	4.3	3.6
Suffolk (Boston), Mass.	140,900	117	1,204	1.3	0.9
Essex, Mass.	102,700	78	1,317	0.9	0.6
Middlesex, Mass.	130,000	180	722	1.2	1.4
Norfolk, Mass.	93,700	87	1,077	0.9	0.7
Providence, R.I.	380,800	461	826	3.5	3.5
Bristol, Mass.	97,000	168	577	0.9	1.3
Worcester, Mass.	151,150	129	1,172	1.4	1.0
Hampshire, Mass.	63,950	36	1,776	0.6	0.3
Hampden, Mass.	120,000	70	1,714	1.1	0.5
Windham, Conn.	93,600	83	1,128	0.9	0.6
Hillsborough, N.H.	46,400	92	504	0.4	0.7
Kennebec, Maine	30,250	167	181	0.3	1.3
Penobscot, Maine	6,900	95	73	0.1	0.7
New York hinterland					
New York complex	2,287,500	2,766	827	20.8	21.3
New York City	1,392,000	1,597	872	12.7	12.3
Putnam, N.Y.	200,000	350	571	1.8	2.7
Passaic, N.J.	607,000	801	758	5.5	6.2
Morris, N.J.	88,500	18	4,917	0.8	0.1
New Haven, Conn.	87,650	89	985	0.8	0.7
Tolland, Conn.	59,008	69	855	0.5	0.5
Hudson Valley					
Dutchess, N.Y.	356,800	319	1,118	3.2	2.5
Columbia, N.Y.	72,500	58	1,250	0.7	0.4
Erie Canal Corridor					
Oneida, N.Y.	193,955	141	1,376	1.8	1.1
Onondaga, N.Y.	53,200	77	691	0.5	0.6
Cayuga, N.Y.	41,900	360	116	0.4	2.8
Wayne, N.Y.	64,500	59	1,093	0.6	0.5
Monroe, N.Y.	62,250	72	865	0.6	0.6
Erie, N.Y.	57,500	39	1,474	0.5	0.3
Philadelphia hinterland					
Philadelphia complex	1,472,114	1,526	965	13.4	11.7
Philadelphia	1,088,864	1,084	1,004	9.9	8.3
Delaware, Pa.	27,000	78	346	0.2	0.6
Chester, Pa.	41,750	65	642	0.4	0.5
New Castle, Del.	314,500	299	1,052	2.9	2.3
Schuylkill, Pa.	92,000	80	1,150	0.8	0.6
Columbia, Pa.	57,895	71	815	0.5	0.5
Baltimore hinterland					
Baltimore complex	286,500	650	441	2.6	5.0
Pittsburgh hinterland					
Pittsburgh (Allegheny), Pa.	443,500	251	1,767	4.0	1.9
Fayette, Pa.	65,700	66	995	0.6	0.5

TABLE 4.8
continued

				Percentage of Nation	
Place/County	Value of Machinery ($)	Number of Employees	Value per Employee ($)	Value (%)	Employees (%)
Total four complexes	4,513,414	5,404	835	41.1	41.6
Total selected counties	7,211,822	8,456	853	65.7	65.0
East (New England, Middle Atlantic)	8,273,059	9,957	831	75.3	76.6
Nation	10,980,581	13,001	845	100	100

SOURCE: U.S. Bureau of the Census, *Compendium of the Sixth Census, 1840* (Washington, D.C.: Blair and Rives, 1841).

NOTE: Counties included in this table had either 0.5 percent or more of the national value of machinery ($54,900 or more) or 0.5 percent or more of the nation's employment in machinery (65 or more). Metropolitan industrial complexes include the core metropolitan counties and nearby counties as listed. For New York City, the counties meeting the criteria for inclusion were New York, Kings, and Richmond.

STOVE FOUNDRIES

At least until the early 1830s rural blast furnaces cast stove plates, and furnaces either assembled the stoves for sale locally, shipped the plates to urban retailers who served the subregional market encompassing the furnace, or sent the plates to merchant wholesalers (some of whom owned blast furnaces) for assembly and sale in metropolitan markets. Iron furnaces did not specialize in stove plates because any furnace could produce these commodity items at low cost as an adjunct to making pig iron and various castings. During the 1820s and early 1830s urban foundries expanded to serve local and subregional markets, and low-cost anthracite coal became increasingly available as a fuel source. These foundries began producing stove plates at prices that competed with the local delivered prices of plates from rural furnaces; consequently, foundries increasingly listed stoves among their product lines. The emergence of the specialized foundry rested on several features of stoves: the demand for them increased with growing urban populations and rising income; heterogeneous stove styles opened opportunities for manufacturers to differentiate themselves and undercut diversified foundries; and batch manufacturing processes and the division of labor generated production economies of scale, thus limiting the number of firms in the industry.

Joel Rathbone, an Albany wholesaler, started dealing in stoves around 1827; he assembled stove plates from New Jersey, and his largest markets consisted of New York City and Philadelphia. Rathbone must have achieved some economies: from his bulk orders for stove plates, large-scale assembly of stoves, and sizable sales in urban markets. Nevertheless, shipping stove plates to Albany and the lack of control over design and quality created inefficiencies, even though Rathbone became one of the largest stove assemblers by the early 1830s. Over time he started to gain control over design and quality. Several designers and pattern makers, such as Samuel Hanley, who began designing and manufacturing ornamental stove patterns in 1830, and James Wager, who started in the pattern business in 1835, worked in nearby Troy. Around 1836 Albany foundries made some castings from Rathbone's patterns, and two years later he erected a foundry and built a cupola furnace to make stove castings. By the late 1840s Albany housed at least four large stove foundries and several smaller ones, producing about seventy-five thousand stoves annually and employing 950 workers. Troy contained at least nine stove foundries that turned out about one million dollars' worth of castings annually and employed about 1,000 workers by the mid-1850s. The Albany-Troy area thus developed a robust community of practice dealing with the design and manufacture of stoves.

During the 1840s some Philadelphia foundries began specializing in stoves, and in 1845 Smith and Brown entered business as designers and manufacturers of patterns. By the mid-1850s the city contained at least five large stove foundries that annually turned out about 125,000 stoves valued at $1.3 million. At the same time, stove foundries operated in other cities, including the Fall River Foundry in Massachusetts; M'Arthur and Company in Hudson in New York; and Hayward, Bartlett and Company in Baltimore.[11]

In 1850, a little more than ten years after the emergence of the first stove foundries, 230 firms produced stoves as a significant part of their product line, and this number rose to 290 ten years later (table 4.9). Over that decade the production share of the top four states—New York, Pennsylvania, Massachusetts, and Maryland—declined slightly, from 81 percent of national value added (value of products minus raw material costs) in stove manufacturing to 77 percent. Firms in New York and Pennsylvania increased their share of stove production from 59 percent to 66 percent, whereas those in Massachusetts and Maryland faced declining sales and lost market share. By 1860 the rest of the East continued to house a small share of national production. The number of firms outside the East had more than doubled over the 1850–60 period, but they did not gain market share from eastern producers.

Firm size, measured by capital, differed markedly across states and regions in 1850 (see table 4.9). New York's firms were much larger than others in the East, and they had the highest value added per employee. Eastern firms located outside the four leading states possessed tiny capitalization and had minuscule value added per employee. In contrast, the average capitalization of firms situated outside the East matched New York's firms, but its firms had over twice the value added per employee. By 1860 the average firm's capitalization had risen significantly throughout the nation. New York's firms remained the largest, and Maryland's more than doubled their capitalization, making them almost as big as New York's. During the 1850s all sections

TABLE 4.9
Stove Manufacturing by State, Region, and Nation, 1850 and 1860

Place	No. of Firms	Value of Products ($)	Capital per Firm ($)	Value Added: Per Employee ($)	Value Added: Percentage of Nation
1850					
East	212	5,158,282	12,872	903	81.3
New York	51	2,222,880	25,600	1,150	37.3
Pennsylvania	97	1,270,142	8,593	697	21.1
Massachusetts	37	972,450	8,032	911	12.1
Maryland	19	665,000	14,289	833	10.7
Total 4 states	204	5,130,472	13,274	914	81.2
Rest of East	8	27,810	2,625	88	0.1
Outside East	18	966,466	25,036	449	18.7
Nation	230	6,124,748	13,824	760	100.0
1860					
East	244	8,779,262	24,766	841	82.8
New York	84	4,450,560	34,747	888	42.9
Pennsylvania	107	2,526,685	19,784	783	23.0
Massachusetts	26	617,470	11,392	791	4.9
Maryland	10	589,000	32,750	1,004	6.0
Total 4 states	227	8,183,715	24,931	855	76.8
Rest of East	17	595,547	22,565	702	6.0
Outside East	46	1,930,710	31,322	815	17.2
Nation	290	10,709,972	25,806	837	100.0

SOURCE: Secretary of the Interior, *Abstract of the Statistics of Manufactures, Seventh Census, 1850* (Washington, D.C.: Government Printing Office, 1859); U.S. Bureau of the Census, *Manufactures of the United States in 1860, Eight Census* (Washington, D.C.: Government Printing Office, 1865).

NOTE: The data for 1850 combine stoves and ranges, and the 1860 data are only stoves. The 1850 data refer overwhelmingly to stoves, because in the 1860 data, hot air furnaces, ranges, and such totaled only about 7 percent of the value of stove products and about 4 percent of the capital invested in stoves.

of the nation converged on the higher national average of value added per employee. Stove manufacturing shifted from small foundries that produced undifferentiated stoves made from commodity plates to large stove foundries that turned out products incorporating innovations in design and style. The capacity of stove foundries outside the East to participate in these advances foreshadowed the extensive stove manufacturing in the Midwest after 1860.[12]

A small number of large stove foundries captured sizable market areas by 1860. As few as twenty-three eastern counties accounted for 91 percent of the East's value added in stove manufacturing and three-quarters of the nation's total, and the six counties with the largest value of production generated almost two-thirds of the East's value added in stoves and slightly over half of the nation's (table 4.10). These consisted of the metropolitan counties of New York, Philadelphia, and Baltimore, New York's satellite (Westchester), and the conurbation of Albany-Rensselaer (Troy). Stove foundries with access to sizable subregional markets had extended their market power to larger territories.

TABLE 4.10
Leading Centers of Stove Manufacturing, 1860

County	No. of Firms	Value of Products ($)	Capital per Firm ($)	Value Added		
				Per Employee ($)	% of East	% of Nation
Top 6 (value of products)	84	5,439,565	45,380	915	63.4	52.5
Philadelphia	28	1,430,765	44,761	804	15.0	12.4
Albany, N.Y	7	1,038,700	144,714	815	12.4	10.3
Rensselaer, N.Y.	8	1,022,250	60,000	1,060	13.5	11.2
Westchester, N.Y.	8	825,250	59,750	860	9.2	7.6
Baltimore	10	589,000	32,750	1,004	7.2	6.0
New York	23	533,600	11,309	1,275	6.1	5.1
Boston hinterland						
Middlesex, Mass.	3	145,200	26,667	968	1.7	1.4
Bristol, Mass.	6	276,200	20,917	653	2.4	2.0
Providence, R.I.	4	259,000	51,250	619	3.0	2.5
Hampden, Mass.	4	59,700	7,000	1,001	0.6	0.5
Hillsborough, N.H.	1	95,000	57,000	778	1.4	1.2
Penobscot, Maine	4	69,160	6,775	737	0.8	0.7
Rutland, Vt.	1	50,000	25,000	614	0.7	0.5
New York hinterland						
Hudson Valley						
Columbia, N.Y.	2	161,800	42,112	1,125	2.1	1.8
Saratoga, N.Y.	4	201,850	17,000	1,879	2.7	2.2
Erie Canal Corridor						
Oneida, N.Y.	6	165,860	28,283	969	1.9	1.6
Onondaga, N.Y.	1	50,000	40,000	300	0.2	0.2
Monroe, N.Y.	2	69,200	25,000	105	0.2	0.2
Erie, N.Y.	4	204,200	24,500	420	1.9	1.6
Philadelphia hinterland						
York, Pa.	2	90,100	37,500	689	0.9	0.7
Pittsburgh hinterland						
Pittsburgh (Allegheny), Pa.	5	381,750	66,000	798	4.5	3.7
Beaver, Pa.	5	47,750	6,360	654	0.6	0.5
Erie, Pa.	2	125,000	32,500	1,218	1.9	1.6
Total selected counties	140	7,891,335	38,366	858	90.8	75.2

SOURCE: U.S. Bureau of the Census, *Manufactures of the United States in 1860, Eighth Census* (Washington, D.C.: Government Printing Office, 1865).

NOTE: Counties included in this table had either 0.5 percent or more of the national value of products ($53,550 or more) or 0.5 percent or more of the nation's employment in stove manufacturing (40 or more).

Like foundries that produced steam engines, stove foundries were concentrated in the prosperous agricultural environs of the metropolises and in other prime farming areas of the East such as the Connecticut Valley (Hampden), the Hudson Valley, the Erie Canal Corridor, and southeastern Pennsylvania (York). The top twenty-three eastern counties contained just under half of the nation's stove firms. Their capacity to capture subregional and sometimes regional markets contrasted with the small firms whose markets remained confined to their immediate vicinity. These foundries benefited from the lower transportation cost of the heavy, bulky iron stoves to their local market, thus allowing them to compete against larger, distant foundries. Nevertheless, these small firms could neither compete on style locally nor overcome their inefficiencies to expand outside their local markets; consequently, they remained marginal producers.

Even as the specialized stove foundry emerged, most foundries remained diversified manufacturers of machines and components, but this mixture did not signify stagnation. Machinists enhanced their skills, and large foundries added to their repertoire of heavy machine tools; thus, these firms developed more sophisticated capabilities to build machines and parts. The two decades following 1840 witnessed the emergence of the large engineering works.

THE LARGE ENGINEERING WORKS COME OF AGE
New York City

The increase in the number of stationary steam engines in New England and the Middle Atlantic from about one thousand in 1840 to almost eleven thousand in 1860 opened avenues for the transformation of the industry (see fig. 4.2). The big engineering works came of age by the late 1840s as leading foundries enlarged their operations. In New York City the capitalization of the Allaire Works reached $300,000; it employed 250 and manufactured $200,000 worth of sugar mills and steam engines annually. Archimedes Iron Works raised its capitalization to $125,000 and its employment to 220 and produced engines, sugar mills, and dredging machines valued at $235,000.

Along the Hudson River north of New York City, the Matteawan Company's machine shop and foundry had almost doubled its employment to 200 since 1830, and it manufactured over $250,000 worth of cotton machinery, sugar mills, and steam engines annually. Its prominent engineer and superintendent, Charles Leonard, had network ties to other premier peers, and the firm's New York City agent, William Leonard, had superb connections to local foundries whose machinists possessed the latest knowledge about innovations in heavy capital equipment. Their network bridges to multiple machinist hubs solidified Matteawan's leadership among the nation's foundries. The West Point Foundry at Cold Spring had employed 360 in 1830; by the late 1840s the number ranged from 300 to 500. Arguably, it ranked as one of the nation's greatest heavy industrial machinery works and produced roughly a half-million dollars of goods annually. Its equipment consisted of three air furnaces, three cupola furnaces, twelve cranes of different horsepower capable of lifting from four tons to fifteen tons, three trip hammers—one of them could work shafts of eight tons—and large machine tools, including a boring mill, fifteen lathes, four planing machines, one slitting machine, and a variety of drilling machines.

Still, by the mid-1840s the machinery works of New York City, as well as those of Philadelphia and of other eastern industrial centers, used unsophisticated machine tools and brute force for most of the manufacturing processes, not withstanding the cranes, trip hammers, and large machine tools. These firms could not build iron steamships of comparable quality to British versions, but between 1833 and 1860 tariff rates of 24 percent to 30 percent on iron manufactures hindered imports of superior British engines. The limited technological sophistication of American machinery works relative to the British, however, should not be construed as a refusal to adopt advanced technology. In the American market of high interest rates, relative to Britain, but abundant resources (including low-cost iron and fuel), machinery users demanded cheap, low-quality engines that depreciated rapidly, and they did not need to economize on fuel. In 1845, amid this technological environment, the United States Congress began subsidies for carrying mail to foreign countries in ocean steamships, and two years later it added subsidies for steamships to carry mail from the east coast to California and Oregon, which lasted until 1859. Most, if not all, of the iron steamships constructed during the time of the subsidies would have been unprofitable to build in their absence.

New York City's machinery works eagerly participated in this new market, and the large marine engines for ocean steamships became another item of their foundry repertoire. Nevertheless, over the twenty years after 1840 the substantial growth of the economy exerted the most impact on the expansion of the large engineering works. The Novelty Works, which had roots around 1830, embarked on its path to fame in 1838, when two machinists and two financiers formed Ward, Stillman and Company, and they undertook to enlarge the works. By 1850 its annual sales reached almost one million dollars, and this consisted of marine engines, over one hundred thousand dollars worth of sugar mill machinery, and other equipment. The firm employed 1,170 workers, three-quarters of them skilled and semi-skilled metal workers—iron founders (248 workers), machinists (359), and boiler makers (242). Like the West Point Foundry, the Novelty Works housed large machine tools (lathes, planers, cutting engines, and bending and punching engines), and big cranes moved the large iron pieces. The scale and complexity of this vast engineering works next to the East River inspired Jacob Abbott's now-famous article on it in *Harper's New Monthly Magazine* in 1851.

The Morgan Iron Works dated from 1838, when T. F. Secor, Charles Morgan, and William Calkin started a foundry, and eight years later, consonant with the growing economy and the lure of mail subsidies for ocean steamships, an expanded facility provided the scale to compete in building large marine engines. In 1847, within three years of marrying Frances Morgan, Charles's daughter, George Quintard entered the firm, and three years later, after Morgan had gained control of the firm, the thirty-year-old Quintard became the top manager of the Morgan Iron Works. He commenced the purchase of large, improved planers, lathes, slotting machines, and steam trip hammers. The combination of his executive skills and Morgan's connections as a steamship magnate vaulted the Morgan Works to a leading position in marine engine construction.

Most foundries produced stationary engines and steamboat/marine engines on order from other firms, but by the mid-1850s, in a hint of future possibilities, New York City's William Burdon used one pattern to build as many as fifty engines at a time. This took advantage of the economies of batch production, and the firm held in stock engines that ranged from three horsepower to forty horsepower. At this time New York City housed seventeen steam engine and boiler foundries with aggregate invested capital of $1.9 million and employment of 3,130, making the city the greatest foundry and machinery center in the nation and its premier community of practice.[13]

Philadelphia

Similar transformations swept through Philadelphia's foundries as older, prominent firms expanded and new, large firms emerged. By the late 1850s the Southwark Foundry of Merrick and Sons, manufacturers of various types of heavy machinery, operated massive equipment, including two cranes, each capable of lifting fifty tons, and three other cranes, each capable of lifting thirty tons. Its boring mill could bore a cylinder eleven feet in diameter and fourteen feet high; its planing machine could plane dimensions of eight feet wide and fifteen feet deep; and it had numerous large and small lathes, planers, and drilling machines.

Around 1847 Isaac Morris, from Levi Morris and Company, led the formation of I. P. Morris and Company. Its Port Richmond Iron Works contained equipment similar to Merrick and Sons. The works' largest planer, even more impressive than that in the Southwark Foundry, weighed over thirty-five tons and could plane an eight foot width and thirty-two foot length; the boring mill could bore a cylinder sixteen feet in diameter and seventeen feet long. Port Richmond supplied equipment for many sugar mills in Louisiana, Cuba, and Puerto Rico and large pumping engines for Philadelphia's and Buffalo's waterworks. Like many of the leading firms, it employed a prominent machinist, Lewis Taws, whose networks provided access to knowledge about the latest technology at other major foundries.

Boston and Providence

Likewise, Boston's foundries enlarged their scale of operations after 1840. The venerable South Boston Iron Company, under the leadership of Cyrus Alger and his sons, continued to expand. Within a decade of beginning production in the mid-1840s, the foundry of Chubbuck and Campbell began producing about fifty stationary engines annually, along with various other machinery. The foundry of Harrison Loring, similarly diversified, reached an employment of 150 and soon added large marine engines to its product line. During the ten years after 1845 Boston's steam engine works increased in number from two to seventeen, and both capital and the value of products increased tenfold, respectively, to $1.3 million and to $1.8 million. By the late 1830s nearby Providence housed nationally prominent steam engine builders, and by 1850 it contained three firms that together employed 240 workers. The Providence Steam Engine Company had become Thurston, Gardner and Company, and its employment rose from about 80 in the mid-1840s to 250 a decade later, at which time its annual production reached $200,000. As one of the nation's most innovative steam engine builders, Robert Thurston acted as a hub of the networks.[14]

THE NETWORK BRIDGES BETWEEN PROVIDENCE AND PHILADELPHIA

Providence's renown machinists operated as network hubs at the local, subregional (Rhode Island and vicinity), New England, and East Coast scale, a pattern dating back to the first decade of the nineteenth century. Their community of practice attracted ambitious young machinists who came to gain training and employment at one of the many successful foundry and machinery works. Willard Fairbanks grew up on a farm in Wrentham, Massachusetts, about fifteen miles north of Providence, and by the early 1820s he moved to that city to work as an apprentice in one of the machine shops. In 1828 the twenty-three-year-old Fairbanks began to manufacture marine and stationary steam engines, and over time his firm and its successors attracted some of the top young machinists in the East, including John Poole, Edward Bancroft, William Sellers, and George Corliss. Each of these extraordinary machinists went on to head their own firms, which ranked among the leading machinery and machine tool producers of the late antebellum.

The Brandywine Valley residents, Poole and Bancroft, epitomized the movement of the sons of the professional and business elite into machinist careers. Poole's father claimed Oliver Evans, the well-known steam engine builder in Philadelphia, as a friend, and Bancroft, Poole's brother-in-law, had an older brother who owned a cotton mill in the valley. In the early 1830s the eighteen-year-olds Poole and Bancroft began trips from the Brandywine to New York state and New England to work in machine shops, thus following the well-worn path by which young machinists acquired technical skills (map 4.1). They shamelessly drew on the financial resources of their elite families and on the network ties of these family members and of their friends and acquaintances which reached to prominent business people. Charles DuPont—successful clothmaker in the Brandywine, member of the wealthy DuPont family of gunpowder manufacturers, and a founder of the Franklin Institute in Philadelphia—provided a loan of one hundred dollars to underwrite Poole and Bancroft's first trip. The

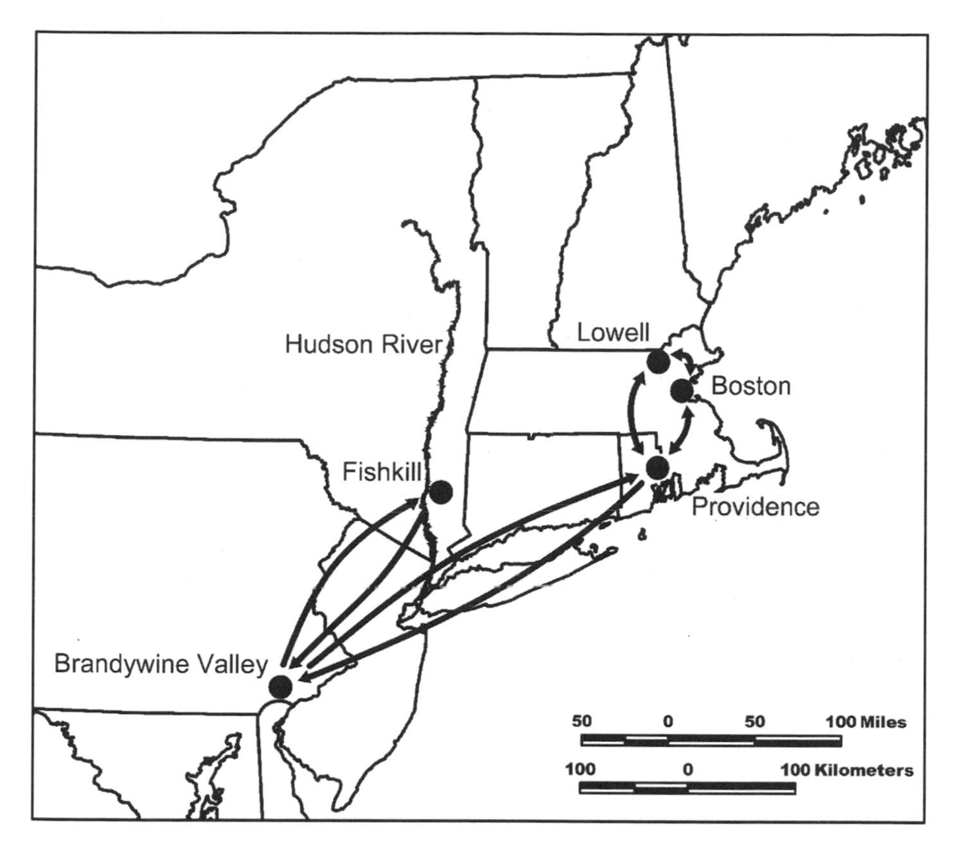

Map 4.1. Travels of John Poole and Edward Bancroft to Machine Shops during the 1830s

loan constituted a princely sum of money which gave them flexibility in seeking job experiences that would provide income and pay for expenses that wages did not cover. For later trips DuPont supplied letters of introduction, and, with his renown and social network connections, the two obtained entrée to the leading foundries, machinery firms, and machine shops.

On their first trip Poole and Bancroft spent about one or two years at the Matteawan Machine Shop in the Hudson Valley, arguably one of the greatest machine shops of that era. They returned to the Brandywine for a short period of time then left for New England around the mid-1830s. They stayed several years in Providence, probably at Fairbanks' steam engine works. Over the next few years they spent some time in the Boston and Lowell areas, and these visits and work experiences must have included major foundries and machine shops because the men used the letters of recommendation from DuPont. Poole returned to the Brandywine, and by 1838–39 he had started a machine shop in the basement of Joseph Bancroft's (Edward's older brother) cotton mill. His shop and subsequent machinery firm became famous in the Brandywine and nearby Wilmington area, and ambitious young machinists came to work in it. Bancroft may have returned to the Brandywine for a short while, but around the late 1830s he had joined a partnership with Fairbanks in Providence. Their firm, Fairbanks, Bancroft and Company, soon acquired a reputation as one of the top foundry and steam engine works, attracting talented youth.

William Sellers came from the wealthy merchant milling family of John Sellers Jr. in Upper Darby, about five miles southwest of Philadelphia. Along with Poole and Bancroft, Sellers drew on his elite connections to gain access to the top machine shops. After finishing private school at the age of fourteen, he began an apprenticeship at the new machine shop of his uncle, John Poole; Sellers stayed about seven years. In 1845 he drew on the friendship between his uncle and Edward Bancroft and went to Providence to work at Fairbanks, Bancroft and Company. Within a year Fairbanks left to start the Taunton Locomotive Manufacturing Company. The Providence partnership became Bancroft, Nightingale and Company, and Sellers stayed there for two or three years before returning to Philadelphia, where he had many family ties, and starting a machine shop to manufacture mill gearing and machine tools. One year later Bancroft left Providence and joined Sellers in Philadelphia, the two men forming Bancroft and Sellers in 1848. Bancroft, in his late thirties, and Sellers, in his mid-twenties, exemplified the ability of talented machinists, especially those from the elite, to vault to the pinnacle of their professions as engineers/machinists while still young. Following the death of Bancroft in 1855, William Sellers and Company went on to become one of the nation's foremost manufacturers of machine tools.

Likewise, the exhilarating opportunities in the emerging career of engineer/machinist attracted George Corliss, another of the ambitious sons of the social, professional, and business elite. Corliss's father, a physician/surgeon in the Hudson Valley town of Easton, north of Albany, moved his family to Greenwich, about a dozen miles farther north. This town housed the cotton textile firm of William Mowry and Son, which had roots back to 1804, when Mowry arrived from the Providence area following his apprenticeship with Slater. George Corliss acquired the trappings of an elite education, remaining in school in Greenwich until he was fourteen, when he entered the employ of William Mowry and Son for four years. His father then sent the eighteen-year-old to Castleton Academy, about fifty miles north of Greenwich, for three more years of schooling. At the age of twenty-one Corliss returned to Greenwich and opened a store, and within a short time he began work on a leather-stitching machine for boots, for which he received a patent in 1842.

Corliss must have become acquainted with the Providence machinist networks through the long-standing ties of the Mowry firm to the mechanics trained by Slater; Corliss moved to Providence in 1844 to market his invention. His experience working for a cotton textile firm and his status as a highly educated member of the professional elite made him a prime candidate for the position of draftsman, one of the most skilled jobs in Bancroft, Nightingale and Company. The firm agreed to hire Corliss provided he drop his idea to build a leather-stitching machine and focus, instead, on steam engines. Providence's community-of-practitioners provided fertile ground for Corliss to gain insight into the key issues of engine construction quickly. Within several years he became a partner; Bancroft soon left for Philadelphia, and the firm became Corliss, Nightingale and Company.

Within two years of arriving in Providence, Corliss developed an automatic variable cut-off engine that provided uniformity of motion, thus allowing close regulation of the engine. Its efficiency in using steam and its economy in fuel consumption added to the engine's attractions. Demand for this type of engine grew as manufacturing requirements became more sophisticated and as the size of firms increased across a range of industries, bolstering the need for large, efficient steam engines. Corliss did not invent the automatic variable cut-off action in isolation; previously, other inventors had experimented with the cut-off action, and patent disputes raged in the courts into the 1860s. His firm's competitive advantages rested on the community of practice of machinery building in the Providence area as well as on the area's machinist networks that reached from the local scale to the rest of the East Coast. These knowledge networks kept his firm abreast of technical skills in building engines and other machinery. Within his firm a community of practice of talented machinists innovated the use of new machine tools to produce engines. During the first decade (1847–57) of building the Corliss engine, the firm's capital soared from $37,000 to almost $300,000, and it built almost 250 engines. Nonetheless, the firm remained a diversified machinery maker even as Corliss engines contributed significantly to profits. As one of the most sophisticated machinery firms, the Corliss Works demonstrated that the equipment market at the close of the antebellum still remained too small to support a high degree of specialization.[15]

THE ACHIEVEMENTS OF MACHINERY MANUFACTURING

The reports of the visiting committee of distinguished commissioners appointed by the British government to attend the New York exhibition of 1853 supply an independent benchmark of the status of the East's large engineering works and iron foundries. The members included Joseph Whitworth, a prominent engineer and the world's leading manufacturer of machine tools, and George Wallis, the headmaster of the Birmingham School of Art and Design. The men did not hesitate to criticize inadequate machine-building techniques and crude machinery, but they expressed favorable views of most foundries and machinery firms, especially those in the leading metropolises. Wallis noted the great progress in iron product manufacturing since the mid-1840s, and Whitworth confirmed that marine engine works, such as those in New York City, remained diversified machinery builders. According to them, the heavy machine tools in the foundries still lagged behind the quality of those used in British foundries, some of them state-of-the-art enterprises. Still, from Whitworth's perspective, the East's foundries were, "for the most part, large and well arranged, and furnished with good powerful cranes."

Machinery manufacturing had enlarged significantly during the two decades of rapid urban-industrial growth following 1840. The nominal value of machinery products more than quadrupled, from eleven million dollars to forty-seven million dollars, and the real value increased even more because the wholesale price level for metal and metal products fell by 25 percent. By 1860 the East housed slightly under 60 percent of the nation's value of products and value added in machinery, but this figure represented a decline in national market share from the three-quarters it had accounted for twenty years earlier (tables 4.8 and 4.11). The decline resulted because most machinery manufacturing remained oriented to local, subregional, and regional markets, and, as a result, the share of this sector in the Midwest increased. Although important advances in machine tools appeared before 1840, most of them remained relatively crude as of the 1850s, especially equipment for high-speed, precision metal cutting and grinding.[16]

Most machinery manufacturing consisted of casting and forging components, and skilled mechanics used metal tools and simple machine tools to make machinery. Economies of scale within firms remained limited, and much of the production involved custom manufacturing of machinery. This required close contact with buyers and the capacity to provide repair services, especially for complex machinery. The large amount of sugar mill machinery constructed for Louisiana and Caribbean sugar mills and the many steamboat engines built by specialized shops such as those in Pittsburgh still remained exceptional. Growing agricultural and industrial areas such as those in the Midwest also gained machinists; some of them acquired training in the region, whereas others migrated from the East. As a result, firms in the East and in the rest of the nation had similar capital per firm and value added per employee (see table 4.11). New York, Pennsylvania, and Massachusetts together accounted for 40 percent of the nation's value added in machinery. The persistent prominence of New York and Pennsylvania reflected their leadership in heavy capital equipment since the early nineteenth century. Their machinists and the investors who underwrote their foundries continued to utilize their networks of know-how trading, which provided access to technological innovations and to markets.

A small number of centers accounted for most of the machinery manufacturing in the East in 1860. As few as thirty-one counties housed 82 percent of the value of pro-

TABLE 4.11
Machinery Manufactured in Iron Foundries, Steam Engine Works, and Machine Shops, by State, Region, and Nation, 1860

				Value Added	
Place	No. of Firms	Value of Products ($)	Capital per Firm ($)	Per Employee ($)	Percentage of Nation
New England	275	7,542,612	22,616	801	16.4
Maine	33	668,345	12,852	527	1.0
New Hampshire	24	414,480	11,850	784	1.0
Vermont	24	501,276	42,717	837	1.0
Massachusetts	126	3,323,751	20,067	793	7.3
Rhode Island	22	924,175	31,655	777	2.1
Connecticut	46	1,710,585	27,413	963	3.9
Middle Atlantic	422	20,082,067	30,262	750	41.5
New York	184	10,037,493	31,741	816	20.2
Pennsylvania	166	6,086,287	24,984	684	12.6
New Jersey	50	1,964,747	23,192	921	4.9
Maryland	16	1,641,000	86,519	535	3.0
Delaware	6	352,540	39,833	655	0.7
East	697	27,624,679	27,246	764	57.8
Outside East	480	19,132,807	30,004	883	42.2
Nation	1,177	46,757,486	28,371	810	100.0

SOURCE: U.S. Bureau of the Census, *Manufactures of the United States in 1860, Eighth Census* (Washington, D.C.: Government Printing Office, 1865).

duction; thus, the level of concentration had declined only slightly since 1840 (tables 4.8 and 4.12). Leadership remained stable: almost half (fifteen) of the top counties in 1860 ranked among the leaders twenty years earlier, and many of the remaining counties occupied the same subregions as the leaders had in 1840. The metropolitan industrial complexes of Boston, New York, Philadelphia, and Baltimore contained over half of the East's machinery manufacturing, testimony to the remarkable dominance of their foundries since the first burst of steam engine production at the start of the century. The communities of practice in these complexes conferred substantial competitive advantages on their firms.

In the hinterland outside each metropolitan complex, the prosperous agricultural-industrial areas and the subregional metropolises remained the major places of machinery manufacturing. These included Providence and Worcester, the Connecticut Valley of Massachusetts (Hampden), and southeastern Maine, all around Boston; Hartford and New London in Connecticut, the Hudson Valley, and the Erie Canal Corridor, all in New York's hinterland; and the agricultural-industrial and mining areas (Berks, Luzerne, and Schuylkill) northwest of Philadelphia. With few exceptions the limited variance across counties in the size of firms, measured by the amount of capital per firm, and in productivity, measured by the amount of value added per employee, suggests that know-how trading through machinist networks diffused technology among the leading centers, creating a broad community of practice.

TABLE 4.12
Leading Centers of Machinery Manufactured in Iron Foundries, Steam Engine Works, and Machine Shops, 1860

				Value Added		
Place/County	No. of Firms	Value of Products ($)	Capital per Firm ($)	Per Employee ($)	% of East	% of Nation
Boston hinterland						
Boston complex	45	1,722,188	34,631	780	6.5	3.8
Suffolk (Boston), Mass.	28	972,650	23,075	880	3.4	2.0
Middlesex, Mass.	17	749,538	53,665	695	3.1	1.8
Providence, R.I.	15	797,975	35,993	831	3.2	1.9
Worcester, Mass.	26	472,893	9,077	796	1.9	1.1
Hampden, Mass.	6	422,000	57,783	819	1.5	0.9
York, Maine	4	366,800	51,150	613	1.5	0.9
Cumberland, Maine	5	439,300	53,720	374	0.8	0.5
Washington, Vt.	6	304,800	146,317	1,089	1.0	0.6
New York hinterland						
New York complex	97	7,920,134	42,693	854	27.7	16.0
New York City, N.Y.	62	5,809,492	51,160	846	18.8	10.9
Rockland, N.Y.	2	311,500	10,000	465	1.1	0.6
Putnam, N.Y.	1	420,000	240,000	792	1.6	0.9
Hudson, N.J.	5	332,000	22,900	1,328	1.5	0.9
Passaic, N.J.	8	286,892	23,688	453	1.1	0.6
Essex, N.J.	19	760,250	21,332	1,572	3.6	2.1
Hartford, Conn.	7	368,485	37,886	761	1.4	0.8
New London, Conn.	12	851,400	51,500	1,178	3.7	2.1
Hudson Valley						
Dutchess, N.Y.	7	331,960	32,143	714	1.4	0.8
Albany, N.Y.	8	205,200	21,875	738	0.8	0.5
Erie Canal Corridor						
Seneca, N.Y.	4	452,500	81,250	1,199	2.1	1.2
Erie, N.Y.	9	374,520	27,844	775	1.7	1.0
Chemung, N.Y.	3	272,400	22,667	570	0.5	0.3
Philadelphia hinterland						
Philadelphia complex	81	3,430,276	30,316	683	12.3	7.1
Philadelphia, Pa.	62	2,466,096	28,352	720	9.0	5.2
Montgomery, Pa.	4	215,300	28,875	472	0.6	0.3
Burlington, N.J.	4	211,700	43,125	743	0.8	0.5
Mercer, N.J.	6	188,680	28,800	539	0.7	0.4
New Castle, Del.	5	348,500	47,400	652	1.2	0.7
Berks, Pa.	5	538,138	95,000	505	2.1	1.2
Schuylkill, Pa.	10	330,444	21,300	694	1.1	0.6
Luzerne, Pa.	10	439,340	36,800	697	1.8	1.0
Baltimore (metropolis, complex), Md.	10	1,492,500	133,580	521	4.7	2.7
Pittsburgh (Allegheny, Pa.)	24	1,031,968	20,688	714	3.4	2.0
Total four complexes	233	14,565,098	40,734	755	51.2	29.6
Total selected counties	394	22,565,221	39,198	756	81.2	47.0
East (New England, Middle Atlantic)	697	27,624,679	27,246	764	100	57.8
Outside East	480	19,132,807	30,004	883		42.2
Nation	1,177	46,757,486	28,371	810		100

SOURCE: U.S. Bureau of the Census, *Manufactures of the United States in 1860, Eighth Census* (Washington, D.C.: Government Printing Office, 1865).

NOTE: Counties included in this table had either 0.5 percent or more of the national value of machinery, steam engines, and such ($233,787 or more) or 0.5 percent or more of the nation's employment in machinery (181 or more). New York City consists of New York and Kings counties, both of which met the criteria for inclusion in the data.

THE RESILIENCE OF IRON FOUNDRY NETWORKS

Foundries played pivotal roles in the agricultural and urban-industrial growth of the East. Although all of the subregions that experienced this growth housed these suppliers of heavy capital equipment, the widespread distribution of machinist skills among the workforce did not translate into strong market positions for most firms. A small share of foundries in the metropolitan industrial complexes and nearby subregional metropolises, along with those in the rich farming areas of the East, dominated production. Machinists in these foundries operated in sophisticated networks that channeled technological knowledge, and investors supplied capital to underwrite foundries and provide access to markets.

Publications such as the *Journal of the Franklin Institute* transmitted increasing amounts of technical knowledge, but their impact remained subdued for many decades. Instead, machinists relied on the robust mechanism of the artifact-activity couple, learning technical skills in foundries by working on machines with other skilled mechanics. Technical knowledge passed among foundries as workers changed jobs, and this mobility within and among the largest complexes of New York and Philadelphia gave their firms competitive advantages. These firms also accessed job seekers in all of the other leading foundry centers. The networks operated effectively because machinists benefited from sharing technical knowledge within a community of practice, and mechanics who did not participate in the networks incurred substantial costs, such as lack of access to technical skills and to innovations.

Hub individuals provided contacts with pivotal members of other networks, but no subset of them controlled a large share of technical knowledge. The machinist networks and the top foundry centers arose early in the nineteenth century, indicating the capacity of machinists and firms to leverage their skills and capital to retain market dominance. Leading mechanics and firms passed from the scene, yet the set of top foundry centers remained relatively stable, confirming that machinist skills were embedded in the networks of know-how trading in communities of practice, rather than in the individuals and firms.

Iron foundry networks also contributed to the development of locomotive manufacturing, and foundries built many of the earliest engines. Within a short time, however, the number of foundries making locomotives declined because these heavy capital goods required strict attention to the changing demands of railroads for better locomotives. Locomotive manufacturing was concentrated in far fewer firms than were stoves, the other major capital equipment sector whose production had shifted to specialized foundries.

CHAPTER FIVE

Networked Machinists Build Locomotives

> We have recently paid several visits to the extensive works of Messrs. Richard Norris & Son in Philadelphia, one of the largest and most perfect in all its appurtenances in this or any other country . . . . The present capacity of the whole works, when fully manned, would enable the delivery of a finished locomotive every second working day in the year.
>
> *"The Transportation of Passengers and Wares, a Visit to the Norris Locomotive Works," United States Magazine*

Locomotives captivated Americans' imagination. These powerful machines symbolized economic dynamism, and they were among the most complicated machinery in existence. By the 1840s a typical locomotive contained four thousand parts, and within a decade this number had risen to six thousand. Once production exceeded a few engines, the manufacturing process required large spaces for the foundry, boiler shop, machine shop, and erecting building. Even if an existing iron foundry, textile machinery firm, or general machine shop carried out the work, these businesses needed to be enlarged to accommodate locomotive building. The Norris Locomotive Works in Philadelphia, which was visited in 1855 by a writer for *United States Magazine*, covered one hundred city lots, and the nearby Baldwin Locomotive Works, as well as the Rogers Works in Paterson, New Jersey, were of a similar scale. They were the "big three" locomotive works of the 1850s.

The glamour and size of locomotives implies that their production provided a special spur to the development of the machinery industry, but that conclusion is mistaken. By 1860, following almost three decades of growth, the value of locomotive production totaled five million dollars, the same as the value of textile machinery. In contrast, the production value of steam engines and related machinery reached forty-seven million dollars, almost ten times greater. Few railroads built their own engines; instead, antebellum locomotive manufacturing had roots in iron foundries, steam engine works, textile machinery firms, and machine shops. Locomotive production joined other sectors in contributing to the development of the heavy capital equipment industry and of the broader category of machinery manufacturing.

As leaders in locomotive engineering, British builders captured about one-quarter of the American market through the 1830s because they possessed the technology and the capacity to supply initial demand. Nevertheless, the tariff of roughly 25 percent on locomotives hurt British firms, which faced difficult markets in the United States. Early British designs did not suit the American terrain, and lower population densities and longer distances placed a premium on cheaper, simpler construction of track and equipment than was customary in Britain. American railroads were risky ventures, and, to meet their diverse needs, locomotive manufacturers needed to coordinate their production closely with railroad management. During the 1830s domestic builders quickly learned the technology through copying, and they assembled capital and skilled labor to produce this new machinery in quantity.[1]

RAILROADS AND THE LOCOMOTIVE INDUSTRY

Because construction of railroad lines created the initial demand for locomotives, the expansion of track mileage indicates the market trajectory that locomotive producers faced. After 1830 the annual mileage built increased significantly, though erratically. It rose during the economic expansion of the 1830s, declined part of the time when the economy contracted (1837–43), and ascended to new, higher levels during the long period of growth from 1844 to 1860 (fig. 5.1). Cumulative mileage built leaped from twenty-three miles in 1830 to eleven hundred miles five years later. Then the pace of growth settled into an astonishing compound annual rate of about 13 percent, reaching almost thirty-one thousand miles by 1860. Cumulative gross investment in railroads closely matched the increase in mileage; investment rose from thirty-three million dollars in 1835 to one billion dollars by 1860, a 14 percent compound annual growth. Consequently, locomotive builders faced an initial spurt in demand and a long-term, rapidly growing market, punctuated by episodic surges and collapses.

Throughout the 1830s and 1840s most railroads in the East built track within 150 miles of Boston, New York, Philadelphia, or Baltimore. Multiple lines radiated into their hinterlands, and their names captured that character: for example, the Boston and Providence Railroad, the New York and New Haven Railroad, the Phila-

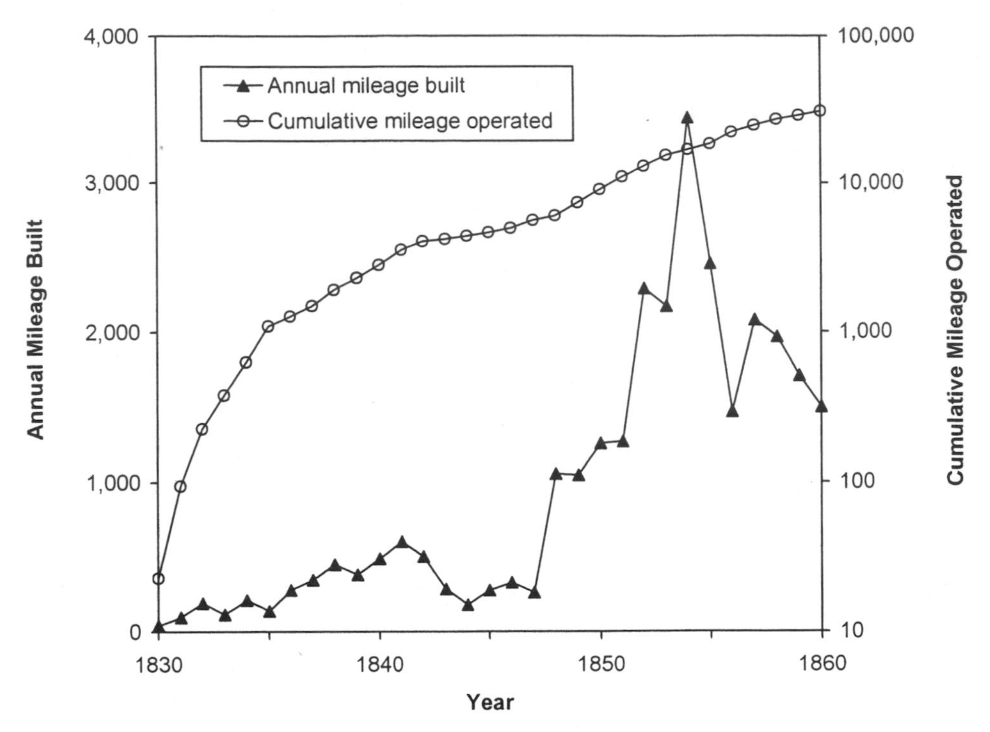

Figure 5.1. Railroad Mileage, 1830–1860. *Source:* U.S. Bureau of the Census, *Historical Statistics of the United States, Colonial Times to 1970, Bicentennial Edition, 2 Parts* (Washington, D.C.: Government Printing Office, 1975), ser. Q321, 329.

delphia, Germantown and Norristown Railroad, and the Baltimore and Ohio Railroad. Smaller cities within this territory built railroad lines such as the Norwich and Worcester Railroad and the Harrisburg, Portsmouth, Mt. Joy, and Lancaster Railroad. Early locomotive markets therefore lay within easy reach of the foundries and machinery works of metropolises and their satellites, and some foundries in smaller hinterland cities accessed nearby railroad markets.

Using annual regional gross investment in railroads as a proxy for locomotive demand, Middle Atlantic firms faced the earliest large market for engines in the 1830s, whereas New England firms confronted much smaller markets that commenced a modest upturn around the mid-1830s (fig. 5.2). During the economic recovery following 1843, New England's gross investment surged, surpassing the Middle Atlantic's investment for the remainder of the decade. After the late 1840s, however, New England's investment declined, whereas the Middle Atlantic's swelled and stayed significantly above New England's during the next decade. The South's railroad investment rose rapidly in the late 1830s but then declined and remained low until the

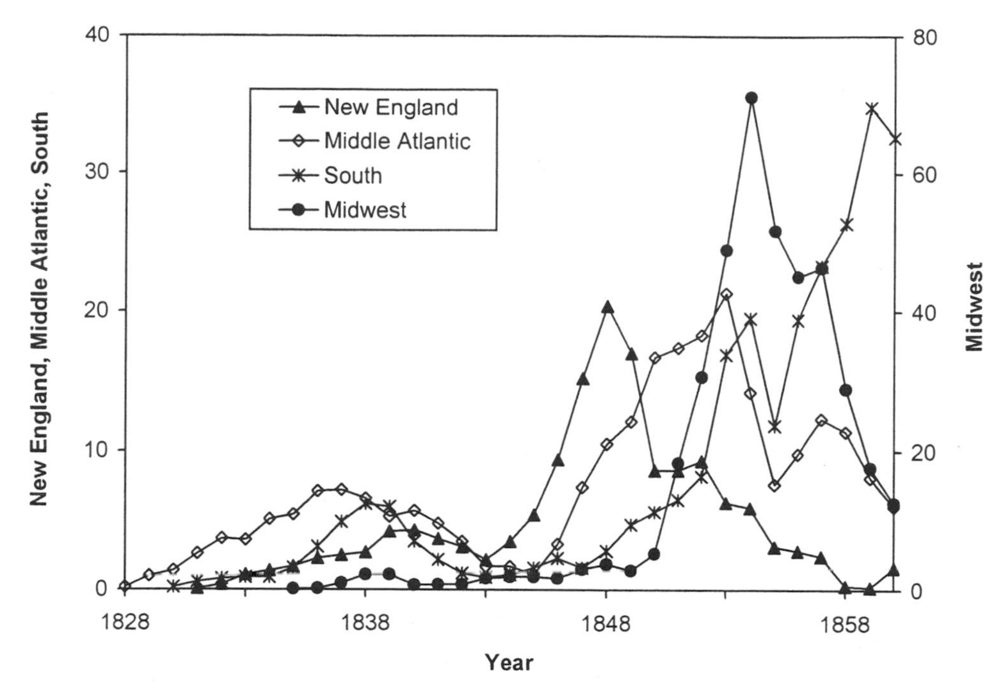

Figure 5.2. Regional Gross Investment in Railroads (in Millions of Dollars), 1828–1860. *Source:* Albert Fishlow, *American Railroads and the Transformation of the Ante-Bellum Economy* (Cambridge: Harvard University Press, 1965), 397, table 53.

late 1840s, after which investment surged to higher levels. In contrast, the Midwest's investment remained low until the late 1840s then skyrocketed to a peak in 1854, far surpassing other regions. The swift collapse still left it above the East, although the South captured the lead at the end of the decade.

Firms that acquired substantial market share during early growth of an industrial sector sometimes gained advantages in technical and organizational skills, production economies, and knowledge of marketing and sales, thus hindering challenges to their leadership by later entrants. Cumulative railroad construction expense by state during the initial stage from 1828 to 1839 maps out the basis of the battle among locomotive builders for market share (fig. 5.3). Pennsylvania and Maryland jumped into the lead while investment stayed small, but by 1833 Pennsylvania's cumulative investment soared. Six years later its twenty-three million dollar investment was over twice that of each of the next largest states—Massachusetts, New York, and Maryland. Consequently, Philadelphia area firms possessed the earliest access to large locomotive markets, possibly benefiting from nearby New Jersey, which achieved a more than quadrupling of railroad construction between 1831 and 1834.

Among the major locomotive builders the Philadelphia firms of Baldwin and Nor-

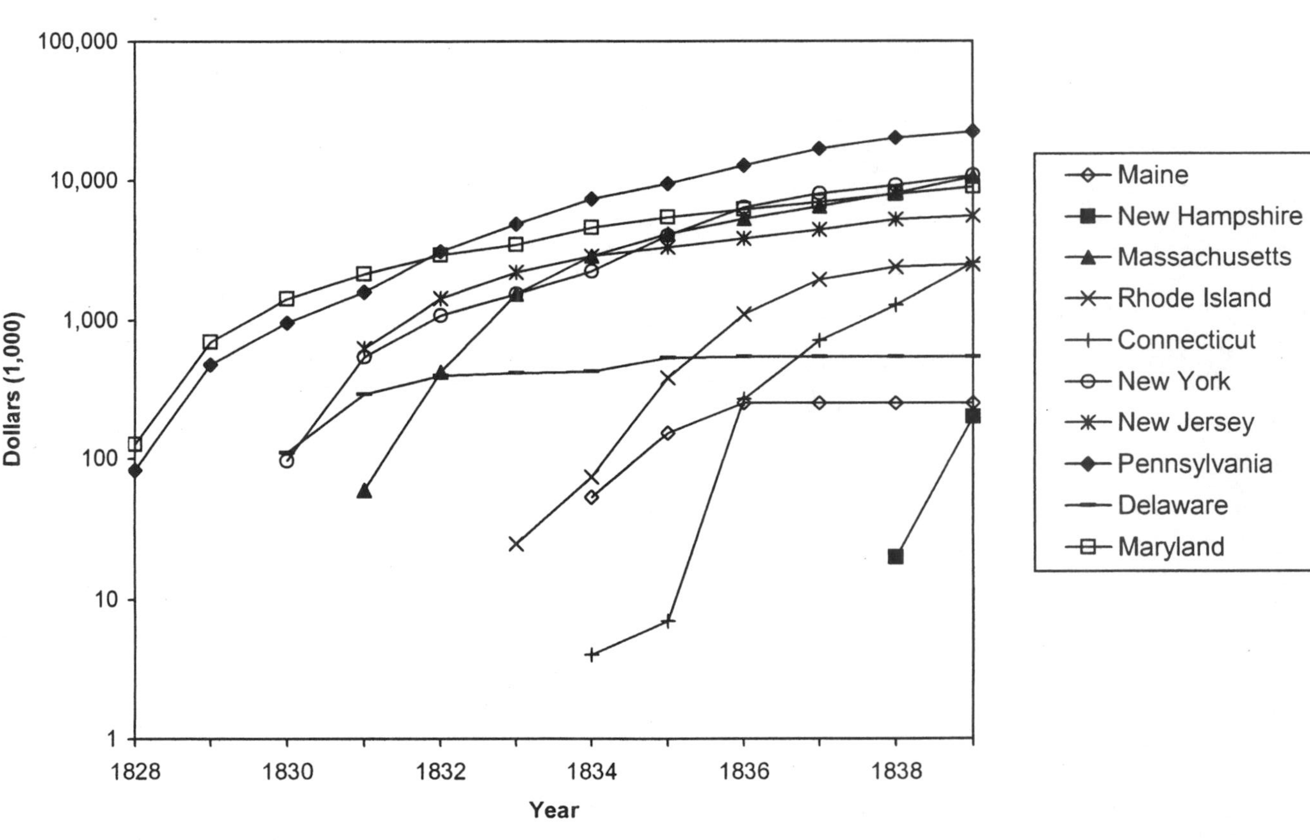

Figure 5.3. Cumulative Railroad Construction Expense by State, 1828–1839. *Source*: Albert Fishlow, *American Railroads and the Transformation of the Ante-Bellum Economy* (Cambridge: Harvard University Press, 1965), 385, table 51.

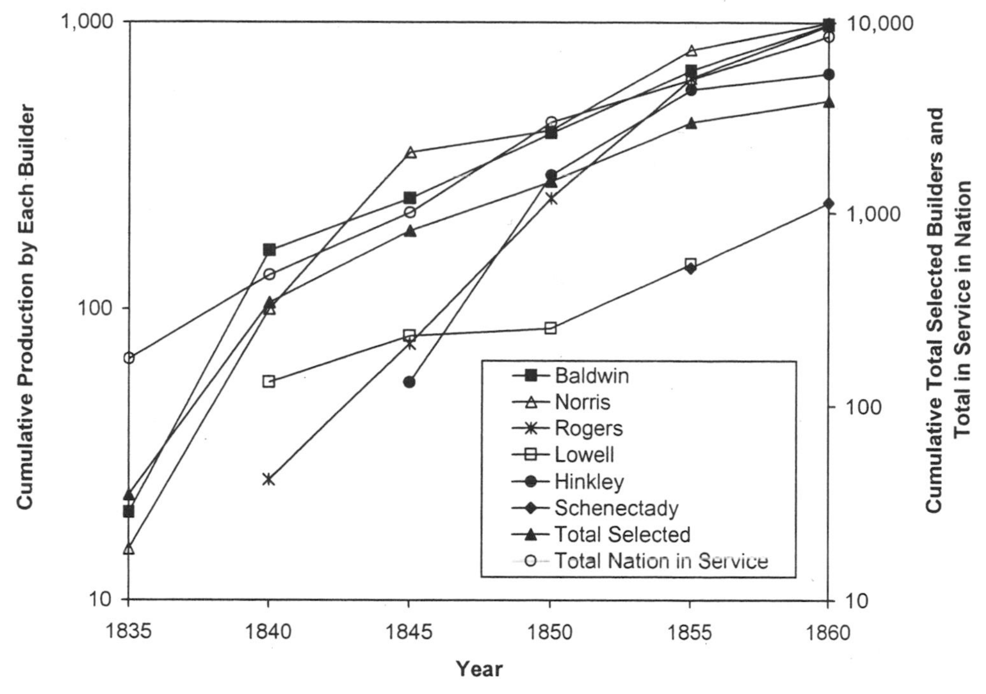

Figure 5.4. Cumulative Locomotive Production of Major Builders and Total Number in Service in the Nation, 1835–1860. *Sources:* George S. Gibb, *The Saco-Lowell Shops: Textile Machinery Building in New England, 1813–1949* (Cambridge: Harvard University Press, 1950), 641, app. 6; John H. White, *American Locomotives: An Engineering History, 1830–1880*, rev. and exp. ed. (1968; rpt., Baltimore: Johns Hopkins University Press, 1997), 20, table 1.

ris ranked as the top producers, and by 1840 they held a commanding position in the cumulative number built (fig. 5.4). The construction of the Baltimore and Ohio Railroad dominated Maryland's investment between 1828 and 1839, thus giving Baltimore firms a market. From 1833 to 1839, however, the railroad built some of its engines through a contracting relation with locomotive builders at its Mount Clare shops near Baltimore, undermining the emergence of independent locomotive firms.

Locomotive builders in New York City and in nearby New Jersey benefited from railroad construction in that state, but Connecticut did not invest in railroads until the late 1830s (see fig. 5.3). Beginning in the early 1830s, New York state railroads, especially those radiating from New York City, invested heavily. By 1836 the state's cumulative total of six and a half million dollars surpassed Maryland's, and it reached eleven million dollars three years later. That construction supported the emergence of the Paterson, New Jersey, firm of Rogers, and it went on to rival Baldwin and Norris (see fig. 5.4). Boston-area firms benefited from the rapid rise in subregional investment in the early 1830s, which proceeded at a similar pace and reached the same level as New York state by 1839. The Lowell Machine Shop (called the Locks and Canals Shop until 1845) started meeting this demand in the mid-1830s, and subsequent railroad construction drew in the Hinkley firm as a major Boston builder in the early 1840s. Rhode Island's investment lagged until 1835, after which it provided modest, supplemental demand for locomotives. The northern New England states witnessed little investment.[2]

ENTERING THE LOCOMOTIVE BUSINESS AND REMAINING IN IT

Although railroad construction in the 1830s set the stage for the emergence of the future dominant firms (Baldwin, Norris, and Rogers), they could not monopolize the market before 1860. They captured about two-thirds of the market in the first half of the 1840s, but their share declined to roughly 40 percent during the next decade (see fig. 5.4). With the backing of well-to-do investors, aspiring locomotive manufacturers who possessed standard machinist skills acquired in foundries, steam engine factories, textile machinery firms, and machine shops faced few barriers to entry before 1860. Early builders demonstrate this trend: Baldwin's (Philadelphia) metalworking experience rested in tool making, hydraulic presses, and stationary steam engines; Rogers (Paterson) started out manufacturing textile machinery; the Stevens Works (Hoboken, N.J.) made steam engines; and the West Point Foundry (Cold Spring, N.Y.) produced heavy castings, steam engines, and sugar mill machinery. The high profits on individual locomotives, production processes that used well-known techniques, and capital requirements below fifty thousand dollars enticed many firms into the business, but most of them turned out only a few engines over a short period.

Large-scale, sustained manufacturing thrust producers into the risky volatility of a heavy capital equipment business with a small number of sellers and buyers, all subject to the investment cycle of railroad construction and the economy-wide business cycle (see fig. 5.2). Annual production of the Baldwin Locomotive Works hit five peaks and four troughs of the locomotive cycle during its first thirty years of business (fig. 5.5). Individual locomotives typically sold for about eight thousand dollars, and during production the builder expended large sums for materials, components purchased from suppliers, and labor. Processes remained labor intensive, and firms employed few machines until the 1850s. If a railroad ordered several locomotives at a time, the builder might have between twenty-five thousand and fifty thousand dollars in sales to one customer. Small producers making as few as five locomotives in a year could face substantial credit risk if a railroad defaulted before paying cash for the locomotives it ordered. If the company paid with its securities (bonds or stocks), a practice that became prevalent by the mid-1840s, the builder faced losses if the securities fell

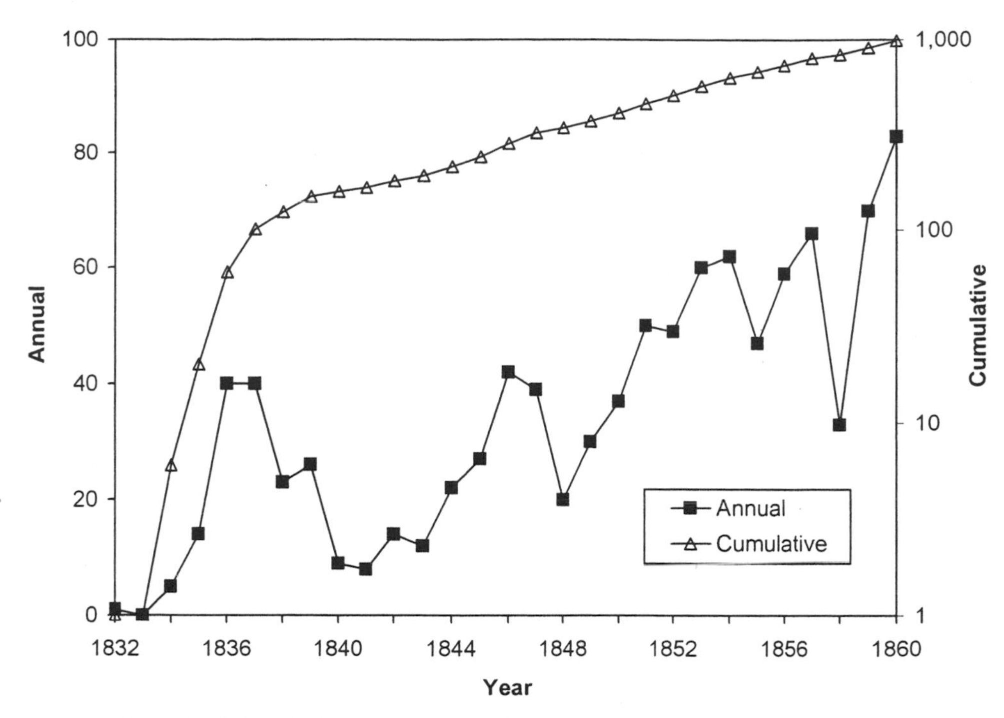

Figure 5.5. Annual and Cumulative Locomotive Production of Baldwin, 1832–1860. *Source:* John K. Brown, *The Baldwin Locomotive Works, 1831–1915* (Baltimore: Johns Hopkins University Press, 1995), 241–42, app. A.

in value or became worthless before they could be sold. The vicissitudes of troughs in the business cycle exacerbated these risks, because the resale value of locomotives plunged. These risks probably explain why few builders sustained meaningful production over long time spans.

Before 1860 locomotive builders did not have sophisticated financiers in their senior management, although some shrewd builders had financiers as partners. The location of production in a metropolis (Boston, New York, Philadelphia, or Baltimore) or in a nearby satellite became a necessary, though not sufficient, condition for long-term success. Locomotive managers could employ local social networks to access senior railway officials—the metropolises housed the headquarters of most major lines—to solicit orders and to assess their railroads' financial strength. They also used these networks to maintain ties with financiers who provided expert evaluations of the railroads' credit worthiness and supplied credit for manufacturing and discounted railroad securities, in the form of bonds and stocks, to provide cash.[3]

NETWORKS OF LOCOMOTIVE BUILDERS

Foundries and machinery works, especially in textiles, provided optimal training sites for locomotive builders because they could build steam engines, and locomotives constituted large, mobile versions of them. Railroad repair shops—in essence, machinery works with foundries—likewise trained engine builders. Machinists moved among these firms and shops, and, as some foundries and machinery works began turning out locomotives, these firms became training grounds for builders, who then moved to other locomotive firms or to railroad repair shops. Several clusters of long-term, successful builders emerged, constituting the hubs of locomotive machinist networks. Each hub possessed internally cohesive networks of know-how trading which generated a community of practice, thus strengthening the viability of local locomotive firms. The mobility of senior managers and machinists from one hub to another created network bridges of contacts over which flowed knowledge about demand for locomotives and about innovations in building, reinforcing the dominance of the hubs.

Outside these hubs, foundries and machinery works, including those with locomotive builders, possessed network ties to the hub firms. The movement of hub machinists to these external firms provided contacts for selling locomotives, and the machinists in turn kept builders in the hubs attuned to the demand for locomotives and to technical innovations in them. The network ties also served as recruitment vehicles to attract skilled machinists to the hubs. The contacts of hub firms ranged widely across the East to foundries, machinery works, locomotive builders, and railway repair shops. These network ties may have been as consequential as the bridges among the hubs in reinforcing their dominance.

PROMINENT TRAINERS YET MINOR BUILDERS

Baldwin, Norris, and Rogers—the top builders—trained large numbers of locomotive machinists, but these firms could not have trained enough of them to meet the needs of every producer of locomotives or every railway repair shop. Many foundries and machinery firms trained mechanics for locomotive firms, and several firms and centers of machinery manufacturing achieved prominence as training grounds, even though they did not engage in innovative locomotive production or they made few, if any, locomotives.

The Lowell Machine Shop

In 1834 the Locks and Canals Shop, arguably the greatest antebellum machine shop and the supplier of textile machinery to the Lowell mills, hired George Whistler to head its locomotive building, paying him the lucrative salary of three thousand dollars and providing a rent-free dwelling. Whistler possessed a West Point engineering education, had studied locomotives in England, and had acquired four years' experience with three American railroads. Locks and Canals completed its first locomotives in 1835 and turned out a total of 143 locomotives over the next twenty years; then its production declined to insignificance, and after 1861 it ceased manufacturing engines (see fig. 5.4). Although not unprecedented, hiring external talent to lead new machinery efforts epitomized the limited commitment of the Lowell Shop to locomotive manufacturing. Patrick Jackson, a leader among the Lowell industrialists, had consulted with Philadelphia's Baldwin about establishing production, and the shop continued to rely on external advice. It operated more as a recipient, rather than as an initiator, of innovative technical knowledge.

The extraordinary role of the Lowell Machine Shop as a training ground for many machinists, managers of machine shops, and superintendents of entire industrial works more than compensated for its deficiencies as an innovative locomotive builder and even as a creative machine shop. It purchased the best machine tools and equipment and built copies for itself; it employed talented machinists; it successively hired Paul Moody, George Brownell, and William Burke, considered among the nation's premier manager/machinists, to head its shop; and it possessed some of the most sophisticated production processes and accounting and management controls for manufacturing anywhere in the East. Aspiring machinists and future shop managers and superintendents, along with individuals who already had experience who came to work at the Lowell Shop, acquired superb technical skills and managerial knowledge that they could transfer to other firms.

Before working at the Locks and Canals Shop in 1835–36, Parley Perrin had worked in other New England machine shops. At Lowell he built calico printing presses and worked on locomotives under George Whistler. Following this training, Perrin moved to Seth Boyden's engine works in Newark, New Jersey, and built locomotives during 1836–37. Ten years later the thirty-four-year-old took a position at the new Taunton Locomotive Manufacturing Company in Massachusetts. With his extensive experience he started as foreman, draftsman, and locomotive designer and later became superintendent. While in his twenties, George Griggs learned the machinist trade at Lowell, and in 1834 the Boston and Providence Railroad appointed the twenty-nine-year-old as master mechanic (the premier machinist position). He worked mainly at the Roxbury (adjacent to Boston) repair shops, completing his first locomotive in 1845. During his career at the railroad he built about thirty engines and received various patents for locomotive designs and components. In 1846 Griggs became a shareholder of the Taunton Locomotive Manufacturing Company, and he sent one of his top mechanics to the firm to help the firm build its first locomotive.

Wilson Eddy started as an apprentice machinist at the Lowell Shop when it commenced locomotive production, putting him into contact with Whistler, the department head. In 1840 Whistler—now superintendent of the Western Division of the Boston and Albany Railroad—hired the twenty-seven-year-old Eddy as a foreman at the Springfield, Massachusetts, repair shops. By the end of the decade Eddy had advanced to the position of master mechanic for the railroad. He began building locomotives and, over the next thirty years, built more than one hundred. After learning the machinist trade, Aretas Blood went to the Lowell Shops around the late 1830s and stayed for seven years. By 1849 he had moved to the Essex Machine Shop in nearby Lawrence, where he produced locomotive parts as a subcontractor. Following four years of experience there, he used his savings to purchase a share in the new Manchester Locomotive Works in New Hampshire. After another four years, the forty-three-year-old Blood had become superintendent of the works. These alumni of the Lowell Machine Shop are emblematic of the many machinists who received training there and later worked on locomotives. The top mechanics became pivots of machinist networks in New England by their early twenties, and most had attained prominent positions in firms by their thirties.

The Matteawan Company

As locomotive production accelerated around the mid-1830s, the Matteawan Company already had experience building steam engines, sugar mill machinery, and other equipment. This outstanding firm possessed close ties to New York City's foundries and machinery works and to those in its satellites, especially in Jersey City, Hoboken, Newark, and Paterson. It supplied several locomotives for the Hudson River Railroad, but its total production remained small.

Matteawan served as a training ground for so many Paterson machinists that a nineteenth-century author of the city's industrial history felt compelled to note that "it is wonderful how many of the great pioneer mechanics and successful manufacturers came to Paterson by way of Matteawan." Several prominent machinists who worked at Matteawan did not meet there, but their careers intertwined in Paterson. In 1821 the twenty-four-year-old Charles Danforth arrived in Matteawan following his experience at cotton mills in the Taunton area. After serving as a foreman for four years, he made various career moves that positioned him in 1831 to become a partner in Paterson's

Godwin, Clark and Company. When the firm failed eight years later, Danforth took over the works and eventually added a partner in 1848. Within four years the Danforth, Cooke and Company began to manufacture locomotives and other machinery.

Danforth's partner John Cooke might have received early apprenticeship training at Matteawan. John's father, Watts, along with all of his sons, had moved to Matteawan around the mid-1830s, and the family relocated to Paterson at the end of that decade. Watts joined Rogers, Ketchum and Grosvenor around 1842 and made patterns for mill gearing, part of the firm's diversified machinery work, which also included locomotives. At the same time, his son John started learning the pattern-making trade in the firm. After William Swinburne, the leading engine builder at the Rogers firm, left in 1845, John Cooke became superintendent of locomotive construction. He left the firm in 1852 to become a partner, as a builder of engines, in the firm of Danforth, Cooke and Company. Cooke's job changes, from being an apprentice as a teenager to superintendent of a machine shop by the age of twenty to being a partner in a firm before the age of thirty, exemplifies the rapid upward career mobility of talented machinists.

Swinburne arrived at Matteawan in 1827 at the age of twenty-seven and made patterns for machinery. Within six years he moved to Paterson to work for Rogers, Ketchum and Grosvenor, which at that time still produced mostly textile machinery. In 1835, when the firm started to manufacture locomotives, Swinburne participated briefly, but he spent most of the next seven years in charge of patterns for mill gearing. He shifted to locomotive building around 1842 and three years later left Rogers and joined with partners in Swinburne, Smith and Company to engage in general machinery business; by 1848 the firm focused on locomotives.

The Matteawan Company thus provided an extraordinary venue for training machinists, and the job mobility of its trainees, which took them to other machinery centers, transmitted technical skills. The community of practice of Paterson's locomotive works owed a substantial debt to Matteawan, whose technical skills were embedded in machinist networks. This expertise continued in Paterson (and elsewhere) long after the Matteawan Company collapsed in the late 1850s.

Providence Machine Shops

In contrast to Lowell and Matteawan, the machine shops in Providence and its immediate vicinity produced few, if any, locomotives. Nonetheless, this hub of machinist networks—which had bridges throughout New England, into the Hudson and Mohawk valleys, to New York City's New Jersey satellites, and to Philadelphia—served directly and indirectly as a training ground for many locomotive machinists. Aside from its textile and general machine shops, Providence's steam engine firms constituted superb venues for learning skills that fit locomotive building. In 1826 the twenty-one-year-old Willard Fairbanks lived in Providence, and during the years that followed he became a skilled engineer. By the late 1830s he became a partner in steam engine firms, including Fairbanks, Clark and Company, and a few years later he partnered in Fairbanks, Bancroft and Company. In 1846 Fairbanks helped found the Taunton Locomotive Manufacturing Company, and Parley Perrin, one of the top machinists at Fairbanks, Bancroft and Company, became foreman, draftsman, and locomotive designer. Edward Bancroft, Fairbanks's partner, and William Sellers, another leading machinist in the steam engine works, left for Philadelphia in 1847. Within a year their machine tool firm, Bancroft and Sellers, began production; by the 1850s its diverse market included locomotive builders such as Philadelphia's Baldwin.[4]

The Lowell Machine Shop, the Matteawan Company, and the Providence machine shops focused on textile machinery and industrial equipment. Nonetheless, they contributed to the cross-fertilization of technical skills among machinist sectors and to the strengthening of a larger community of practice of locomotive builders. These trainers operated as network hubs in their machine sectors, but the leading centers of locomotive manufacturing—Philadelphia, Paterson, and the Boston region—functioned as the pivotal hubs of locomotive machinists and the homes of the most important communities of practice.

THE HUBS OF LOCOMOTIVE BUILDERS

Philadelphia

As the headquarters of the Baldwin and the Norris Works, two of the three largest antebellum locomotive firms, Philadelphia held the mantle as the foremost hub of locomotive machinist networks. When engine production began in Philadelphia—Baldwin started in 1831 and Norris a year later—this sector possessed instant membership within a local cohesive network of foundries and machine shops. During the early years most network bridges reached from Philadelphia's locomotive builders to railroad repair shops and to the other railroad units because the builders constituted the chief source of skilled machinists, engineers, and shop superintendents. By the mid-1840s, however, railroads trained sufficient numbers of these employees to supply skilled machinists regularly to the locomotive builders; thus, workers moved both ways across the network bridges.

In 1831 Andrew Vauclain worked on Baldwin's first engine, and during the early years of the firm's locomotive building Vauclain took engines to their purchasers and assembled them. Interspersed with his employment at the Baldwin Works, his subsequent career included service as general superintendent of the Ohio and Mississippi Railroad, where he oversaw the establishment of their repair shops in East St. Louis,

Illinois; positions with the Reading Railroad, including master mechanic of the Port Richmond repair shops; and, following 1856, positions in the Altoona repair shops of the Pennsylvania Railroad.

Enoch Lewis started as an apprentice with Philadelphia locomotive builders Eastwick and Harrison in 1836. In a subsequent career move he served as superintendent of railroad car construction in Russia for the St. Petersburg and Moscow Railroad between 1844 and 1846. After returning to the United States, he worked for four years as a foreman in machine shops in Trenton, New Jersey, and then at Ballardvale, Massachusetts, where he supervised locomotive building and machine tool making. In 1850 the twenty-nine-year-old moved to the Pennsylvania Railroad, where he held various supervisory positions for seven years, including work at the repair shop at Mifflin and at the Middle Division headquarters at Altoona.

Baldwin's personal networks bridged to Morristown, New Jersey, a leading iron forge and foundry center, and in 1839 he temporarily drew in Stephen Vail, owner of the Speedwell Iron Works, as a partner. Baldwin's networks also bridged to the New York state railroads, including the Mohawk and Hudson. In 1842 the fifty-year-old Asa Whitney, former superintendent of that railroad, became Baldwin's partner. Starting in his late teens, Whitney had worked in various machine shops in Massachusetts, and through his twenties he worked in New Hampshire. Then he moved to New York state, where he acquired diverse experience, including building textile machinery. In 1830 the Mohawk and Hudson Railroad hired the thirty-nine-year-old to supervise the erection of machinery on the inclined planes at Albany and Schenectady and to build railroad cars. Within three years the railroad named him superintendent, a position he held for six years. His success in this position and the contacts he acquired situated him to be appointed by Governor Seward as the canal commissioner of New York state in 1839. This position provided recognition throughout the state and the East and opened up opportunities to deepen his contacts with railroads whose projects paralleled the Erie Canal. At the same time, Whitney turned his attention to innovative activities and in 1840 acquired a patent for a locomotive engine.

When Whitney became Baldwin's partner, the firm gained a top manager and senior machinist who brought efficient and effective network contacts that included bridges to New York state machinist networks and railroad companies as well as instant name recognition in the social and political networks of the East Coast elite from his position as New York's canal commissioner. Whitney stayed with Baldwin four years and strengthened the firm's management. Within a year after resigning, he acquired two patents for railroad car wheels, and, along with his sons, he organized a firm near the Baldwin Works to manufacture car wheels; a year later Whitney added another patent for car wheels. His firm became a leading manufacturer of car wheels for railroads and enhanced Baldwin's contacts with railway companies.

By the 1840s no builder could make all of the four thousand parts in the typical locomotives of that period. Philadelphia's extensive variety of local metalworking firms supplied numerous components to the Baldwin Works, and suppliers engaged in know-how trading with Baldwin, enhancing its access to technical knowledge. This exchange reached more widely because supplier networks of large builders such as Baldwin also spanned the East. It obtained iron plate from Pennsylvania forges, files from Pittsburgh, axles from Baltimore, brass from Boston, and wrought-iron tires from Connecticut. By 1850 the Baldwin Locomotive Works employed roughly four hundred workers, and at the end of that decade the Baldwin and the Norris firms each had around six hundred employees. This ranked them among the nation's largest employers of machinists.

Philadelphia locomotive networks also reached to hinterland places that housed railroad repair facilities. In Reading, about sixty miles northwest of Philadelphia, the repair shops of the Philadelphia and Reading Railroad housed one of the nation's greatest concentrations of machinists. By 1850 they had a capital investment of eight hundred thousand dollars and employed 643 workers, although not all of them labored as machinists. The Baldwin and the Norris locomotive works supplied engines to the railroad, thus creating a direct network link to its repair shops, which were headed by James Millholland, one of the nation's top machinists. During his early twenties New York City's Allaire Works employed him, thus giving him access to its extensive local machinist networks. In 1838 the Baltimore and Susquehanna Railroad appointed the twenty-six-year-old as master mechanic, and ten years later he became master of machinery of the Philadelphia and Reading repair shops. Millholland built his first locomotive around 1849, and over the next two decades he completed almost one hundred engines. This talented machinist contributed numerous innovations to locomotive design and manufacturing and operated as a pivot of know-how trading in locomotive technical knowledge, thus directly benefiting Philadelphia's locomotive works.[5]

Paterson

The locomotive firms of Paterson emerged in a cohesive local milieu of firms and machinists (figs. 5.6 and 5.7). Their professional contacts drew in enterprising mechanics who came to make their fortunes in the booming industrial center, and local locomotive producers utilized these contacts to supply machinists to the railroads. Thomas Rogers and his firm occupied the hub position in Paterson's network of locomotive builders. As with many locomotive machinists, he had gotten his start in the textile machinery industry. Within four years of his arrival in Paterson in 1812, he joined with John Clark Jr. to found the textile machinery firm of Rogers and Clark.

In 1822 the firm obtained additional capital from Abraham Godwin Jr., a local hotel owner, and Rogers became a partner in the reconstituted firm of Godwin, Rogers and Company. A new partnership of Godwin and Clark formed after Rogers left in 1831 and continued to manufacture textile machinery and other machine products. Immediately prior to Rogers's departure, Charles Danforth, the inventor of a spinning frame, arrived and infused the new partnership with talent and strengthened its network bridges to top machinist centers. His career had included employment at the Matteawan Company textile mill and machinery works (1821–25); building textile machinery in Newburgh and Ramapo, New York; and work in the textile machinery and metalworking city of Taunton and relocation to Paterson (1829–31), during which time he also went to England (1831). By 1840 the partnership of Godwin and Clark had been reconstituted as Charles Danforth and Company.

In 1832, when Thomas Rogers joined with New York City financiers Morris Ketchum and Jasper Grosvenor to found Rogers, Ketchum and Grosvenor, they embarked on textile machinery production, but railroad equipment manufacturing quickly attracted them. Their large works contained a machine shop and foundry, and they soon made ironwork for railroad bridges, supplied wheels and axles to a South Carolina railroad, and turned out wrought-iron car wheels. By 1835, aiming to embark on locomotive manufacturing, they added buildings to their plant and finished their first locomotive two years later. Between the late 1830s and 1847 they produced fewer locomotives than Baldwin, but the Rogers Works surpassed the Philadelphia firm from 1848 to 1857 (see figs. 5.5 and 5.6).

Shortly after the founding of the Rogers firm, it quickly became a magnet for talented machinists, who transmitted technical skills and innovations from other machinist centers and, in turn, provided network bridges back to these places (see fig. 5.7). William Swinburne, who worked at Matteawan from 1827 to 1833, joined Rogers and served in a variety of design capacities, especially as a pattern maker for machinery parts. He worked on the firm's first locomotive and continued in various supervisory positions, including service as superintendent of the locomotive works (1842–44). The Cooke family members arrived at Rogers around 1840, following their work at Matteawan, and John Cooke became superintendent of locomotive production in 1845; he held that position for seven years.

William Hudson's arrival in 1852 affirmed the prominence of Rogers, Ketchum and Grosvenor. Around 1830 he apprenticed in England at the famous locomotive works of Robert Stephenson and Company then migrated to the United States and served as engineer for various northern railroads and as engineer for the Auburn State Prison in New York. From 1849 to 1852 Hudson served as master mechanic of the Attica and Buffalo Railroad, during which time he visited Rogers to consult on locomotive production. Hudson acquired these deep technical and managerial experiences

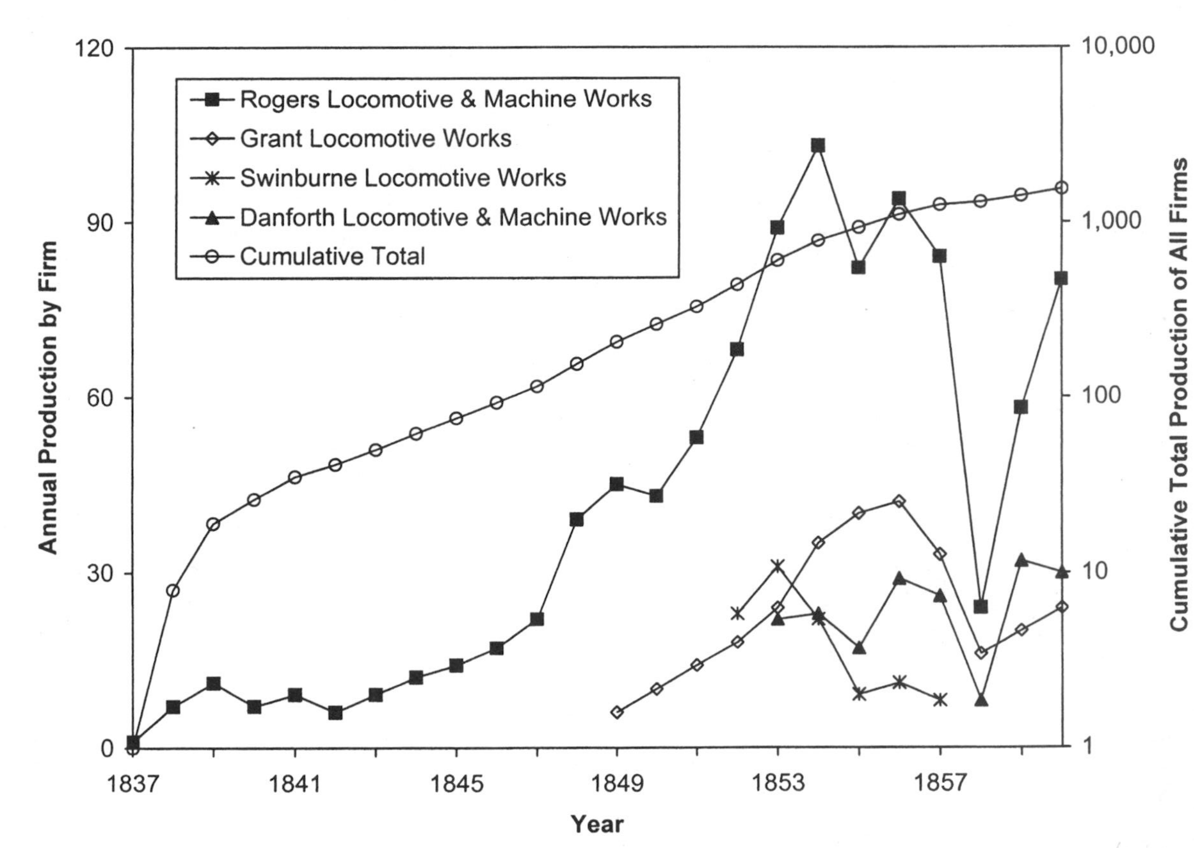

Figure 5.6. Locomotive Production in Paterson, New Jersey, 1837–1860. *Source:* L. R. Trumbull, *A History of Industrial Paterson* (Paterson, N.J.: Carleton M. Herrick, 1882), 124, 131, 144, 147.

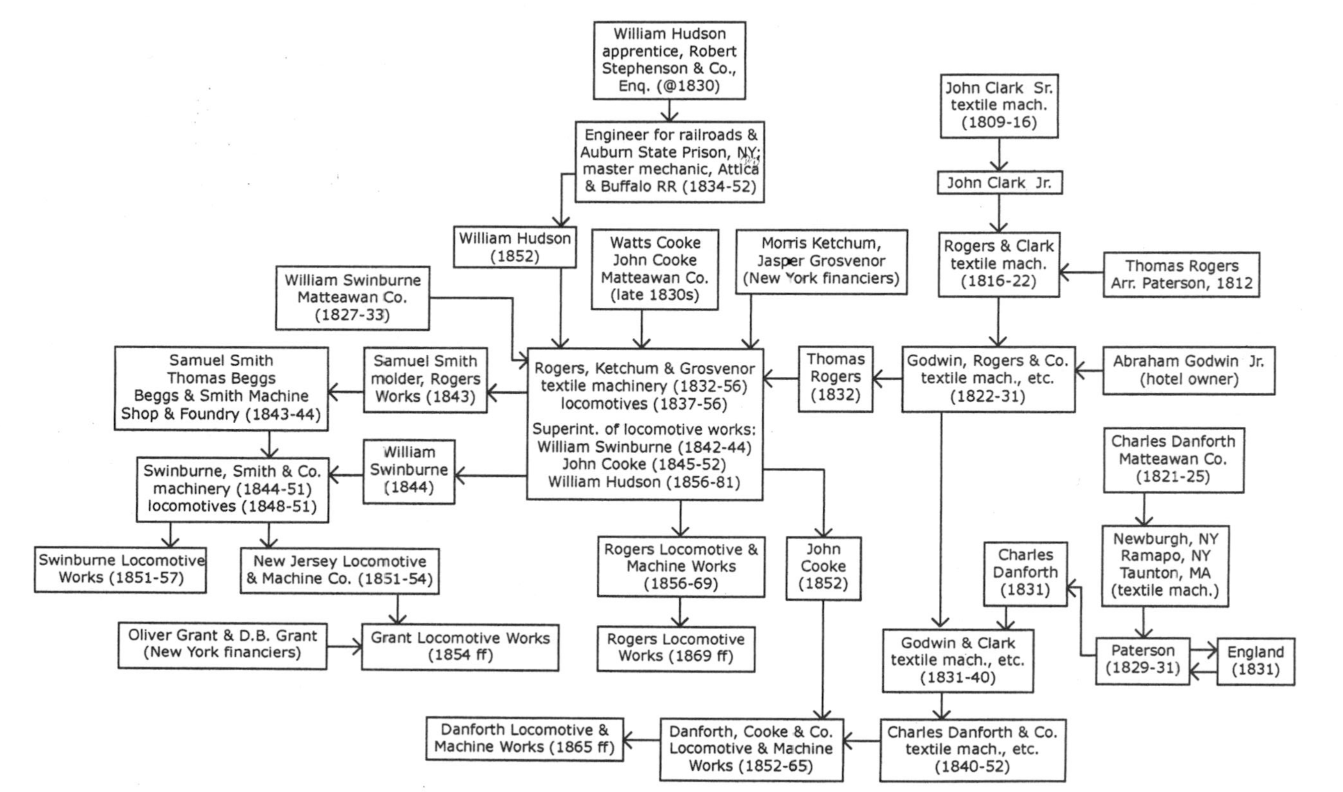

Figure 5.7. Networks of Paterson's Locomotive Builders. *Sources:* L. R. Trumbull, *A History of Industrial Paterson* (Paterson, N.J.: Carleton M. Herrick, 1882); John H. White, *American Locomotives: An Engineering History*, 1830–1880, rev. and exp. ed. (1968; rpt., Baltimore: Johns Hopkins University Press, 1997), 541.

by the age of thirty-two; then he started work at Rogers. The firm had thus obtained one of the nation's youngest, most experienced locomotive builders, who possessed network bridges to many leading hub individuals and firms. Rogers promoted him to superintendent four years later, a position he held for a quarter-century.

As the hub of the locomotive machinist networks of Paterson, the Rogers Works spurred locomotive production in other firms (see fig. 5.7). Samuel Smith, a molder at Rogers, left in 1843 and joined with Thomas Beggs to found Beggs and Smith Machine Shop and Foundry. This firm was soon transformed into Swinburne, Smith and Company, when William Swinburne resigned as Rogers's superintendent of locomotive production in 1844. The Swinburne firm began building locomotives in 1848, but the partners ran into financial difficulties three years later. This poor footing led to a disagreement over strategy, and the firm subsequently reincorporated as the New Jersey Locomotive and Machine Company. It continued to face financial problems until the New York City financiers Oliver Grant and his son D. B. Grant took it over and founded Grant Locomotive Works, placing the firm on a solid footing for many years. When New Jersey Locomotive incorporated in 1851, Swinburne left and started Swinburne Locomotive Works, but it succumbed to the financial crisis of 1857. John Cooke, superintendent of locomotive production at Rogers, Ketchum and Grosvenor from 1845 to 1852, received a lucrative annual salary of eighteen hundred dollars, but this did not dissuade Charles Danforth and Company from trying to recruit him. It intended to begin locomotive manufacturing, and Danforth offered the twenty-seven-year-old Cooke an even better package—a partnership. Following 1865, this firm became the Danforth Locomotive and Machine Works.

Paterson's community of practice of machinists had developed a broad base in foundries, textile machinery manufacturing, and machine shops over a period of almost three decades prior to the beginning of locomotive building in the late 1830s. Many of its mechanics had moved to Paterson from leading machinist centers, and their firms made the city a hub of machinist networks. As the locomotive producers expanded output, the network bridges of the mechanics and their firms made the city's works a magnet for other talented machinists. These arrivals infused the firms with technical skills learned elsewhere, especially at the Matteawan Company's premier machine shop, which in turn strengthened the network bridges of Paterson's locomotive builders. The city's development as a hub of locomotive manufacturing during the quarter-century leading up to 1860 therefore rested on a vigorous community of practice whose members, both machinists and firms, engaged in extensive know-how trading locally and with leading machinist centers. Paterson had more locomotive firms than Boston, yet its firms and those in its satellites made the Boston region a third hub of locomotive manufacturing.[6]

The Boston Region

Compared to the Middle Atlantic, New England's railroad investment began several years later and increased at a slower pace (see fig. 5.2). Nonetheless, locomotive manufacturing quickly emerged, and it drew on the Boston region's many skilled machinists who had expertise in building steam engines and textile machinery. The Mill Dam Foundry of Boston, the first to produce locomotives in New England, finished two in 1834, and the Locks and Canals Shop, under the supervision of the talented engineer George Whistler, completed seven in 1835, becoming the nation's top builder. During Lowell's first five years of production, New England railroads—most of which either radiated from Boston or ran within about one hundred miles of the metropolis—purchased three-fourths of the fifty-six locomotives that the machine shop finished.

Under the leadership of Patrick Jackson, the Boston Associates controlled the Lowell Machine Shop. They employed their cohesive network and their wide-ranging links to the rest of the region's business elite in order to persuade New England railroads to purchase locomotives from their shop. Nevertheless, it did not sustain a big commitment to locomotive production during the 1840s, even while New England railroad investment surged. From 1851 to 1855 the machine shop participated in the frenzy of production for the Midwest's rail boom. It sold twenty locomotives to the region's railroads as part of a total output of fifty-seven then retreated from engine manufacturing (see figs. 5.2 and 5.4). This failure to challenge the leading builders—Baldwin, Norris, and Rogers—for market dominance seems puzzling. The Lowell shop possessed superb business network ties to railroad owners in the East and the Midwest and numerous network links to locomotive machinists because it had trained substantial numbers of them. Its lack of full commitment and subsequent withdrawal probably reflected a strategic decision to avoid the high risk of locomotive manufacturing, especially when carried out at a large scale. The Lowell shop continued building textile machinery as well as various other machines and foundry products.

Although Boston housed several major foundries and machine shops, no significant locomotive production followed the Mill Dam Foundry's incipient effort, begun in 1834, until Holmes Hinkley, Boston's leading steam engine builder, began to manufacture locomotives in 1840. He associated with Gardner Drury in the firm of Hinkley and Drury, and within two years they had reached an annual output of ten locomotives. By 1845 their firm had completed a total of fifty-six locomotives, a successful start considering that New England railroad investment had declined from 1840 to 1843 (see figs. 5.2 and 5.4). Hinkley's extensive machinist networks had arisen from his many sales of steam engines in New England and probably gave the firm credibility as a locomotive supplier. The firm's production soared during the late 1840s, along with the jump in New England railroad investment, and reached a cumulative total of 291 locomotives by 1850. To meet demand from the midwestern railroad boom, the firm completed as many engines during the next five years as in the previous decade.

By 1860 Hinkley had completed 660 locomotives, about two-thirds as many as Baldwin, Norris, and Rogers, making him the nation's fourth largest producer. As a large-scale, successful steam engine and locomotive manufacturer, Hinkley attracted and trained numerous machinists. After working for Hinkley for seven years, John Souther left in 1846 to begin building locomotives with J. Lyman in Boston. By the mid-1850s this firm became the Globe Locomotive Works; its other senior machinists included Souther's brother George and D. N. Pickering, the former master mechanic for the Boston and Worcester Railroad.

Other Boston region locomotive firms began production during the boom in New England railroad investment in the late 1840s. The Springfield Locomotive Works originated in 1847, and J. M. Blanchard, one of the owners, had previous experience building engines for New York railroads. Springfield offered a prime location for manufacturing locomotives because the city and the surrounding area constituted one of New England's largest concentrations of skilled machinists, and extensive know-how trading occurred among the major employers. The city housed the repair shops of the Western Division of the Boston and Albany Railroad, a source of skilled machinists for the Springfield Locomotive Works. Machinists also could be recruited from the Springfield Armory and the Ames Manufacturing Company. The Portland Company Locomotive Works in Maine started production in 1848 by building five locomotives for the Atlantic and St. Lawrence Railroad. Within five years it had completed sixty-two locomotives and employed about 360 workers. These Boston region machinist networks reveal the strong ties between locomotive firms and the railroads' repair shops, whose skilled machinists maintained and occasionally built engines.

Manchester

The refusal of the Lowell Machine Shop to make a substantial commitment to locomotive manufacturing did not dissuade other textile machinery firms from moving into engine work. By the early 1840s Manchester, New Hampshire, had become a large textile factory city, and its most prominent firm, the Amoskeag Manufacturing Company, housed a sizable machine shop with a foundry. In 1849 an enlarged shop began to produce locomotives amid the surge in New England railroad investment (see fig. 5.2), and within three years the company added other buildings, including a boiler shop, tank shop, forge room, paint shop, and pattern house. By the mid-1850s Amoskeag's machine shop could turn out sixty locomotives annually and, at the same time, continued to produce textile machinery and turbine wheels. The cohesive network ties of Amoskeag's investors and management to Boston's businesses—the pivots of New England's railroads, finance, and wholesaling—kept the machine shop attuned to the demand for locomotives. Furthermore, the firm's importance as a cotton textile machinery producer meant it employed experienced machinists and attracted novice mechanics to its shops, facilitating its rapid rise as a locomotive manufacturer.

Although the surge in New England railroad investment had run its course, the level remained above the annual investment prior to 1846, thus providing a regional market. The boom in Midwest investment, however, probably sent a bigger market signal to the founders of the Manchester Locomotive Works, started around 1853 by several individuals connected with Amoskeag (see fig. 5.2). O. W. Bayley, one of the investors, had headed Amoskeag's machine shop for many years and became superintendent of locomotive production of the Manchester Works. Aretas Blood, another of the investors, had worked at the Lowell Machine Shop for seven years and then at the nearby Essex Machine Shop, before coming to Manchester in 1853, where he took over as superintendent in 1857. Within a few years of starting production, Manchester Locomotive employed as many as two hundred workers, thus underscoring the extent of the network bridges among machinist centers which enabled new locomotive works, headed by leading machinists, to attract skilled workers.

Taunton

The mid-1840s upturn in New England railroad investment also lured investors to start locomotive production in Taunton. Its firms drew on the large concentration of textile and steam engine machinists there and in Providence. The founding of Taunton Locomotive Manufacturing Company in 1846 strengthened the city as the hub of an extraordinary network of machinists. One partner, Willard Fairbanks, came from the prominent Providence steam engine firm of Fairbanks, Bancroft and Company, and he brought along Parley Perrin from his firm. Another partner, George Griggs, master mechanic at the Roxbury repair shops of the Boston and Providence Railroad, had recently started his locomotive building career there. He sent Benjamin Slater, one of his top machinists, to Taunton Locomotive, and Slater stayed until his death in 1854. Both Perrin and Griggs had worked at the Lowell Machine Shop, thus providing a bridge to that machinist center, and the work of Griggs and Slater at the railway repair shops in Roxbury and Slater's subsequent move to Taunton epitomize the machinists bonds between railway repair shops and locomotive works.

The job mobility of Perrin during his thirteen-year career before arriving at Taunton created network bridges to a substantial number of premier machinist hubs (table 5.1). The twenty-one-year-old Perrin began his career in 1833 in a key machinist position, building textile equipment at the Pawtucket shop of Brown and Clark. Within a year or so, he moved a little over twenty miles northeast to Bridgewater, Massachusetts, to work at a saw manufacturing firm, and from there he went to the Lowell Machine Shop. By 1836 Perrin had moved to Newark, New Jersey, to work with Seth Boyden, the well-known machinery builder. This move exemplified the network ties between New England's machinist centers and the vast complex of foundries, steam engine firms, and machinery firms in New York City and nearby New Jersey (see table 4.6). Within two years he returned to Pawtucket to work at Walcott Manufacturing Company, repairing mule-spinning machines. His ties to the hub of machinists in and around Providence were thus reinforced.

The subsequent move of Perrin to Philadelphia to work at Levi Morris and Company, one of that city's leading iron foundries, enhanced his experience with steam engines and gave him access to a highly networked community of practice, arguably almost as large as the machinist networks of New York City and vicinity (see tables 4.4, 4.7, and 5.1). While in Philadelphia in 1839–40, Perrin seems to have attended mechanics classes at the Franklin Institute and made use of their library. Because the institute functioned as a hub of machinist knowledge through its activities and its prominent board members, this access gave Perrin valuable information about the latest innovations in machinery technology. In 1841 he returned to the Providence (and Pawtucket) machinery hub and ran his own shop for a short time; subsequently, he spent a few years tending to a relative's farm. In 1845 he entered the premier firm of Fairbanks, Bancroft and Company, from which he moved to Taunton Locomotive as draftsman and designer. Perrin's frequent job changes over the thirteen-year period, which included employment in top firms located in the leading hubs of machinists, provided him unparalleled network bridges for his position at Taunton Locomotive between 1846 and 1888. By the 1850s machinists attracted to the firm came from the same hubs where Perrin had worked, confirming the effectiveness of network ties as job recruitment mechanisms.

Taunton Locomotive's success in selling engines to New England railroads from 1846 to 1849, when they purchased forty-two of the firm's forty-four locomotives, may have encouraged Taunton's William Mason to consider entering locomotive manufacturing. He owned half of William Mason and Company; and Mills and Company, a Boston textile selling house that also owned mills and textile machinery shops, owned the other half. By the late 1840s the Mason company ranked as one of the nation's largest textile machinery companies, with a fully equipped foundry, but sales fluctuated significantly and failed to increase. In 1852 Mason built boiler and erecting shops, enlarged the blacksmith shop and the foundry, and added new machine tools. He completed his first locomotive the following year. Although his timing coincided with the last leg down in the New England railroad investment cycle, the Midwest's

TABLE 5.1
Parley Perrin's Machinist Career, 1833–1888

Date	Employer	Location	Type of Work
1833–34	Brown and Clark	Pawtucket, R.I.	textile machinery
1834–35	Manassah Andrews	Bridgewater, Mass.	saws
1835–36	Lowell Machine Shop	Lowell, Mass.	calico printing presses, locomotives
1836–37	Seth Boyden	Newark, N.J.	machine tools, locomotives
1838–39	Walcott Manufacturing Co.	Pawtucket, R.I.	repair mule spinning machines
1839–40	Levi Morris and Co.	Philadelphia, Pa.	steam engines, textile machinery
1841–42	Perrin's Machine Shop	Pawtucket, R.I.	machinery
1845–46	Fairbanks, Bancroft and Company	Providence, R.I.	steam engines, machine tools
1846–88	Taunton Locomotive Manufacturing Co.	Taunton, Mass.	locomotives

SOURCE: John W. Lozier, *Taunton and Mason: Cotton Machinery and Locomotive Manufacture in Taunton, Massachusetts, 1811–1861* (New York: Garland, 1986), 403–5.

railroad boom had begun (see fig. 5.2). Mason and Taunton sold about half of their locomotives in the Midwest during the 1850s.

As a leading textile machinery builder, Mason possessed bridges to major machinist hubs, and, along with the network ties of Mills and Company, he easily drew on locomotive talent locally and from outside Taunton. Mason hired Charles Thomas, the draftsman who designed locomotives for the Amoskeag Manufacturing Company from 1849 to 1852, to design his early locomotives. The network links of Mason to forges, foundries, and machine shops across the East formed the conduits through which to purchase the numerous specialized components that went into a locomotive.[7]

THE SHAKE-OUT OF THE LATE 1850S

The Baldwin firm's decision in 1854 to drop its diversified foundry and machinery business and to specialize in locomotives signaled a shift in strategy by the largest locomotive producers, and by 1860 Baldwin, Norris, and Hinkley had mostly made that change. Rogers Locomotive, however, maintained a sizable machinery business until 1869. Some smaller producers such as Swinburne Locomotive mostly built engines, but this firm collapsed in 1857. The second tier of firms, including Taunton, Mason, Danforth, Grant, and Manchester, manufactured various machinery even as they turned out locomotives, and most smaller firms also remained diversified.

During the railroad construction boom of the early 1850s, lines in the Midwest contributed the most to the boom, but those in the Middle Atlantic and the South also made sizable investments; New England railroads, however, significantly reduced their efforts (see figs. 5.1 and 5.2). At this time foundries, textile machinery firms, and machine shops in the East and Midwest, along with a few foundries and machine shops in the South, plunged into locomotive production, but they quickly retreated or went bankrupt during the financial contraction of 1857, when railroad investment declined precipitously and many lines became insolvent. Railroads often failed to pay for locomotives they had ordered and for which production had commenced, and they did not pay for engines that had been delivered; therefore, locomotive firms could not recover their expenses. Even leading builders such as Philadelphia's Baldwin and Paterson's Rogers witnessed wrenching collapses in production around 1857, but they quickly recovered. Likewise, smaller firms such as Paterson's Grant and Danforth righted themselves following the dive in engine building (see figs. 5.5 and 5.6).[8]

YOUNG JOB-HOPPING MACHINISTS

Locomotive builders in the early centers of iron foundries, textile machinery firms, and machine shops in the vicinities of Boston, New York, and Philadelphia dominated production during the antebellum. They possessed technological skills and maintained close contacts with the leading buyers of locomotives—the headquarters of railroad companies in the metropolises and in their major inner-hinterland cities—and relied on metropolitan financiers to rate the credit worthiness of railroads. As machinists, locomotive builders had extensive local network ties, and these linkages reached their greatest extent in the leading hubs of builders—Philadelphia, Paterson, and the Boston region. Locomotive firms trained many machinists, but engine building mainly required standard mechanic skills. Consequently, some machinist firms or centers of machinists—the Lowell Machine Shop, the Matteawan Company, and the Providence machine shops—which did not sustain production or produced few, if any, engines, served as training grounds for many locomotive machinists.

Railroad companies integrated with these machinist networks through their chief engineers and mechanics, most of whom operated from the railway repair shops, and machinists moved back and forth between railways and locomotive builders. Because skilled machinists continued to be in scarce supply relative to demand during the antebellum, they could be highly mobile during their twenties and into their early thirties. During their careers the most able among them worked in several of the leading centers of foundries, textile machinery firms, and machine shops. Talented, job-hopping machinists commanded premium wages. Some of them reached foreman or even superintendent positions during their twenties, and many of them attained that level by their early thirties. Subsequently, they changed jobs less frequently, probably as a consequence of family obligations, attainment of partnership in the firm, and high wages.

The job-hopping of top machinists early in their career transmitted technical skills and created strong network bridges among the leading machinists and their firms, facilitating know-how trading about the construction of engines and sustaining a community of practice of locomotive building. Consequently, even the big three of Baldwin, Norris, and Rogers could not exert structural constraint on other producers to prevent them from acquiring top machinists and technical innovations in locomotive building. A large number of talented machinists created technical innovations in locomotive design and components, and the networks of know-how trading constituted the conduits through which to transmit the innovations. A few builders such as Baldwin, Norris, and Rogers maintained production for at least two decades, but others began production and then ceased. Nevertheless, locomotive machinists skills continued, and production grew significantly because the skills were embedded in networks that endured long after individuals and firms had changed direction or departed entirely.

Locomotive manufacturing owed many debts to the textile machinery industry. Some firms in that industry were among the first to manufacture locomotives, and several, including the Lowell Machine Shop, the Matteawan Company, and the Providence textile machine shops, trained many machinists for the locomotive builders. The textile machinists established a networked community from 1790 to 1820, which continued to expand and strengthen its networks during the following decades.

CHAPTER SIX

Resilient Cotton Textile Machinist Networks

> The Lowell Machine Shop . . . can furnish machinery complete for a mill of 6,000 spindles, in three months, and a mill can be built in the same time.
>
> *"Statistics of Lowell Manufactures," Hunt's Merchants' Magazine*

> The machine shop [of the Matteawan Company] is one of the most complete in the country, and its several departments are well arranged to produce in the most perfect manner, every part of a cotton mill, from the water wheel or steam engine, to the card, spindle and loom.
>
> *"American Machinery—Matteawan," Scientific American*

Before 1820 the Providence core spawned textile machinist hubs in nearby areas of Massachusetts and Connecticut, in southern New Hampshire, in the Upper Hudson Valley, and in Utica, New York. Other hubs arose in eastern Massachusetts, the Connecticut Valley, New York City's environs (the Lower Hudson Valley and nearby New Jersey), and Philadelphia. In each hub mechanics built machinery for the cotton mills that employed them, but, after that initial effort, only a few—Providence and vicinity, Philadelphia, and Boston's environs—became major nests of textile machinery builders by 1820. At that time the employment in cotton mills totaled only twelve thousand. Yet textile machinist networks retained unusual resilience over the subsequent forty years, as employment in the cotton textile industry soared tenfold.

With the exception of New York City's environs, no other important nest of textile machinery builders emerged before 1860.

CONCENTRATED MARKETS FOR TEXTILE MACHINERY

Over the 1820–60 period textile machinery markets remained concentrated, and New England cotton mills purchased much of the equipment. At the beginning of the period the region housed half of the nation's capital investment in cotton textiles; its share rose to about 70 percent by the early 1830s and stayed at that level until 1860. The growth of cotton cloth production in New England, therefore, serves as a proxy for the expansion of the textile machinery industry. From a base of forty-six thousand yards in 1805, annual cloth production exploded, with the exception of the brief contraction after the War of 1812, to fourteen million yards in 1820 (fig. 6.1). From that base cloth output soared to seventy million yards within five years, doubling over the next five years. Nevertheless, retardation in the growth rate appeared by the late 1820s and continued until 1860. Consequently, the 1820s may have been a time when the textile machinery industry was transformed, as machinists met surging demand for their equipment.

Spinning machines (with their spindles) and power looms constituted the largest set of equipment in cotton mills; therefore, the distributions of the numbers of spindles and of power looms serve as a reliable measure of the textile machinery market. By 1820 New England made up over half of the national market, somewhat exceeding the Middle Atlantic's share, but within a decade New England's portion rose to two-thirds, reaching three-fourths by 1860 (table 6.1). In contrast, the Middle Atlantic's share slipped to one-third by 1831 and fell to one-fifth by 1860. Massachusetts, Rhode Island, and Connecticut accounted for much of New England's market, housing roughly half of the national market for textile machinery over the 1820–60 period. In the Middle Atlantic region New York and Pennsylvania represented the largest markets. Their national share remained close to one-fourth during the 1820s but fell below one-fifth by 1860.

A few places in the East accounted for most of the textile equipment market. In 1840 only eighteen counties had over $750,000 of capital investment in cotton-textile manufacturing; collectively, they made up 72 percent of the nation's capital investment (table 6.2). Their typical firms had much greater capital ($63,392) than the national average ($37,328). Several counties, including York in Maine, Rockingham and Strafford in New Hampshire, and Hampden and Middlesex in Massachusetts, contained textile firms that had triple the amount of capital as the nation's average firm. Most of the eighteen leading counties corresponded to the nests of textile machinery builders which emerged before 1820. As a share of the nation's textile capital, Providence and vicinity housed just over one-fifth of the capital, and Boston's envi-

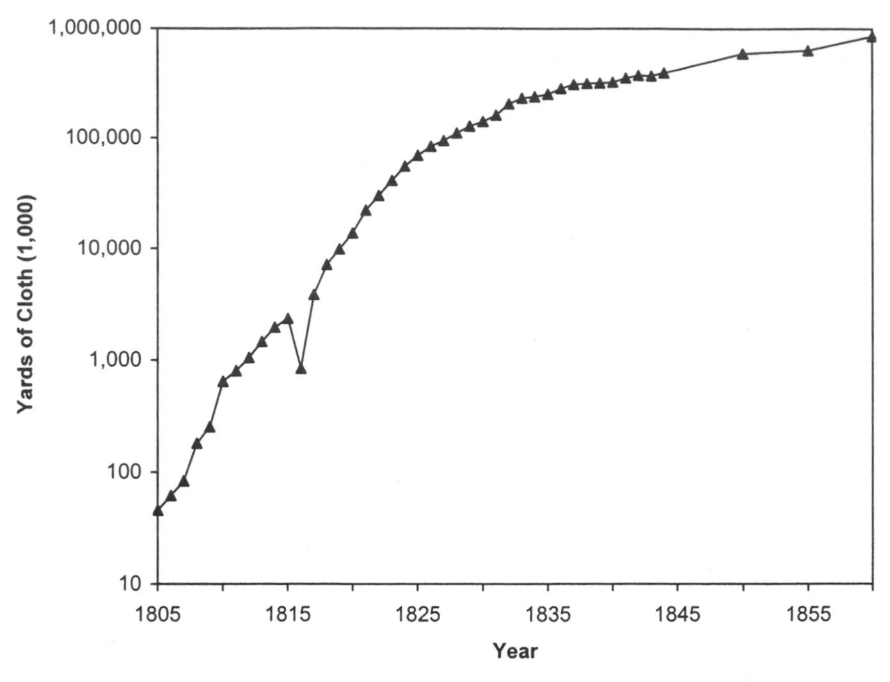

Figure 6.1. Annual Production of Cotton Cloth in New England, 1805–1860. *Source:* Robert B. Zevin, "The Growth of Cotton Textile Production after 1815," in *The Reinterpretation of American Economic History*, ed. Robert W. Fogel and Stanley L. Engerman (New York: Harper and Row, 1971), 123–24, table 1.

rons contained about one-third. Philadelphia's capital investment was much smaller, however, indicating that its nest had not kept pace with the New England builders. New York City's environs accounted for 5 percent of the nation's textile investment, thus hinting at a fourth nest of machinery builders.[1]

DECLINING PRICES OF TEXTILE MACHINERY

As cotton mills increased production, the demand for textile machinery surged, yet mills faced a dilemma (see fig. 6.1). Cloth prices continued plunging during the 1820s, just as they had immediately following the War of 1812, and relentlessly pressuring cotton mills to cut production costs (fig. 6.2). Falling prices of raw cotton accounted for only about 17 percent of the decline in real cloth prices from 1815 to 1833; therefore, mills had to look to other sources to significantly reduce total costs. Although mills cut their labor costs, real wages did not decline and may in fact have risen, which suggests that falling wages did not contribute to lower cloth prices. The rate of return to capital did not exhibit a clear trend; thus, investors did not accept declining returns

TABLE 6.1
Number of Spindles and Power Looms in the Cotton Textile Industry by Region and State as a Share of the Nation, 1820–1860

	Number of Spindles as Percentage of Nation				Number of Power Looms as Percentage of Nation		
Region/State	1820	1831	1840	1860	1820	1831	1860
New England	51.5	66.1	69.9	73.7	60.0	63.8	73.9
Maine	1.0	0.5	1.3	5.4	0.7	0.3	5.4
New Hampshire	5.5	9.1	8.5	12.2	7.4	10.6	13.7
Vermont	1.3	1.0	0.3	0.3	2.3	1.1	0.3
Massachusetts	12.7	27.3	29.1	32.0	19.8	26.9	33.9
Rhode Island	20.6	18.9	22.7	15.6	19.4	17.3	13.7
Connecticut	10.5	9.3	7.9	8.3	10.4	7.8	6.9
Middle Atlantic	43.5	33.1	21.3	19.9	39.7	35.9	19.9
New York	19.2	12.6	9.3	6.7	20.6	10.9	6.2
Pennsylvania	6.6	9.7	6.4	9.1	1.4	18.8	10.3
New Jersey	6.2	5.1	2.8	2.4	4.9	2.4	1.2
Maryland	7.1	3.8	1.8	1.0	11.6	3.0	1.3
Delaware	4.4	2.0	1.1	0.7	1.2	0.7	0.8
Rest of nation	5.0	0.8	8.7	6.4	0.2	0.3	6.2
Nation	100	100	100	100	100	100	100
Total	325,680	1,246,703	2,284,631	5,235,727	1,623	33,433	126,313

Sources: David J. Jeremy, *Transatlantic Industrial Revolution: The Diffusion of Textile Technologies between Britain and America, 1790–1830s* (Cambridge, Mass.: MIT Press, 1981), 277–78, app. D, tables D.2, D.4; U.S. Bureau of the Census, *Compendium of the Sixth Census, 1840* (Washington, D.C.: Blair and Rives, 1841); U.S. Bureau of the Census, *Report on the Manufactures of the United States at the Tenth Census, 1880* (Washington, D.C.: Government Printing Office, 1883).

to make their mills competitive on prices. The large absolute decline in cloth prices from 1815 to the mid-1820s coincided with the revolutionary restructuring of the cotton industry—the adoption of power loom weaving and the integration of the entire production process using new or improved machinery. This one-time restructuring was the most plausible cause of the sharply falling cloth prices.

A decline in the cost of textile machinery might have contributed a small portion of the dip in cloth prices. Even as demand for textile machinery surged, the growing ranks of machinists who built that equipment and efficiencies from building more machines may have increased competitive pressures and caused an initial drop in machinery prices. This interpretation comports with the declining profit rates of machine shops: between 1818–20 and 1822–23 gross profits as a share of machinery sales of the Boston Manufacturing Company fell from about 50 percent to 30 percent. The successor machine shop, the Locks and Canals Company in Lowell, maintained a gross profit rate of 30–35 percent on some sales by the late 1830s, whereas on other sales profit rates may have dropped to as low as 5 or 10 percent. The relentless decline of dividend rates of the Locks and Canals Company and of its successor, the Lowell Machine Shop, from 18 percent during 1827–36 to 2 percent during 1857–60, suggests that profit rates of machine shops fell over the longer period.

TABLE 6.2
Counties with over $750,000 of Capital Investment in Cotton Textile Manufacturing in the East, 1840

Nest of Textile Machinery Builders/County	No. of Firms	Capital ($)	Capital/Firm ($)	Capital as Percentage of Nation
Boston environs	168	17,557,600	104,510	34.4
Middlesex, Mass.	35	8,952,500	255,786	17.5
Essex, Mass.	9	773,000	85,889	1.5
Worcester, Mass.	71	1,507,400	21,231	2.9
Hillsborough, N.H.	25	2,205,700	88,228	4.3
Rockingham, N.H.	6	816,300	136,050	1.6
Strafford, N.H.	17	2,015,700	118,571	3.9
York, Maine	5	1,287,000	257,400	2.5
Providence and vicinity	284	10,719,499	37,745	21.0
Providence, R.I.	130	4,977,000	38,285	9.7
Kent, R.I.	45	1,504,000	33,422	2.9
Windham, Conn.	54	1,711,500	31,694	3.3
Bristol, Mass.	55	2,526,999	45,945	4.9
New York City environs	42	2,691,300	64,079	5.3
Columbia, N.Y.	11	893,300	81,209	1.7
Dutchess, N.Y.	11	865,000	78,636	1.7
Passaic, N.J.	20	933,000	46,650	1.8
Philadelphia	45	1,923,600	42,747	3.8
Four nests of builders	539	32,891,999	61,024	64.4
Other	43	4,001,950	93,069	7.8
Hampden, Mass.	17	2,097,000	123,353	4.1
Baltimore, Md.	13	1,099,200	84,554	2.2
Oneida, N.Y.	13	805,750	61,981	1.6
Total selected counties	582	36,893,949	63,392	72.2
Nation	1,369	51,102,359	37,328	100

SOURCE: U.S. Bureau of the Census, *Compendium of the Sixth Census, 1840* (Washington, D.C.: Blair and Rives, 1841).

Nevertheless, the limited supply of skilled machinists capable of building textile machinery for integrated spinning-weaving mills initially hindered spinning mills from converting to the integrated form and retarded the start of new, integrated mills. Because few machine shops acquired the capacity to meet large textile machinery orders, those with that capability did not have to reduce their prices much, even though they might have achieved lower costs from technical change. Big, integrated textile companies such as the Boston Associates' mills possessed competitive advantages because they owned sizable machine shops that equipped their mills. Their shops ranked among the few that could supply machinery for other mills, and, therefore, they could charge high prices. As more large mill shops, both those of the Boston Associates and those of other textile companies, entered production by the mid-1830s, their profit rates came under pressure as shops competed on price to win orders.

Textile machinery cost per spindle dropped from 1827 to 1860, with the biggest absolute drop occurring before the mid-1830s (fig. 6.3). More shops entered produc-

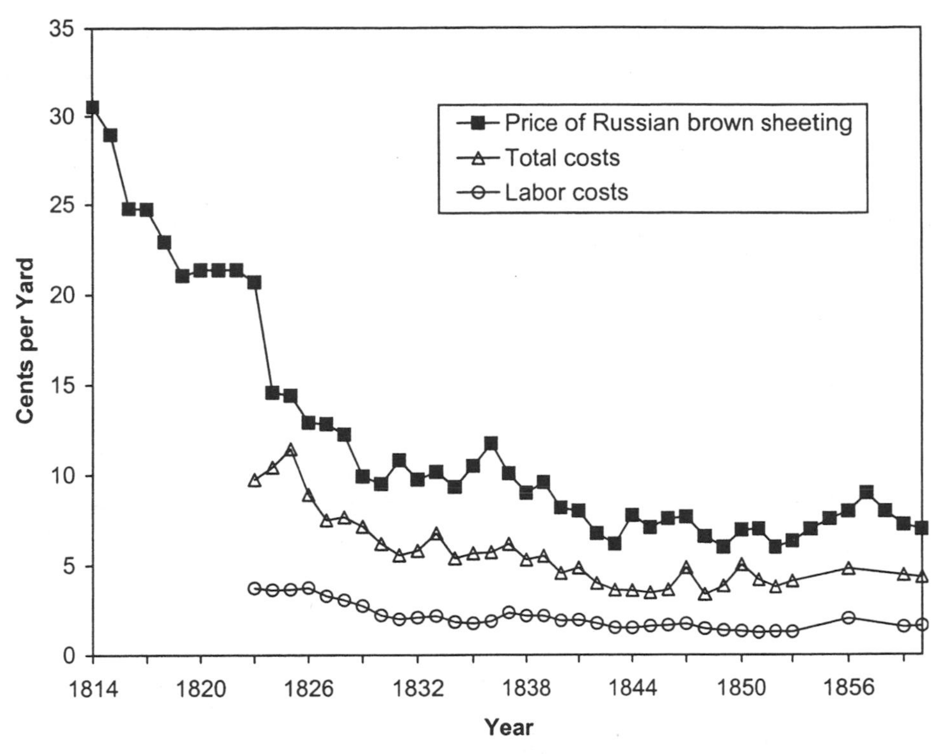

Figure 6.2. Selling Price, Total Cost, and Labor Cost for Cotton Cloth (Cents per Yard), 1814–1860. *Source:* Robert B. Zevin, "The Growth of Cotton Textile Production after 1815," in *The Reinterpretation of American Economic History*, ed. Robert W. Fogel and Stanley L. Engerman (New York: Harper and Row, 1971), 134, table 3.

tion, which pressured profit margins, and this probably accounted for an important share of the initial decline in machinery cost. Other factors also may have caused this decline in cost per spindle. As the volume of orders surged during the 1820s, machinists may have gained efficiencies in constructing textile machinery: it became a larger share of the shop's business; batch production techniques may have been used; and machinists' skills may have improved. Better machine tools, however, did not contribute to the initial decline in machinery costs because their improvement came slowly, and sophisticated ones did not appear until after 1850. Throughout the antebellum machinists engaged in labor-intensive efforts of cutting, grinding, and filing metal. After the mid-1830s rising competition among shops, gradual gains in efficiency, and slow improvement in the quality of machine tools probably accounted for the measured decline of textile machinery cost. All of this evidence, therefore, points to the 1820s and early 1830s as the time when the textile machinery industry underwent transformation.[2]

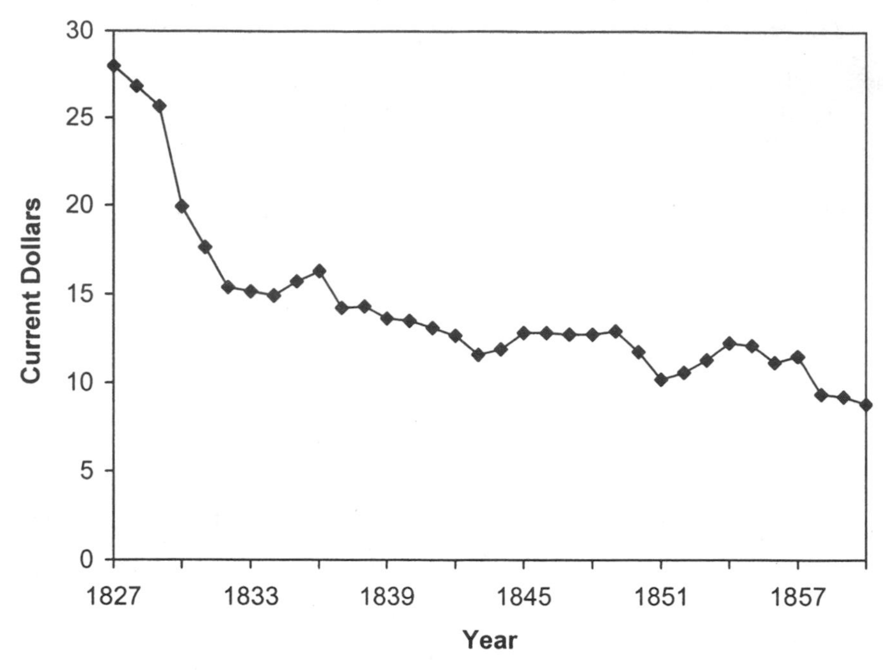

Figure 6.3. Textile Machinery Cost (Current Dollars) per Spindle, 1827–1860. *Source:* Paul F. McGouldrick, *New England Textiles in the Nineteenth Century: Profits and Investment* (Cambridge: Harvard University Press, 1968), 240–43, table 46.

FROM COTTON MILL SHOP TO INDEPENDENT FIRM

During the first thirty years of cotton textile manufacturing, from 1790 to 1820, machinists built 326,000 spindles, and during the first five years of power loom weaving, from 1815 to 1820, they built 1,600 looms. Yet over the next decade they produced about 900,000 spindles and 30,000 power looms, an indication of the prodigious demands on machine shops to supply equipment to support the tenfold rise in cotton cloth output (see fig. 6.1 and table 6.1). Machine shops also had to adjust to dramatic technological changes. With the exception of spooling and warping, all of the processes of cotton cloth manufacturing—opening and cleaning cotton, carding, drawing, roving, spinning, spooling, warping, and weaving—had been mechanized by 1830. The sophistication of much of this machinery significantly exceeded the simple equipment in the earliest integrated mills constructed around 1815.

Paul Moody, the brilliant machinist employed by the Boston Manufacturing Company and, later, the Locks and Canals Company, developed an astounding number of new and improved machines between 1814 and 1824. Over the next two decades numerous inventors and machinists added many improvements, often related to controlling the motion of parts of the machines. These technical innovations magnified the complexity of machinery, raising the level of the skills required of machinists, and the shear number of parts in some of the new types of machines increased the amount of time needed to build them. By 1830 existing cotton mills that materially expanded production could not readily make all of the new machinery, and many of the new mills could not afford the expense of a full-line machine shop to build their equipment. The lengthy time that a small mill shop required to build the machinery would put off the start of production, raising the cost of the mill because returns on investment in land, water rights, the dam, and buildings would be delayed.

By the late 1830s equipment in integrated spinning-weaving mills differed considerably from machines in spinning mills around 1814, testimony to the transformation of machine building (table 6.3). The size of the typical mill, as measured by the number of spindles, doubled; weaving required equipment to prepare the yarn, including spooling, winding, and dressing machines; and power looms completed the weaving stage. The dressing frames constituted sizable expenditures of as much as 40–50 percent of the total cost of power looms for a mill. All of this new machinery, along with the greater number of spindles, enlarged capital requirements by nearly 100 percent. Yet, indicative of the plunging price of machinery, the cost per spindle of all equipment in an integrated spinning-weaving mill of the late 1830s barely exceeded the cost per spindle in a spinning mill of 1814. The jump in the cost of machinery per spindle from the adoption of power loom weaving, still evident in the high cost during the late 1820s, had been sharply reduced by the early 1830s (see fig. 6.3).

The largest new mills of the 1815–20 period became the typical new ones of the late 1830s. The Lowell Machine Shop, named the Locks and Canals Shop before 1845, offers clues about the task of equipping these big mills. During the mid-1830s three hundred workers took about four months to equip a mill of five thousand spindles fully, and during the late 1840s the shop's six hundred employees (not all of whom built textile machinery) completely equipped a mill of six thousand spindles within three months. Extrapolating these figures backward in time suggests that equipping the new mill of three thousand to four thousand spindles in the 1820s required two hundred workers for three to four months of full-time work making the machines.

Nevertheless, the limited supply, relative to demand, of machinists in the 1820s prevented most new mills from attracting several hundred machinists for a short time, because they readily acquired alternative, steadier employment. Even fifty machinists would take about one year to finish the machinery. This labor force would be a scale, however, which exceeded the size of most machine shops, with the exception of the largest shops in the East Coast metropolises and their satellites. During the rapid cotton textile expansion of the 1820s, therefore, mills of three thousand to four thousand spindles may have hired, at most, thirty to forty machinists for as long as eighteen

TABLE 6.3

Machinery for Cotton Mills, 1814 and 1839

Spinning Mill (2,160 Spindles), 1814				Typical Mill (4,992 Spindles), 1839				Lowell Mill (4,096 Spindles), 1839			
No.	Type	Unit Price ($)	Total Cost ($)	No.	Type	Unit Price ($)	Total Cost ($)	No.	Type	Unit Price ($)	Total Cost ($)
1	Picker	300	300	1	Willow	100	100	2	Willows	100	200
24	Cards	320	7,680	1	Scutcher	600	600	2	Pickers	500	1,000
2	Drawing frames	240	480	40	Cards	210	8,400	2	Lap winders	210	420
4	Roving frames	210	840	1	Lapper	250	250	20	Breaker cards	270	5,400
4	Stretching frames	600	2,400	6	Drawing frames	200	1,200	2	Grinders	60	120
2,160	Throstle spindles	5.00	10,800	6	Speeders	660	3,960	20	Finisher cards	245	4,900
	Reels, cans, etc.		1,500	7	Extensors	900	6,300	5	Drawing frames	225	1,125
				4,992	Throstle spindles	4.50	22,464	4	Speeders	545	2,180
				10,000	Rove bobbins	.06	600	8	Stretchers	760	6,080
				12,000	Spinning bobbins	.01	120	2,048	Warp spindles	3.59	7,352
				6,000	Spool bobbins	.03	180	2,048	Filling spindles	3.86	7,905
				6,000	Skewers	.015	90	6	Warpers	130	780
				6	Spoolers	70	420	8	Dressing frames	510	4,080
				6	Warpers	150	900	120	Power looms	70	8,400
				9	Dressing frames	400	3,600				
				128	Power looms	75	9,600				
Total			24,000				58,784				49,943
Per spindle			11.11				11.78				12.19

SOURCES: George S. Gibb, *The Saco-Lowell Shops: Textile Machinery Building in New England, 1813–1949* (Cambridge: Harvard University Press, 1950), Press, 632, app. 1B, 1C; James Montgomery, *A Practical Detail of the Cotton Manufacture of the United States of America* (Glasgow: John Niven, 1840), 115–17.

months to build machinery. Subsequently, most of them moved on to another new mill because the rapid mill expansion meant they could easily find jobs. After completing the machinery, the mill shop reverted to a small operation and mostly engaged in repair work.

During the 1820s the substantial increase in the scale and complexity of textile machinery manufacturing for new mills spelled the demise of the mill's machine shop as the supplier of its new equipment. To compete in a market of collapsing prices of cotton cloth, mills faced intense pressure to acquire the best machinery, but the limited quality control resulting from using roving bands of thirty to forty machinists offered a poor way to build it. Henceforth, new mills ordered most, if not all, of their machinery from external suppliers, and the small mill shops repaired machinery. On the other hand, large mills (such as at Taunton, Mass.) and the great cotton mill complexes of the Boston Associates at Lowell and Chicopee in Massachusetts, at Saco in Maine, and at Manchester in New Hampshire continued to combine extensive machine shops and cotton mills at least until 1840. The owners expanded existing mills and added new ones, thus providing ample work for the shops and enabling them to keep a large number of skilled machinists busy building and repairing machinery.[3]

INDEPENDENT AND PSEUDO-INDEPENDENT SHOPS

Before 1820 Providence and vicinity and Boston's environs housed the majority of the small independent machine shops; some of them rented space from cotton mills. Most, if not all, of these shops built textile machinery as one line of a diversified machine business. Mills with small shops might build machinery for other cotton mills, but they could not meet large orders. The large-scale cotton textile expansion of the 1820s, coupled with the greater complexity of machinery, created opportunities for existing and for new independent machine shops to meet the surging demand for equipment. This demand also supported the emergence of "pseudo-independent" shops owned by cotton mills which hired talented machinists to make machinery for sale to other mills. The big machine shops of the Boston Associates did not compete much for this swelling machinery market because they had to meet the huge equipment orders from their own mill complexes. Nevertheless, independent (including pseudo-independent) shops did not have guaranteed markets for their production.

Independent shops competed with existing mill shops that might supply their own mill's expansion needs and with roving crews of thirty to forty machinists who built machinery for new mills. To be successful, independent shops needed to participate in textile machinist networks to stay abreast of rapid technical changes in machinery, and they had to be accessible to machinery markets to acquire orders for equipment. Only the largest shops could supply most of the equipment for a mill, but the shear number and variety of machines taxed them. The growth of textile production during the 1820s created opportunities for machinery firms to specialize. Cotton mills typically ordered their machinery from multiple suppliers; therefore, machine shops that participated in networks of suppliers possessed competitive advantages in acquiring orders. Consequently, most of the independent textile machinery firms concentrated in the early nests of textile machinists, and much of the cotton mill expansion of the 1820s and 1830s also occurred in these areas.

Providence and Vicinity

Changes in the nest of textile machinists in Providence and vicinity after 1820 demonstrate that technological knowledge and skills were embedded in networks of machinists, firms, and clusters of individuals and firms; they were not solely the property of any single entity. Individuals and firms operating as network hubs before 1820 faded from the scene. Samuel Slater trained a veritable army of leading machinists who went on after 1820 to serve as network hubs, but his direct influence waned as he focused on textile mill projects in nearby Massachusetts and Connecticut. After his death in 1832, the Slater family continued to enlarge their textile holdings and for more than a decade afterward maintained foundry and machinery companies, yet they were merely one set of actors among many others.

David Wilkinson's machine shop, which had been a technological leader, collapsed in 1829. He moved to Cohoes (near Albany, New York), a site of textile mill expansion, but his business struggled. Perez Peck and his brother, leading textile machinery builders since 1810, continued for another four decades. Nevertheless, the technological change in textile machinery and the substantial expansion of the 1820s prevented individuals and firms from attaining dominance; new leaders emerged, and many firms began production. By the early 1830s the nest of textile machinery builders in and around Providence constituted a formidable complex of firms (table 6.4). A large share of Rhode Island's ten iron foundries and thirty machine shops built textile machinery.

Larned Pitcher and Ira Gay formed their partnership around 1819, but these talented machinists went their separate ways five years later. Gay left the firm for a position in New Hampshire, and James Brown, who had trained with Wilkinson, teamed up with Pitcher. They capitalized on the 1820s textile boom to became one of the largest builders of cotton textile machinery (see table 6.4). Brown also turned his attention to improving and innovating machine tools, including turning lathes and gear cutters. In 1830 the firm of Fales and Jenks began manufacturing textile machinery in Central Falls, immediately north of Pawtucket, and, like Pitcher and Brown, this firm remained prominent after 1860. Fall River's textile roots dated from around

TABLE 6.4
The Providence and Vicinity Nest of Textile Machinery Builders, 1832

City/State	Firm (date founded)	No. of Emp.	Value of Products	Products	Market Area
Rhode Island	10 iron foundries, 30 machine shops	1,242	—	—	—
Pawtucket, R.I.	Larned Pitcher (1810)	45	$57,540	cotton machinery	—
Fall River, Mass.	1 machinery firm, 2 cotton mills (1823)[a]	—	$46,000	cotton machinery	local
Taunton, Mass.	Taunton Manufacturing Co. (?)	100	—	textile machinery	local, vicinity
Attleborough, Mass.	Daniel Read (1810)	15–20	$12,250	cotton machinery	R.I., Mass., Conn.

SOURCE: Louis McLane, *Documents Relative to the Manufactures in the United States Collected and Transmitted to the House of Representatives, 1832, by the Secretary of the Treasury*, House Doc. No. 308, 22nd Cong., 1st sess. (Washington, D.C.: Duff Green, 1833).
[a] Brayton Slade and Co. (machinery firm) and machine shops of Anawan Manufacturing Co. and Fall River Manufacturing Co.

1813, and some of its leading entrepreneurs had worked in Rhode Island textile mills, including service in one of Slater's mills. The textile machinery firm of Hawes and Marvel started production in 1821, and Brayton Slade and Company also emerged around this time. These firms supplied equipment for the large expansion of cotton textile manufacturing in Fall River during the 1820s, and Hawes and Marvel went on to further success.

Taunton stayed bound to the machinist networks of Providence and vicinity as workers moved back and forth over the twenty-mile distance separating them. This job mobility enhanced know-how trading in this community of practice of building textile machinery. John Thorp and Silas Shepard had worked together on power looms in the machine shop of one of Taunton's cotton mills. Around 1816–19 Thorp moved to Providence and spent the next decade as a prolific textile machinery inventor. In 1823 a group of wealthy Taunton residents joined with members of the Boston Associates to form the Taunton Cotton Manufacturing Company, combining several mill shops that had been selling machinery in the surrounding area. During the textile boom of the 1820s this pseudo-independent shop expanded its sales, and by 1832 it employed as many as one hundred workers (see table 6.4). The next year the shop added an iron foundry, one of the earliest textile machine shops to do so.

During the 1830s its market area enlarged to include sales outside the Taunton area to the rest of New England and, perhaps, to the Middle Atlantic. The shop attracted notable machinists such as George Danforth, who had worked on a speeder around 1824, and the loom innovator William Crompton. He arrived from England in 1836 and spent several years in Taunton, including a return trip to England. Then, from 1839 to 1841, he moved to Lowell to build looms, and then afterward settled in

Worcester to continue power loom manufacturing. His career after leaving Taunton reinforced the connections between the Providence and Boston nests of textile machinery firms.[4]

Boston's Environs
Ira Gay

When Gay left Pawtucket in 1824, he became the resident agent of the Nashua Manufacturing Company. Along with Slater and Pitcher, Gay also acquired an ownership stake in the Amoskeag Manufacturing Company in nearby Manchester. Members of the Boston Associates joined them, strengthening the network links between the Providence and Boston nests of builders. Gay's departure for southern New Hampshire seems puzzling. The previous year the first of the Merrimack Manufacturing Company's mills had commenced operations in Lowell, fifteen miles south of Nashua, and plans existed to expand the mills significantly. The machinist and merchant networks connecting Providence and Boston communicated this information to Gay. Furthermore, he would have been apprised of plans already under way to move the machine shop of the Boston Manufacturing Company at Waltham to Lowell. Its mills would generate few orders to external shops for machinery because the state-of-the-art, five-story machine shop that opened in 1825 would equip them.

Nevertheless, Gay made an astute move. He certainly received a lucrative pay package from the Associates, who generously rewarded their senior management. The mill of the Nashua Manufacturing Company under construction in 1823 would hold twenty thousand spindles, and it provided ample vent for Gay's considerable machine-building skills. By 1832 the company had $550,000 of fixed capital investment, employed about four hundred workers, and turned out 2.6 million yards of cloth annually. At that scale Gay would have been even more richly remunerated by the time he left the firm.

Yet additional considerations motivated his move to Nashua. Within a radius of about fifty miles a veritable boom in cotton mill building was under way in New Hampshire which would generate extensive demands for textile machinery. After completing the Nashua mill's equipment, its pseudo-independent shop started selling textile machinery to other mills, a practice that was unusual for the Associates' big complexes before 1840. Gay contributed innovations to the self-acting mule which the Nashua shop sold widely in the late 1820s. It also authorized a machine shop in Whitestown, near Utica, at the core of the cotton textile region of central New York, to build them, reinforcing the textile machinist networks between that region and the Providence area and Boston's environs. From the late 1820s to the early 1830s the Nashua shop sold a full range of textile machinery.

In 1833 Gay established Ira Gay and Company in North Chelmsford, near the Lowell complex, but the local market did not attract him because the Locks and Canals shop equipped the mills. Gay's firm, which later became Gay and Silver and then the North Chelmsford Machine and Supply Company, built textile and other machinery; it became one of the nation's foremost firms. Although this company looked to distant markets, huge opportunities loomed in three nearby New Hampshire counties which accounted for 10 percent of the nation's cotton textile capital investments by 1840. Within a hundred miles of his shop a large Massachusetts textile industry, excluding the Associates' complexes, also existed (see table 6.2).

Otis Pettee

The Elliot Manufacturing Company in Newton Upper Falls, within ten miles of Boston's merchant core, jumped into the fray of the 1820s textile boom. It hired Otis Pettee, a twenty-eight-year-old machinist who had trained in textile machinery building at his father's blacksmith shop in Foxborough, southwest of Boston. Before he reached the age of twenty, Pettee had already set up a local thread mill, and by 1819–20 a cotton factory in Newton Upper Falls hired him. Although this position soon ended, it provided contacts that culminated in the Elliot Company hiring him in 1823 to build a mill and its machinery, a task he completed within a year. Over the next several years the mill shop filled small orders for machinery from nearby factories, cementing its status as a pseudo-independent shop.

Merchant networks altered machinery building at the Elliot Company. In 1828 Samuel Cabot, a member of the Associates who had taken a sizable position in the Elliot Company, took charge of the Lowell Manufacturing Company. This new cotton textile endeavor of the Boston Associates ranked below a much larger project at Lowell, the Appleton Cotton Manufacturing Company, which began at the same time. Because the Appleton project preoccupied the Locks and Canals machine shop, Cabot gave the Lowell Company's order to Pettee. The successful completion of this order—the investment in tools and machinery came to $100,000—transformed Pettee's business. The merchant networks soon led to an even larger machinery order totaling as much as $150,000 from the Jackson Company (incorporated in 1830), another sizable enterprise of the Associates. The Elliot Company directors, however, had no intention of transforming their shop into a large textile machinery producer. Pettee ended up receiving financing on his own, probably from the Jackson Company in Nashua, and he took full responsibility for completing the order.

Indicative of the strong bridges that connected various local textile machinist networks in the East, news of Pettee's breakup with the Elliot Company quickly generated invitations from the Nashua Manufacturing Company and from manufacturers in western Massachusetts and eastern New York either to relocate his machine shop or to take a position as a senior mechanic. He decided to remain nearby and build his own machine shop, purchasing much of the equipment of the Elliot Company's shop. Members of the Associates, including a stockholder in the company, provided mortgage loans on Pettee's property to help fund his shop. By 1833 he had a forge and machine shop, which he enlarged several times over the next few years. Within four years he added a foundry, which had become necessary because it allowed greater coordination and quality control of the machine-building process.

Pettee established his firm in an increasingly challenging textile machinery environment. He faced strong competitors, including Ira Gay and Company and the machine shops of Providence and vicinity, many of which dominated their local markets. The large shops of the Associates' mill complexes had begun to look externally for orders because they confronted a slowing growth rate of their cotton mills. Consequently, Pettee looked mostly elsewhere for markets, and from the late 1830s until his death in 1853, his machine shop sold a full range of equipment to new mills in the South and in Mexico. Members of the Associates with whom Pettee maintained close ties probably supplied entrée to these markets. Their mercantile contacts provided intimate knowledge about southern mill developments, and they also may have supplied the Mexico connection. One of the cotton mill developers in Mexico in the late 1830s had come from Dover, New Hampshire, the site of one of the Associates' largest projects (the Cocheco Manufacturing Company). In 1850, shortly before Pettee's death, his machine shop employed about three hundred workers—a high point for a long time because the shop declined for several decades.

In 1832, immediately prior to the founding of Gay's and Pettee's machine shops, Boston's inner environs housed a sizable number of independent textile machinery firms (table 6.5). Most of them probably emerged during the textile boom of the 1820s, and other shops entered production during the slower pace of textile expansion in the early 1830s. Typically, these firms listed cotton and/or woolen machinery as their product line, but their degree of specialization cannot be determined. Firms specialized in cards and bobbins, and several, including William Whittemore's in Cambridge, produced substantial amounts. Bobbins sold for just pennies, and companies such as two in southern New Hampshire could have manufactured as many as two hundred thousand each, indicative of the enormous scale of textile manufacturing by the early 1830s. Some firms' market areas, such as Crosby and Bliss's in Milford, New Hampshire, remained restricted to their vicinity, whereas others sold to several New England states, especially Massachusetts, New Hampshire, and Maine—the sites of extensive textile manufacturing. A few, albeit atypical, firms sold throughout New England and outside of it. The absence of sales to Rhode Island and Connecticut cotton mills reveals the limited success that textile machinery firms in Boston's inner environs had in invading markets of other nests of textile machinery firms.

TABLE 6.5

The Boston Environs (Excluding Worcester County) Nest of Textile Machinery Builders, 1832

State/City	Firm (date founded)	Number of Employees	Value of Products ($)	Products	Market Area
Massachusetts					
Cambridge	William Whittemore (1821)	14	40,000	cards	1/3 Mass.; 1/16 export; rest Middle Atlantic, South, West
Cambridge	A. Whittemore Jr., Co.	3	5,000	cards	South, West
Lexington	D. Holmes and J. White (1831)	12	15,000	cards	Mass., N.H.
Dedham	Golding's Machine Shop (recent)	15	12,000	cotton and woolen machinery	New England
Canton	Copeland and Kinsley (not recently)	8	6,000	cotton and woolen machinery	East
Andover	Sawyer and Phelps	30	30,000	woolen machinery	N.H., Vt.
Andover	John Smith	35	30,000	cotton machinery	Mass., N.H.
Methuen	Cotton mill machine shop (1823)	12	4,000	cotton machinery	
South. New Hampshire					
Exeter	Unknown	45	45,000	cotton machinery	N.H., Maine, Mass., Md.
New Market	Unknown	13	4,500	bobbins	N.H., Maine, Mass.
Hillsborough County					
Dunstable	Unknown	3	—	power looms and such	vicinity
Mason	Loami Chamberlin (1817)	6	1,606	cotton and woolen machinery	N.H.
Milford	Crosby and Bliss (1831)	10	6,000	cotton machinery	N.H.: Milford, Jaffrey, New Ipswich
Wilton	Unknown (1824)	8	4,000	bobbins	
Total		214	203,106		

SOURCE: Louis McLane, *Documents Relative to the Manufactures in the United States Collected and Transmitted to the House of Representatives, 1832, by the Secretary of the Treasury*, House Doc. 308, 22nd Cong., 1st sess. (Washington, D.C.: Duff Green, 1833).

TABLE 6.6

Worcester County, Massachusetts, in the Boston Environs Nest of Textile Machinery Builders, 1832

City	Firm (date founded)	Number of Employees	Value of Products ($)	Products	Market Area
Worcester	W. M. Bickford	10	10,700	woolen machinery	East
Worcester	Hobart, Goulding and Smith	30	48,000	textile machinery	East
Worcester	Rice and Wheelock	12	11,800	woolen machinery	East
Worcester	J. Simmons and Co.	5	6,335	woolen machinery	New England
Worcester	Wheelock and Phelps	18	18,000	woolen looms	mostly New England, few N.Y.
Worcester	White and Boyden	14	10,010	woolen machinery	mostly East
Fitchburg	Samuel Hawes	2	1,000	bobbins	Worcester County
Holden	A. Dryden	20	12,000	cotton and woolen machinery	Mass.
Leicester	Unknown	1	672	bobbins	U.S.
Leicester	Eight machine card shops	59	106,782	machine cards	U.S.
Leicester	Unknown	2	1,500	shuttles	U.S.
Mendon	Paine and Ray	30	24,000	cotton and woolen machinery	½ Mass., ½ R.I.
Milford	Unknown	9	4,000	shuttles	East
Millbury	Samuel Brown (1826)	10	6,182	spindles	Mass. (mostly), Conn., N.H., N.Y.
Millbury	Leland and Sabin (1826)	14	10,950	woolen looms (mostly)	Mass., N.H., Vt.
Millbury	Hervy Waters (1831)	24	19,048	cotton machinery	Mass., Conn., Vt., N.Y., Pa.
Northborough	Unknown	11	4,000	cotton machinery	New England
Northbridge	Two shops	20	16,500	cotton and woolen machinery	New England
Oxford	W. Chatman	10	6,600	cotton machinery	Mass.
Spencer	Two shops	3	1,372	bobbins	Mass.
Sutton	J. Farnum and M. Ruggles	6	2,800	shuttles	New England, N.Y.
Sutton	Origen Harbach	4	4,000	shuttles	New England
Sutton	Asa Woodbury	6	3,200	spindles	New England
Uxbridge	John White	9	7,200	shuttles	Mass., R.I.
West Boylston	Samuel Flagg	23	23,310	cotton machinery	Mass.
Winchendon	E. and W. Murdock	9	5,000	bobbins	vicinity
Total		361	364,961		

SOURCE: Louis McLane, *Documents Relative to the Manufactures in the United States Collected and Transmitted to the House of Representatives, 1832, by the Secretary of the Treasury*, House Doc. 308, 22nd Cong., 1st sess. (Washington, D.C.: Duff Green, 1833).

Worcester County

The textile machine shops in Boston's outer environs of Worcester County operated at larger scales and reached greater market areas than shops in the inner environs (tables 6.5 and 6.6). Most of the county's firms claimed to market their machinery throughout New England, about five reached the entire East, and three said they reached the national market—probably the East and perhaps the South. Although firms that claimed to market in New England may have sold some equipment in Rhode Island, only Paine and Ray in Mendon and John White in Uxbridge explicitly identified that state as a market. They operated immediately north of Rhode Island's border in towns whose textile developments had close ties to Providence area investors and machinists. Mendon, for example, housed the Blackstone Manufacturing Company which Providence merchants, including the prominent firm of Brown and Ives, started in 1808–9. Paine and Ray also owned a small cotton mill that sold goods mostly in Rhode Island, cementing their business ties there. Like Boston's inner environs, Worcester County specialized in cards and bobbins, and its firms also made shuttles. By the early 1830s an extreme focus of textile machinery firms remained limited primarily to simple equipment.

Worcester County sprawled across central Massachusetts almost forty miles from east to west and almost fifty miles from north to south, yet four-fifths of its textile machinery firms concentrated in the city of Worcester and within a ten-mile radius of it (see table 6.6). Before 1820 the city of Worcester had housed several small machine shops, including some that turned out textile machinery for the mills in nearby towns. The village of Millbury housed as many as forty small workshops and factories that used waterpowered machinery. During the 1820s several iron foundries began operating in Worcester, foreshadowing its growth as a metal manufacturing and machinery city, and this milieu supplied fertile ground for the enlargement of textile equipment production. In the early part of the decade at least four firms turned out such machinery, and most, if not all, of the six firms that operated in 1832 began business sometime after the mid-1820s, in response to the textile boom.

The city's machinists constituted a networked group that fluidly formed and dissolved partnerships in various metalworking businesses. Their cohesiveness made these associations possible, but they did not form an exclusive network. They readily incorporated new machinists into the networks, and many of the leading mechanics maintained contacts outside the county, thus helping its textile machinery firms reach large market areas, and their success contributed to a vibrant, growing industry during the 1830s and afterward. The know-how trading that flourished within this community of practice of textile machinists also extended to other machine sectors, which strengthened all of them because various metalworking industries required many of the same technical skills.

Network bridges linked the machinists of Worcester County with those of Providence and vicinity, and one of several connecting points included the cluster of cotton mills and machine shops in Northbridge, fifteen miles southeast of the city of Worcester. The Northbridge Cotton Manufacturing Company, which was organized in 1809, had united a Providence group—investors and a machinist, William Howard—with Northbridge investors, including Paul Whitin Sr., a blacksmith, and with the Earle brothers who manufactured cards in Leicester, immediately west of Worcester. Whitin probably learned textile machinery skills from Howard. Following changes of ownership and the death of Whitin in 1831, his sons formed the firm of Paul Whitin and Sons, consisting of two mills and a machine shop. Northbridge had at least three cotton mills, one woolen mill, and two machine shops that produced textile machinery (see table 6.6). John, one of Paul's sons, had previously worked as a machinist for the Northbridge Cotton Manufacturing Company. He took over Paul Whitin and Sons' machine shop and developed a picking machine to remove foreign particles in cotton, one of the first steps in textile manufacturing. In 1832 John acquired a patent on it, and the brothers' two mills used the first three machines the firm produced.

News of the new picker must have circulated through elite networks connecting patent recording in Washington, D.C.—where members of the legislative and executive branches included merchants, industrialists, and lawyers from the Boston and Providence areas—to the owners of cotton mills, some of whom were in Washington. Mill owners also acquired information through publications such as Philadelphia's *Journal of the Franklin Institute*, which published patent specifications and commentary on the patents, including Whitin's patent in one of its issues. Leading mechanics magazines, including the *Boston Mechanic, and Journal of the Useful Arts and Sciences* and the New York *Mechanics' Magazine and Register of Inventions and Improvements*, reprinted the Franklin Institute's patent articles. Elite cotton mill investors living in Boston and Providence knew about these publications and would undoubtedly have passed along information about Whitin's patent to their mill managers. Within a year of his patent cotton mills must have placed orders because the Whitin machine shop enlarged production capacity to twelve pickers annually in 1834. Within two years production surged to over seventy pickers; after that the number turned out each year settled into a range between forty and sixty.[5]

New York City's Environs

In contrast to the areas in and around Providence and Boston, textile machinery manufacturing within New York City's environs remained subdued before 1820, consonant with the small scale of the textile industry. The subsequent growth of textile machinery production diverged from that of the New England nests of builders, which had a modest number of large firms and many small builders, whereas most of the textile equipment in the nest of builders in New York City's environs originated in a few large machinery firms.

The Matteawan Company

Wealthy New York City merchants backed the Matteawan Company at Fishkill, in Dutchess County. It had made cotton textiles since 1814, and its machine shop sold equipment to other mills in the Hudson Valley because its owners possessed ample capital to fund cotton manufacturing and production of machinery for sale. After 1820 the valley's textile mills continued to expand, and by 1840 Dutchess County and its northern neighbor, Columbia County, contained over 3 percent of the nation's capital investment in textiles and over one-third of New York state's (see table 6.2). Under the leadership of Superintendent William Leonard, Matteawan increased its textile machinery sales in the 1820s to serve the valley's growing market. This mechanical genius developed innovations in textile machinery, and the company became one of the leading importers of English textile innovations; it also stole them. Matteawan added a foundry and in 1832 built a separate machine shop; together, these operations employed two hundred workers at that time. The machinery works manufactured various products, and textile machinery remained an important part of the business.

Although Matteawan's cotton mill had six thousand spindles by 1846 machinery production from the pseudo-independent shop dominated. It totaled $262,000, whereas the output of cotton goods amounted to $174,000. Four years later the machinery works, considered one of the best in the nation, employed about four hundred mechanics and reputedly could fully equip a cotton mill. The connections of Matteawan's prominent mechanical and financial leaders made it a pivot of know-how trading of technical knowledge about machinery. Numerous apprentice and experienced machinists eagerly sought work at this outstanding machine shop because the mechanism of the artifact-activity couple provided a basis for superb training. Many of them left for other firms and spread the skills they had learned there. This enhanced Matteawan's nodal position in machinist networks and made it one of the leading contributors to communities of practice in textile machinery, locomotives, and industrial machinery. Nevertheless, technical prominence did not ensure financial success; Matteawan failed in the Panic of 1857, partly due to its owners' financial problems. Its achievements and demise demonstrate that technical skills and knowledge were embedded in the networks of machinists and firms, and these assets did not disappear when individuals or firms dropped out of the networks.

Paterson

For years the Great Falls of the Passaic had attracted industrialists. Nonetheless, only two of the sixteen cotton mills in Paterson with known founding dates (one mill cannot be dated) and still in production in 1832 had started manufacturing before 1820. The total capital investment of one million dollars in its seventeen mills amounted to over half of New Jersey's cotton textile investment in 1832, and the city housed six of the state's ten largest mills. The aggregate capital of its mills, however, fell below the $1.3 million dollars invested in the Merrimack Cotton Manufacturing Company, the signature firm of the Boston Associates' Lowell project. By 1840 Passaic County, including Paterson, still housed over half of New Jersey's capital investment in textiles (see table 6.2).

Thomas Rogers and John Clark Jr., who had formed their partnership in 1816, turned out textile machinery as one of their diversified product lines because the limited demand for mill equipment did not support specialization. In 1822, when Abraham Godwin Jr. partnered with Rogers and Clark, the firm obtained additional capital that they used to build a larger facility and add a foundry. The latter represented a sizable investment and signified that the machinery business would be expanded. Following that date, textile machinery orders increased significantly, judging from the number of cotton mills that survived until 1832; thirteen mills started up in Paterson and twenty-five elsewhere in New Jersey. In 1831 Rogers left the partnership and established a machine shop with a foundry.

Prior to Rogers's departure, the firm acquired the services of Charles Danforth, and the new partnership became Godwin and Clark. Danforth started his machinist career in the Taunton, Massachusetts area, and by 1821, when he went to work as a foreman at the Matteawan Company, the twenty-four-year-old already was a well-known textile machinist. He left after four years and worked in several cotton mills in the Hudson Valley, where he developed his cap-spinning frame which he patented in 1828. Over the next several years he spent time in Taunton with his brother George, who had developed what came to be called the "Taunton speeder." This machine transferred roving to bobbins in preparation for spinning, and George patented it in 1824. Between 1829 and 1831 Charles spent time in Paterson and in England then moved to Paterson permanently to become a partner in Godwin and Clark. His cap-spinning frame became a signature product of the company's line of textile machinery.

Thus, by 1832, Paterson housed two textile machinery firms that ranked among the

TABLE 6.7
New York City Environs and Philadelphia Area Nests of Textile Machinery Builders, 1832

City/County/State	Firm (date founded)	No. of Emp.	Value of Products	Products	Market Area
New York City environs					
Paterson, N.J.	Godwin, Clark and Co. (1821)	200	$200,000	cotton machinery	U.S. (mostly), Mexico
Paterson, N.J.	Rogers, Ketchum and Grosvenor	120	—	cotton machinery	U.S. (mostly), Mexico
Philadelphia area					
Philadelphia, Pa.	Jenks	110	—	textile and other machinery	—
New Castle County, Del.	Jacob Alrichs and Son (1808)	25	$24,000	cotton and woolen machinery	within 160 miles
New Castle County, Del.	McLary and Bush (1828)	21	—	textile and other machinery	within 100 miles

SOURCE: Louis McLane, *Documents Relative to the Manufactures in the United States Collected and Transmitted to the House of Representatives, 1832, by the Secretary of the Treasury*, House Doc. 308, 22nd Cong., 1st sess. (Washington, D.C.: Duff Green, 1833).

largest at that time (table 6.7). The local market did not provide enough business to keep them busy, and they claimed the nation as their sales area (primarily the East, with some sales in the South) and exported small amounts to Mexico. Both firms went through various ownership changes, but they continued to produce textile machinery for the remainder of the antebellum, even as locomotives became increasingly significant for the Rogers's firm after 1837 and for Danforth's firm after 1852 (see fig. 5.7).

Like Paterson's locomotive machinists, its prominent textile mechanics and their firms operated as network hubs. Many of its mechanics received training at the Matteawan Company, one of the leading innovators in textile machinery, and they had close ties to areas in and around Providence, especially with Taunton. Paterson's textile machinists also possessed strong connections to their peers in Boston's environs. Within a year of Danforth's move to Godwin and Clark, Kirk Boott, one of the leaders of the Boston Associates, negotiated with the Paterson firm for rights to manufacture Danforth's cap-spinning frame in the Locks and Canals machine shop. The Lowell mills would install them, and the Lawrence Manufacturing Company's mills were the first to get the machinery. At the time their production began in 1834, they contained over twenty-four thousand spindles, and two years later the spindle count rose almost one-third. With four cotton mills and a capital investment of $1.2 million, clearly this mammoth installation boosted the reputation of Danforth's machine.

The networks of know-how trading transferred textile machinery innovations from

Paterson to the other nests of textile machinists and kept its firms abreast of technical advances generated elsewhere. The multifaceted business of the city's machinery firms fueled their capacity to remain vibrant hubs in machinist networks because techniques in seemingly different industries had substantial complementarities. Firms could build textile machinery and locomotives as well as other equipment; therefore, the city's mechanics and firms contributed to communities of practice in several machinist sectors. By 1850 Paterson had six full-line machinery firms with foundries, a sign of significant production capabilities, which together employed almost fifteen hundred workers.[6]

The Philadelphia Area

A robust cotton textile industry of small firms had emerged in Philadelphia before 1820, but, like in the New York City area, its industry remained small by 1840 (see table 6.2). The counties of Philadelphia, Delaware, and Montgomery in Pennsylvania and of New Castle in Delaware housed almost 6 percent of the national capital investment in cotton textiles. This figure matched the 5 percent share for the counties in New York City's environs. The scale of the Philadelphia area's textile industry paled next to New England's concentrations: Middlesex County, the heart of the Associates' efforts, contained 18 percent of the nation's investment; Providence County, the center of the Providence merchants' endeavors, had 10 percent; and other large textile clusters surrounded each of these. Philadelphia's small cotton mill sector impacted its textile machinery industry. Several major textile machinery firms operated by 1832, including the firm of Jenks in Philadelphia, but few small ones. In contrast, Providence and vicinity and Boston's environs had a plethora of small firms as well as a number of large ones (see tables 6.4–7). For the most part, the firms in the Philadelphia nest of textile machinery builders sold their equipment locally.

The firm of Alfred Jenks and Son in the Bridesburg section of Philadelphia had started making textile machinery in 1810, when the father arrived from Pawtucket following training under Slater. He had even deeper ties to Rhode Island because the Jenks family members there operated machine shops, such as Fales and Jenks which began manufacturing textile machinery in Central Falls in 1830. Alfred and his son, therefore, constituted a major bridge to the nest of machinists in Providence and vicinity. Their firm's more than one hundred employees in 1832 placed it among the larger machine shops in the East (see table 6.7). The talented father and son obtained many important patents for textile machinery as well as for other types of equipment, making them hubs of know-how trading of technological innovations and technical skills, and the son's prominence continued after his father's death around 1844.

By the mid-1850s Jenks's Bridesburg Works constituted a substantial industrial enterprise that included a large foundry, a shop housing eighteen forges and four trip hammers, and a machine shop; its mechanics made many of their machine tools. When the works operated at capacity, employment totaled as many as four hundred workers. The machinists in the Jenks Works participated as key actors in the ongoing functioning of communities of practice of textile machinists and of other industrial mechanics in Philadelphia, and the company served as a pivotal hub of machinist networks throughout the East.[7]

Outside the Nests of Textile Machinery Shops

Independent textile machine shops located away from the nests of machinery builders could capture their local markets for mill equipment in competition with shops in the nests. Some of the firms outside the nests competed in subregional and even in regional markets (table 6.8). Like Boston's environs, specialized producers of simple components such as cards, bobbins, and shuttles coexisted along with manufacturers of a diverse range of machinery.

The Connecticut Valley and nearby areas from southwestern New Hampshire to central Connecticut accounted for the majority of the shops outside the nests, and the largest number occupied the Massachusetts portion of the valley. These machine shops emerged to serve the cotton mills that proliferated during the 1820s boom. By 1832 cotton mills ranged from small ones such as the Woodbridge Manufacturing Company in South Hadley (started in 1828), which had thirteen thousand dollars of fixed capital investment, to large mills such as the Chicopee Manufacturing Company (1822), whose capital investment reached almost a half-million dollars after the Boston Associates took it over. Although the Chicopee firm had its own machine shop, many of the smaller mills constituted markets for the independent machinery firms. Because the trade links of the valley north of the Connecticut border focused on Boston, machinist migrations followed well-worn paths between the valley and the metropolis's environs. Furthermore, the cluster of machine shops in and near Worcester operated less than fifty miles to the east. The valley's machinist networks also included private firearms manufacturers and the Springfield Armory, and these networks reached to Providence and vicinity. During the 1830s the market for textile machinery continued to expand and by 1840 Hampden County, Massachusetts, housed seventeen mills whose total capital amounted to 4 percent of the nation's, close to Philadelphia County's share (see table 6.2).

Cotton mills in the Hudson Valley around Albany and in the nearby hills in Berkshire County, Massachusetts, had emerged before 1820, but the subsequent cotton mill boom provided the bigger spur to textile machinery manufacturing. By 1832 the towns of Adams, Lee, and North Bennington housed a textile machine shop cluster

TABLE 6.8

Manufacturers Outside the Nests of Textile Machinery Builders, 1832

Area/City/State	Firm (date founded)	Number of Employees	Value of Products ($)	Products	Market Area
Connecticut Valley					
Claremont, N.H.	R. Elmes and Co. (1830)	30	2,500	looms and carding machines	¾ N.H., ¼ N.J.
Winchester, N.H.	Unknown	8	5,000	looms and spinning frames	New England
Winchester, N.H.	Cotton mill machine shop	8	5,000	cotton machinery	vicinity
Enfield, Mass.	J. Wood and Co. (1818)	21	62,500	cards	½ Mass., ½ (Middle Atlantic, Va., Ohio, Mexico)
Northampton, Mass.	E. Jewett (1828?)	—	350	bobbins	—
Northampton, Mass.	J. Stedman (1830)	9	1,500	bobbins and shuttles	Mass.
Springfield, Mass.	Springfield Card Manuf. Co. (1826)	13	24,000	cards	—
Wales, Mass.	Willimansett Card Manuf. Co. (1829)	25	40,000	cards	local
Ware, Mass.	Hampshire Manuf. Co. (1829)	40	20,000	cotton and woolen machinery	New England
Williamsburgh, Mass.	J. and S. Hannum (1820)	6	2,000	carding machines	½ Mass., ½ outside Mass.
New Britain, Conn.	Stanley, Watrous and Co.	50	—	cotton and woolen machinery	—
Hudson Valley					
Adams, Mass.	Three machine shops	100	85,000	cotton machinery	½ New England, ½ New York
Lee, Mass.	Unknown	16	12,000	cotton and woolen machinery	Mass, N.Y:, elsewhere
N. Bennington, Vt.	Doty and Loomis (1831)	20	—	cotton and woolen machinery	—
Other					
Pittsburgh, Pa.	Asa Waters	30	25,000	cotton machinery and metal tools	—
Pittsburgh, Pa.	F. A. Bemis and Co.	25	14,000	cotton machinery	—
Watertown, N.Y.	George Goulding (1827)	15	10,000	cotton and woolen machinery	N.Y.

SOURCE: Louis McLane, *Documents Relative to the Manufactures in the United States Collected and Transmitted to the House of Representatives, 1832, by the Secretary of the Treasury*, House Doc. 308, 22nd Cong., 1st sess. (Washington, D.C.: Duff Green, 1833).

consisting of five medium-size firms. This area also had other machine shops, some of them connected to the rapidly growing paper industry (see table 6.8). The Hudson Valley and the Berkshire Mountains constituted the main sales area of the textile machine shops, and this area continued to thrive as a market through the 1830s. By 1840 the counties of Albany and Rensselaer in New York and of Berkshire in Massachusetts together contained forty-eight cotton mills with an aggregate capital of about one and a half million dollars; this amounted to almost 3 percent of the nation's investment. Several other textile machinery firms emerged in widely separated areas, including one in Watertown, New York, which probably served local mills and those to the south in Oneida County, and two shops in Pittsburgh served the small number of cotton mills in western Pennsylvania.[8]

CHANGES IN THE MARKET FOR TEXTILE MACHINERY

Capital Investment

During the 1820s textile machine shops equipped cotton mills that dramatically grew in size, as measured by capital investment per establishment, because existing spinning mills added power looms and new mills started manufacturing as large, integrated spinning-weaving mills (table 6.9). The greater size of new mills in the main areas of cotton textile growth—the southern portions of New Hampshire and of Maine, Massachusetts, and Rhode Island—spelled the demise of the mill's machine shop as a supplier for the new mill. The one-time construction of a large machine shop, outfitting it with metalworking machinery, and assembly of a large number of machinists to equip a big mill with complex machinery proved too costly. Overall, the capital investment per mill in most states in the East during the 1830s did not increase significantly, but concentrations of huge mills became evident in some counties by 1840 (see tables 6.2 and 6.9). Small independent shops had difficulty filling orders from these large mills. The size of cotton mills in New England continued to rise after 1840, which intensified pressure on the large independent textile machinery firms. To meet sizable equipment orders, these firms required large numbers of mechanics and extensive capital investment to purchase big buildings and to enlarge foundries or add new ones. Similar pressures, but at a much reduced scale, impacted the Middle Atlantic textile machinery firms that confronted larger mills after 1840.

The Number of Customers

Textile machine shops needed to adapt their approach to selling equipment to changes in the number of customers (table 6.10). During the cotton textile boom of the 1820s Massachusetts accounted for about 40 percent of the net increase in the

TABLE 6.9
Capital Investment per Establishment in Cotton Textiles by Region and State, 1820–1860 (dollars)

Region/State	1820	1831	1840	1850	1860
New England	24,851	53,369	48,115	95,448	121,509
Maine	34,000	95,625	155,333	277,475	316,754
New Hampshire	19,368	132,500	89,084	248,875	286,065
Vermont	10,064	17,382	16,871	22,500	33,900
Massachusetts	26,599	50,355	58,047	133,595	155,321
Connecticut	30,570	30,053	25,836	32,962	51,372
Rhode Island	24,142	53,986	32,416	42,247	65,701
Middle Atlantic	48,330	45,567	31,650	36,711	55,292
New York	49,487	32,763	37,990	48,569	68,145
Pennsylvania	24,157	56,097	22,777	21,774	49,746
New Jersey	36,214	39,758	30,764	70,643	30,013
Maryland	121,418	93,217	54,350	93,167	112,725
Delaware	47,778	38,450	30,045	38,342	52,955
Rest of nation	11,848	41,429	16,560	43,481	58,137
Nation	30,205	50,703	37,328	68,100	90,362

Sources: David J. Jeremy, *Transatlantic Industrial Revolution: The Diffusion of Textile Technologies between Britain and America, 1790–1830s* (Cambridge, Mass.: MIT Press, 1981), 276, app. D, table D.1; U.S. Bureau of the Census, *Compendium of the Sixth Census, 1840* (Washington, D.C.: Blair and Rives, 1841); U.S. Bureau of the Census, *Report on the Manufactures of the United States at the Tenth Census, 1880* (Washington, D.C.: Government Printing Office, 1883).

number of mills in the East, which spurred the formation of many machine shops in Boston's environs. Similarly, machinists in Providence and vicinity also received substantial new business because Rhode Island and nearby eastern Connecticut, which accounted for most of that state's new mills, together generated about one-fifth of the East's net gain in cotton mills. These changes in number of mills understate the impact on orders for equipment, because numerous tiny spinning mills collapsed. The net gain in the East fell from 476 mills in the 1820s to 298 in the next decade, but, like the 1820s, this probably underestimates the addition of new mills. Many marginal integrated mills collapsed during the 1830s because they could not sustain operations under the regime of low selling prices of cotton cloth which became a permanent fixture of the market (see fig. 6.2). During 1820–40 the Middle Atlantic states added far fewer mills, and the number of their mills remained about half that of New England's.

After 1840 textile machinery manufacturers faced a transformed market. Even though individual orders for equipment rose, they had to compete with one another over fewer customers; lost orders threatened a shop's survival (see table 6.10). During the 1840s the total number of cotton mills in the East shrank by 16 percent, and New England's machinery firms were impacted the most as the number of its mills declined by 22 percent. Shops in Boston's environs faced even greater pressure as the mill count in Massachusetts, New Hampshire, and Maine fell by 27 percent.

TABLE 6.10
Number of Establishments in the Cotton Textile Industry by Region and State, 1820–1860

Region/State	1820	1831	1840	1850	1860
New England	215	531	726	564	570
Maine	5	8	9	12	19
New Hampshire	31	40	62	44	44
Vermont	11	17	7	9	8
Massachusetts	58	256	300	213	217
Rhode Island	69	116	226	158	153
Connecticut	41	94	122	128	129
Middle Atlantic	103	263	366	351	339
New York	41	112	129	86	79
Pennsylvania	28	67	146	208	185
New Jersey	14	51	56	21	44
Maryland	11	23	24	24	20
Delaware	9	10	11	12	11
Rest of nation	39	7	277	179	182
Nation	357	801	1,369	1,094	1,091

Sources: David J. Jeremy, *Transatlantic Industrial Revolution: The Diffusion of Textile Technologies between Britain and America, 1790–1830s* (Cambridge, Mass.: MIT Press, 1981), 276, app. D, table D.1; U.S. Bureau of the Census, *Compendium of the Sixth Census, 1840* (Washington, D.C.: Blair and Rives, 1841); U.S. Bureau of the Census, *Report on the Manufactures of the United States at the Tenth Census, 1880* (Washington, D.C.: Government Printing Office, 1883).

Although the Middle Atlantic numbers remained stable overall during the 1840s, the distribution of mills changed; New York and New Jersey lost the most, whereas Pennsylvania's number soared by 42 percent. Textile machine shops in New York City's environs, especially in Paterson, confronted many fewer customers. This probably motivated the decision of Rogers, Ketchum and Grosvenor, the largest machine shop in the city, to give increasing attention to locomotive manufacturing. Philadelphia's textile machinery manufacturers, especially Jenks's Bridesburg Works, benefited the most. The relative distribution of cotton mills in the East changed little during the 1850s, suggesting some stability in customers.

Textile machinery markets outside the East surged during the 1830s, but they shrank after 1840; the South housed most of these mill customers. The sharp decline during the 1840s, followed by stability over the next decade, however, indicates that this region's importance as a market for the East's textile machinery firms would come after the Civil War.

The Market for Spindles and Power Looms

The largest financial outlays of cotton mills for machinery went to purchase spinning machines and power looms. The growth in the amount of this equipment, therefore, measures the markets that machine shops faced (table 6.11). By 1820 New England and the Middle Atlantic had a similar installed base of spindles and power looms.

After the decade's mill boom, however, the nests of textile machine shops in Boston's environs and in Providence and vicinity had installed twice as many spindles and about three-quarters more power looms than shops in New York City's environs and in the Philadelphia area. This divergence widened; by 1840 New England's machine shops had installed over three times as many spindles as the Middle Atlantic's shops.

Builders in Boston's environs and in Providence and vicinity participated fully in this machinery construction boom. In 1840 the textile mills in Massachusetts, New Hampshire, and Maine which shops in Boston's environs served contained almost nine hundred thousand spindles, and the mills in Rhode Island and Connecticut which shops in Providence and vicinity served had seven hundred thousand. Yet these nests of builders were more similar than these figures indicate. In Boston's environs the captive shops in the mill complexes of the Boston Associates installed large amounts of equipment in their own mills. Removing this equipment from the calculations for 1820–40 significantly reduces the spindle demand in Massachusetts, New Hampshire, and Maine which the independent shops could serve. These shops in Providence and vicinity installed many of the spindles in Massachusetts mills that bordered Rhode Island. In aggregate, therefore, these shops did even more business than the independent shops in Boston's environs.

By 1860 the differences in the scale of textile machinery production had widened among the East's nests of builders. The installed base of spindles and power looms of New England's machinery firms was almost four times the amount of that of the Middle Atlantic firms (see table 6.11). A sharp divergence appeared between New England's nests of builders. Firms in Boston's environs had an installed base of about 2.6 million spindles and sixty-seven thousand power looms in Massachusetts, New Hampshire, and Maine, whereas firms in Providence and vicinity had about half as much of an installed base in Rhode Island and Connecticut (1.3 million spindles and twenty-six thousand power looms). That large gap between the nests could not result from inadequate measurement, such as wrongly assuming that market areas of machinery shops corresponded to these state groupings, failing to adjust for captive machine shops of the Boston Associates (most were independent by 1860), or ignoring Massachusetts markets on the border of Rhode Island which shops in Providence and vicinity served. Machinery firms in Boston's environs benefited from the much greater expansion of cotton mills within their market area. This expansion had important consequences: these firms accumulated more retained earnings, which augmented their capital bases, and they invested in more and better capital equipment for manufacturing textile machinery, thus dramatically boosting their competitive advantage relative to firms in Providence and vicinity.

TABLE 6.11
Number of Spindles and Power Looms in the Cotton Textile Industry by Region and State, 1820–1860

	Spindles				Power Looms		
Region/State	1820	1831	1840	1860	1820	1831	1860
New England	167,806	823,726	1,597,394	3,858,962	974	21,336	93,344
Maine	3,370	6,500	29,736	281,056	12	91	6,877
New Hampshire	17,811	113,776	195,173	636,788	120	3,530	17,336
Vermont	4,102	12,392	7,254	17,600	38	352	362
Massachusetts	41,286	339,777	665,095	1,673,498	321	8,981	42,779
Rhode Island	66,934	235,753	518,817	814,554	315	5,773	17,315
Connecticut	34,303	115,528	181,319	435,466	168	2,609	8,675
Middle Atlantic	141,517	413,133	487,571	1,039,920	645	12,006	25,102
New York	62,566	157,316	211,659	348,584	334	3,653	7,885
Pennsylvania	21,652	120,810	146,494	476,979	23	6,301	12,994
New Jersey	20,071	62,979	63,744	123,548	80	815	1,567
Maryland	23,024	47,222	41,182	51,835	188	1,002	1,670
Delaware	14,204	24,806	24,492	38,974	20	235	986
Rest of nation	16,357	9,844	199,666	336,845	4	91	7,867
Nation	325,680	1,246,703	2,284,631	5,235,727	1,623	33,433	126,313

Sources: David J. Jeremy, *Transatlantic Industrial Revolution: The Diffusion of Textile Technologies between Britain and America, 1790–1830s* (Cambridge, Mass.: MIT Press, 1981), 277—78, app. D, tables D.2, D.4; U.S. Bureau of the Census, *Compendium of the Sixth Census, 1840* (Washington, D.C.: Blair and Rives, 1841); U.S. Bureau of the Census, *Report on the Manufactures of the United States at the Tenth Census, 1880* (Washington, D.C.: Government Printing Office, 1883).

The Amount of Machinery per Mill

The customers of textile machinery firms also changed. After 1820 the increased amount of capital investment in each cotton mill generated larger orders for equipping new mills (see table 6.9). Coincidentally, from the late 1820s to the early 1830s the cost of textile machinery plunged, after which the decline in costs slowed for the remainder of the antebellum (see fig. 6.3). For a given amount of capital, investors could build cotton mills with ever-larger numbers of spinning machines and power looms. New England's experience illustrates this development: between 1831, when the switch-over to integrated spinning and weaving had mostly run its course, and 1860, the capital investment of the typical cotton textile establishment more than doubled, whereas the numbers of spindles and of power looms more than quadrupled (tables 6.9 and 6.12).

The average number of spindles or of power looms per mill understates the typical size of orders from new mills because the large number of old, small mills pulls down the averages; nevertheless, the measure indicates the relative change. During the cotton mill boom of the 1820s average orders of spindles for new mills doubled in the core states of Massachusetts and Rhode Island (see table 6.12). The large mills built in New Hampshire during the 1820s and in Maine during the 1830s foreshadowed changes

TABLE 6.12
Number of Spindles and Power Looms per Establishment in the Cotton Textile Industry by Region and State, 1820–1860

	Spindles per Establishment				Power Looms per Establishment		
Region/State	1820	1831	1840	1860	1820	1831	1860
New England	780	1,551	2,200	6,770	5	40	164
Maine	674	813	3,304	14,792	2	11	362
New Hampshire	575	2,844	3,148	14,472	4	88	394
Vermont	373	729	1,036	2,200	3	21	45
Massachusetts	712	1,327	2,217	7,712	6	35	197
Rhode Island	970	2,032	2,296	5,324	5	50	113
Connecticut	837	1,229	1,486	3,376	4	28	67
Middle Atlantic	1,374	1,571	1,332	3,068	6	46	74
New York	1,526	1,405	1,641	4,412	8	33	100
Pennsylvania	773	1,803	1,003	2,578	1	94	70
New Jersey	1,434	1,235	1,138	2,808	6	16	36
Maryland	2,093	2,053	1,716	2,592	17	44	84
Delaware	1,578	2,481	2,227	3,543	2	24	90
Rest of nation	419	1,406	721	1,851	0	13	43
Nation	912	1,556	1,669	4,799	5	42	116

Sources: David J. Jeremy, *Transatlantic Industrial Revolution: The Diffusion of Textile Technologies between Britain and America, 1790–1830s* (Cambridge, Mass.: MIT Press, 1981), 277—78, app. D, tables D.2, D.4; U.S. Bureau of the Census, *Compendium of the Sixth Census, 1840* (Washington, D.C.: Blair and Rives, 1841); U.S. Bureau of the Census, *Report on the Manufactures of the United States at the Tenth Census, 1880* (Washington, D.C.: Government Printing Office, 1883).

over the last two decades of the antebellum. The number of spindles in the typical mill in Rhode Island and Connecticut more than doubled, those in Massachusetts more than tripled, and the ones in New Hampshire and Maine grew between four- and fivefold. The number of power looms in these mills probably doubled or tripled. Thus, throughout the 1820–60 period textile machinery firms in both nests of builders in New England had to adjust to ever-larger orders from cotton mills, and machine shops in Boston's environs faced even greater adjustment than those in Providence and vicinity. Bigger individual orders came from a shrinking number of customers as the number of cotton mills declined after 1840; this drop occurred mainly in the 1840s.

In contrast, by 1860 textile machinery firms in New York City's environs and in the Philadelphia area faced individual orders for spindles and power looms of less than half the size of those received by New England firms (see table 6.12). These smaller orders provided less opportunity to achieve economies from the division of labor or from batch production, and the gradual increase in the use of machine tools in the 1850s did not allow these firms to make as efficient use of the new equipment through spreading their fixed costs over more units of output.[9]

THE TEXTILE MACHINERY INDUSTRY TRANSFORMS

By the end of the 1830s textile machinery firms had enjoyed almost two decades of robust markets. Many shops were drawn into production, and they could sell their equipment without undue concern about competition from the large captive machine shops housed at the mill complexes of the Boston Associates or from captive ones at other mill complexes. Because their mills grew considerably, these shops had plenty of business turning out new machinery, replacing obsolete equipment, and repairing machines. Nonetheless, the annual growth rate of the market for machinery, as proxied by production of cotton cloth, declined sharply during the 1820s, and this growth rate continued falling, though not as fast, during the next decade (fig. 6.4). Thus, by the end of the 1830s, textile machinery firms confronted altered competitive conditions, and the volatility during the economic contraction from 1837 to 1843 challenged firms to find appropriate new strategies. The independent shops faced slower growth compared to the heady expansion of the previous years, and their sales could be impacted by new shops or by established ones that tried to gain market share.

Over the twenty years following 1840, pressures on the machinery industry intensified as the size of cotton mills more than doubled (measured by capital investment), the number of customers fell by 16 percent, the volume of machinery ordered more than doubled, and the orders from individual mills tripled (see tables 6.9–12). These larger orders, combined with the increasing complexity of machinery, placed small shops in difficult positions. If they had operated as diversified suppliers of equipment for mills, they could no longer meet the large orders from new mills. These shops had to increase the size of their operations significantly, which required substantial new investment in land, buildings, and equipment (such as an iron foundry). If they had specialized in simple equipment, such as bobbins and shuttles, they might continue these lines, but the ease of entry into these businesses meant that competition drove prices to low levels and offered miniscule profit margins.

The other alternative—specialization in a line of major equipment such as pickers, spinning machines, or power looms—required significant capital investment. The big machine shops at the large mill complexes, such as those controlled by the Boston Associates, loomed over the independent machinery shops. By 1840 these shops could no longer rely on orders from their captive markets to maintain their scales of production, and they increasingly looked externally for sales. Their entry into the market for equipment, while its growth slowed considerably, challenged the entire textile machinery industry.

Figure 6.4. Compound Annual Percentage Change in Yards of Cotton Cloth Produced in New England, 1820–1860. *Source:* Robert B. Zevin, "The Growth of Cotton Textile Production after 1815," in *The Reinterpretation of American Economic History*, ed. Robert W. Fogel and Stanley L. Engerman (New York: Harper and Row, 1971), 123–24, table 1.

The Machine Shops of the Boston Associates

The Lowell Machine Shop

In the early 1830s the Locks and Canals shop at Lowell completed machinery for four mills of the gigantic Lawrence Manufacturing Company, whose total spindles came to over twenty-four thousand and whose bill for machinery and furniture amounted to $350,000. By 1835 the shop employed about three hundred workers and could complete all the machinery for a mill of five thousand spindles in about four months. Over the following three years the shop finished the machinery for three mills of over six thousand spindles each for the Boott Cotton Mills, and the bill came to $250,000. The last major new project of the shop in Lowell commenced in 1839, when it began to manufacture equipment for the Massachusetts Cotton Mills. Within two years it completed the machinery for the four mills, and the equipment totaled twenty-four thousand spindles and cost $290,000. The immensity of the orders from the three firms, whose aggregate number of spindles (sixty-six thousand) equaled the total for all fifty-six cotton mills in the state of New Jersey in 1840, indicates the scale of one of the nation's premier machine shops, and filling these orders did not exhaust its work schedule (see tables 6.10–11).

The manufacture of a four-color printing machine for a customer in Fall River and of a calendar for a Lowell bleachery in 1833 constituted the first consequential diversification of the Locks and Canals machine shop beyond standard textile machinery. The following year the shop added locomotives to its line of equipment, which ran as a separate department from textile machinery. Nevertheless, workers could move between departments because they required similar metalworking skills.

After 1838 the Locks and Canals shop leveraged its long experience making its own tools and machines for building textile machinery, and it increased the sale of small planing machines, steam boilers, grinding wheels, mill shafting, window weights, and castings for pumps and waterwheels. Its large iron foundry, built at a cost of thirty thousand dollars in 1840, provided the internal capacity to cast many of these metal components and those for its textile machinery. Coincident with the last major new mill project in Lowell, the Massachusetts Cotton Mills, the shop expanded its sales of textile machinery to other large cotton mill complexes. Customers included P. Whitin and Sons and the Blackstone Manufacturing Company, both in the Blackstone Valley, and the mill complexes owned by the Associates, such as the Nashua Manufacturing Company, the Salmon Falls Company, and the Great Falls Manufacturing Company, all of them in New Hampshire. The Lowell shop also sold machinery to other textile mills in the East, but they purchased a minority of its equipment sales. It possessed no competitive advantage selling textile machinery based on its innovations. Since the early 1830s the shop had purchased licenses from other machinery innovators to manufacture equipment.

During the slowdown of the early 1840s the directors of the Locks and Canals Company decided to spin off the shop as a separate corporation, and in 1845 the Lowell Machine Shop commenced business. The Associates who invested in the shop followed their standard procedure for hiring top management, appointing William Burke, one of the most talented, young machinists in the East, as superintendent. This thirty-four-year-old had accumulated a wealth of experience, including having gotten his start in 1825–26 in the machine shop of the Nashua Manufacturing Company, where the fifteen-year-old Burke trained under Ira Gay, the resident agent of the company. From 1826 to 1828 Burke worked at the Locks and Canals machine shop then returned to Nashua. In 1833 he helped Gay set up his new machine shop near the Lowell mills in North Chelmsford.

Burke held the position of master mechanic at the Boott Cotton Mills in Lowell from 1836 to 1839. Then he became superintendent of the new machine shop of the Amoskeag Manufacturing Company in Manchester, and five years later he moved to the Lowell Machine Shop. Burke's network ties made him a hub of machinist networks and a bridge to other prominent mechanics such as Ira Gay. When Burke left Amoskeag for the superintendent's position at Lowell, he brought along William Bement to design a line of basic machine tools for sale by the Lowell shop. In 1851 Bement left for Philadelphia to commence his career as an owner of one of the nation's premier machine tool firms, which in turn made Burke a bridge to the machinists in the Philadelphia community of practice.

The Lowell Machine Shop began its independent existence during an upturn in the economy. Leading members of the Boston Associates, including Patrick Jackson, Nathan Appleton, John Lowell, and Abbott Lawrence, invested in it, hiring a premier manager/machinist as superintendent. Nevertheless, this pedigree did not guarantee success over the next fifteen years. The shop could not alter broader structural problems with the textile machinery market and faced tough competitors in all of its business lines. As a member of the capital goods sector, it experienced large fluctuations in sales: between 1846 and 1860 annual sales swung between two hundred thousand and seven hundred thousand dollars, and they trended down; its declining stock price indicated that problems existed. Breaking out of reliance on the Lowell mills for sales, proved more difficult than anticipated; in 1847 these mills accounted for 88 percent of the shop's annual sales of six hundred thousand dollars. Other of the Associates' mill complexes—including the Boston Manufacturing Company at Waltham, the Essex Company at Lawrence, and the Great Falls Manufacturing Company and the Nashua Manufacturing Company in New Hampshire—purchased the remainder of the equipment. The Lowell mills continued to be large buyers during the ten years following 1845, as the firms doubled their number of spindles.

Although the shop's sales lines and market area broadened somewhat, as late as the 1854–60 period, New England customers accounted for 95 percent of the sales. By all accounts textile machinery dominated (especially power looms), locomotives had declined as a line of business, and the remainder of the sales consisted of various machinery and machine tools. Instead of being a fierce rival of many independent textile machine shops, the Lowell shop had not departed much from sales to its traditional customers in the mill complexes controlled by the Boston Associates. It faced strong competitors, including other shops of the Associates' complexes and independent firms such as Fales and Jenks, the Providence Machine Company, the Fall River Iron Works, and Mason (in Taunton), all in and around Providence, as well as Whitin in Massachusetts. The Lowell shop confronted the limits on the expansion of the textile machinery market—the growth of cotton cloth production remained stuck in a tight range of approximately 5 percent compound annual growth from 1840 to 1860. This quandary did not mean it faced demise; in 1860 its five hundred employees ranked it as one of the nation's largest machinery firms.

The Amoskeag Machine Shop

By the late 1820s the Amoskeag Manufacturing Company controlled three cotton mills and two small machine shops. Within a decade it had gained control of all the waterpower along the Merrimack River around Manchester, purchased land, and began building a dam and canals; by 1838, with plans for a new city in place, it proceeded to develop a mill. At the end of 1839 the Amoskeag Company, which acted as a waterpower owner and land developer, had completed two mills for the Stark Manufacturing Company. In 1840 it started two mills for its own account and signaled at the same time that cotton manufacturing would expand dramatically by building a large, three-story machine shop and hiring William Burke, the top master mechanic of the Boott Cotton Mills in Lowell, as superintendent. Within two years the shop added a separate iron foundry, indicating it intended to operate as a large-scale machinery manufacturer with the capacity to produce diverse equipment.

Beginning in the mid-1840s, the construction of a mill for the Manchester Mills Company pointed to a substantial enlargement of cotton manufacturing; it housed 30,000 spindles and 708 power looms. In 1848 Amoskeag built an even larger foundry and another machine shop as large as the first one, but other motives directed this expansion. The firm started producing locomotives, and over the next decade it made 232 engines. Then it exited the industry and sold its interest to the Manchester Locomotive Works. Amoskeag's shop, like Lowell's, faced difficulties competing with the increasingly specialized locomotive builders. Yet its sophisticated machine-building skills could be transferred to another market: building steam fire engines. It produced 5 fire engines in 1859 and 20 the following year and continued that line until 1876, when it had completed 504 fire engines.

Like the Lowell shop, the Amoskeag shop diversified its manufacturing to include products such as stationary and portable steam engines, boilers, tools, turbine waterwheels, and all types of castings. Local textile machinery orders remained too episodic and lumpy—a big order had to be completed in a short period—to keep a large shop fully occupied, and it could not count on a continuous stream of textile machinery orders from outside Manchester. Large machine shops in other mill complexes belonging to the Boston Associates, as well as both small and large independent shops in and around Boston and Providence, competed for these orders.

The New, Large Competitors

The Whitin Machine Shop

In contrast to the Amoskeag Manufacturing Company, the Northbridge, Massachusetts, firm of P. Whitin and Sons looked to textile machinery markets outside the local area. In 1847 the company built an immense brick machine shop, arguably the largest in the nation at that time. The Whitin company kept adding cotton mills to its empire, reaching a total of six mills by 1860, but they took only a small share of its shop's production. The picker business continued to do well into the early 1850s, and Whitin and Sons competed nationally for that market; nevertheless, improved pickers produced by others undercut Whitin's market dominance during that decade.

Fortuitously, the Whitin machine shop had access to nearby eastern Connecticut. The counties of Windham and New London experienced a boom in cotton mill building and in expansion of existing mills; together, they more than doubled their capital investment in cotton manufacturing from 1845 to 1860. These mills needed a full line of textile machinery, and the Whitin shop possessed the capacity to supply most of that equipment. During these years the shop received close to one order annually for a complete line of machinery, which took six months to fill. Between 1854 and 1859 customers placing three or more orders accounted for almost half of the value of Whitin's sales. These satisfied customers confirm that the shop produced superior equipment. Its success continued after 1860 as it became nationally prominent as a full-line manufacturer of textile machinery.

As owners of cotton mills, P. Whitin and Sons possessed network ties to other mill owners within the subregion of southern Massachusetts, eastern Connecticut, and Rhode Island, and John Whitin, the head of its shop, provided direct links to machinists. Nevertheless, the Whitin shop created its most important network ties following the patenting of John's picker in 1832. As orders came in, the shop widened its contacts with textile machinists throughout the East, and it deepened links within its subregion when it diversified the machinery line to fill orders. In contrast, William Mason, whose Taunton firm became one of the leading textile machinery builders beginning in the early 1840s, initially developed his networks ties, as did many of his famous peers, by moving among machinist centers before arriving in Taunton.

The Mason Machine Shop

William Mason's early career mobility reveals the network mechanisms that moved a talented mechanic among positions within the nest of machinists in and around Providence and between it and other nests. The fourteen-year-old Mason, who called the eastern Connecticut shore town of Mystic his home, took a job in 1822 at a small spinning mill in the nearby Quinebaug Valley. The valley had become a hotbed of cotton mill development during the decade following 1807, the date the Wilkinson family of machinists founded their mill in Pomfret. Mason thus acquired experience with machinists trained directly or indirectly by the Slater and Wilkinson cadre of mechanics, which integrated him into the machinist networks of the Providence area.

Within a year of starting work in the mill, several investors hired Mason to start a spinning mill in the Connecticut Valley town of East Haddam. After finishing the project, he returned to the Quinebaug Valley to serve a three-year apprenticeship in a mill's machine shop. In 1828 the twenty-year-old left for the Utica, New York, area to work in a machine shop, a move that had rich precedence. Machinists from Providence and vicinity had maintained close network bonds with Utica mechanics since 1808–9, when Benjamin Walcott Sr. and Jr., premier machinists who had set up cotton mills with Slater in Rhode Island, established mills in Utica. The son had remained and operated as a hub of the Utica area machinists networks, and his father had returned to Rhode Island, thus cementing network ties between these cotton textile centers.

Within six months Mason returned to the Quinebaug Valley, and over the next several years he developed various types of specialized power looms, painted, and built violins. By 1832 he had returned to power loom work and received a contract for looms from John Hyde, a Mystic mill owner. The extensive network ties among owners of textile mills and machine shops in the valley and in adjacent areas provided the mechanism for Hyde to recommend Mason the next year to Asahel Lanpher, owner of a machine shop in the Quinebaug Valley town of Killingly. Lanpher probably profited from the cotton textile boom of the 1820s. He sold a full line of cotton machinery to mills in New England and the Middle Atlantic. Mason worked on a ring frame, and within a year he took over the machine shop for creditors because Lanpher's business had failed; for the next two years, Mason ran the business.

In 1836 the shop received an order to install ring-spinning machinery in a valley cotton mill. The owner, Edmund Smith, had connections to Samuel Crocker and Charles Richmond, who owned cotton mills, an iron foundry, and machine shops in Taunton. Upon Smith's recommendation Crocker and Richmond hired the twenty-eight-year-old Mason as a foreman to supervise the production of ring-spinning frames. In keeping with the practice of generously compensating young, talented machinists, they gave him an annual salary of one thousand dollars. Mason thus moved laterally across the network of machinists in the Providence area, a smooth relocation in this highly integrated nest of textile machinists.

From 1836 to 1842 Mason contributed a variety of innovations to textile machinery, and, under various ownership structures of the machine business, he retained responsibility for building machinery and supervising sales. During 1838–42 the Taunton machinists' Edwin Keith and James Leach ran two textile machine shops that Crocker and Richmond controlled; the shops employed a total of fifty to seventy machinists. Mason remained under contract to them even while he worked with Keith and Leach. His innovative designs and technical skills contributed to the shops' success in selling a broad line of equipment to cotton mills along the East Coast, and these sales provided him with valuable future contacts.

When Keith's and Leach's operations failed in 1842, the pervasive machinist ties between the nest of builders in and around Providence and the nest in Boston's environs quickly generated an offer to Mason to move to the Lowell Machine Shop. James Mills, the Boston dry goods and commission merchant who had owned shares in Taunton mills since 1827, made a counteroffer to Mason. Mills would underwrite the purchase of the assets of Keith and Leach, which included machinery and shops scattered around Taunton. The firm of Mason and Company fortuitously began production in 1842, just as demand for machinery jumped. Within three years the Mills and Company partners agreed to help underwrite a new state-of-the-art machine shop that would be one of the largest in the nation until the Whitin shop was built two years later.

Using profits from previous textile machinery sales, Mason acquired a 50 percent stake in the bigger Mason and Company. At the end of 1845 its huge complex occupied a six-acre site and consisted of a main machine shop, an iron foundry, and a blacksmith shop. Within three years the company added another machine shop. Numerous specialized machine tools, most of them built by the workers, filled the shops. By 1850 the company employed about three hundred mechanics, making it one of the nation's greatest concentrations of machinists in any one company—yet these workers did not fill the vast complex. Although Mason's business had shifted to a higher plane in the years immediately prior to 1845, textile machinery production could not continue expanding at a fast pace because by then the growth rate of the cotton textile industry had leveled off (see fig. 6.4).

Mason ranked as one of the largest suppliers of a full line of textile machinery. Nevertheless, like other manufacturers who possessed specialties that rested on the control of patents, licenses, or particular skills, the majority of his sales consisted of a few types of equipment which buyers purchased in large volumes. His sales of mules and ring frames made him one of the leading installers of spinning machines. Between 1842 and 1853 he received many of his orders through Charles James, a consulting mill engineer and the leading expert on establishing steam-powered textile mills; these mills continued reordering from Mason after 1853. Only nineteen customers accounted for 60 percent of his sales of 750,000 spindles from 1842 to 1861, the majority of them reorders.

Like most textile machinery firms at this time, Mason sold about two-thirds of his equipment within a subregion that surrounded his shop. His market consisted of New England coastal steam-powered mills as well as small mills in Connecticut and Rhode Island. He had built network ties with these mills and with local machinists since the early 1830s. Mills in a corridor from the upper Hudson Valley through New Jersey to southeastern Pennsylvania purchased the next largest share (about 16 percent), and mills scattered across the rest of the East and in the Midwest and the

South bought the remainder. James, the consulting engineer for many textile mills, and Mills, the partner who had wide-ranging merchant connections, helped Mason sell machinery.

From the early 1840s until 1860 Mason's firm installed about 25 percent of all new spindles in the nation. He took market share from many textile machinery manufacturers, especially from small firms that could not compete on quality or which did not have the capacity to supply a large amount of equipment quickly for a mill. He probably hindered the efforts of the big machine shops of the Boston Associates to increase sales outside their complexes. At the same time, he had limited success penetrating the markets in their complexes and the markets near leading textile machinery manufacturers such as those in Paterson and Philadelphia.

Mason's gains in market share, however, did not translate into continuously rising sales. Between 1846 and 1859 annual sales averaged about half of the peak sales in 1845, the year he opened his new machine complex. He could not prevent other firms either from significantly enlarging their existing textile machinery facilities or from building new large ones. During the five years following the opening of Mason's new shop, the Amoskeag Manufacturing Company and the Saco Water Power Company doubled the size of their machine shops, Whitin built the nation's largest shop, and new large enterprises such as the Lawrence Machine Shop opened.

Consequently, like some of the other big textile machine shops, Mason decided to manufacture locomotives. This diversification provided more business for his shop, whereas he faced difficulties increasing textile machinery sales. Furthermore, the midwestern railroad construction boom, the biggest the nation had ever seen up to that time, had started in the late 1840s (see fig. 5.2). In 1852 Mason enlarged his shop complex, and the first locomotive came out the next year. This addition of a machinery line to his firm's production copied other textile machinery peers, and, if they had a foundry, they undoubtedly made castings because textile equipment orders seldom kept a shop continuously busy. Even in the mid-1850s, while the midwestern railroad boom still ran, Mason's machine shop manufactured light and heavy machine tools, blowers, cupola furnaces, gearing, and shafting as well as locomotives and cotton and worsted machinery. He employed close to six hundred workers, almost as many as the greatest machinery firms in New York City and environs and in Philadelphia.

New Firms Maintain Competitive Pressures

While the big machine shops of the Boston Associates such as Lowell and Amoskeag tried to increase their sales outside the complexes and several independent machine shops such as Whitin and Mason built huge plants, other new competitors sprouted during the 1850s. Although sales of textile machinery did not soar, talented, networked machinists could enter the business and have a chance of succeeding. In 1824 C. M. Marvel had moved from Fall River to Lowell to work in the machine shop of the Locks and Canals Company. After twenty-two years of experience, this forty-year-old became the superintendent in charge of building the Lawrence Machine Shop. Within six years he returned to Lowell to establish a machine shop that became the C. M. Marvel and Company. The talented Marvel and his other partners must have had little difficulty raising capital. They built a substantial shop complex that included a three-story stone machine shop, a boiler shop, a foundry, and several other buildings, qualifying them as competitive builders of cotton and woolen machinery. In keeping with the need to stay diversified, they also manufactured stationary steam engines, printing presses, steam and gas fittings, machine tools, and worsted, flax, and hemp machinery.

New specialized textile machinery firms, especially those focusing on the key equipment such as spinning machines or power looms, posed severe competitive challenges for the large diversified manufacturers. These specialists gained advantages from acquiring patent rights to significant improvements in machinery. If they brought together talented machinists and acquired sufficient capital, they could become suppliers for new mills or for those replacing obsolete machinery.

The Crompton Loom Works exemplified these new specialized competitors. In 1851 the twenty-two-year-old George Crompton moved to Worcester, following several years of training at the Colt Patent Fire Arms Factory in Hartford. His experience working under Elisha Root, Colt's superintendent and one of the nation's top manager/machinists, gave Crompton valuable insight into how a sophisticated machine shop operated. He partnered with Merrill Furbush, who trained in the machine shop of Ira Gay and Company in North Chelmsford, and thus likewise brought experience from another of the nation's leading manager-machinists. The efficient networks in Boston's environs which linked mechanics in Worcester and in the Lowell and North Chelmsford area smoothed the way to form this partnership. They founded Furbush and Crompton to manufacture power looms, drawing on patent rights from George's father, a leading designer of looms. Roughly fifty workers built looms, and the partnership continued until 1859, when Furbush moved to Philadelphia to engage in cotton machinery manufacturing. The long-standing network bridges connecting nests of machinists in Philadelphia, the Providence area, and Boston's environs must have provided Furbush with information about career opportunities. Crompton achieved continued success making looms; by the mid-1860s his Crompton Loom Works employed 375 workers and produced about 125 looms per month.

Hopedale, a village about twenty miles southeast of Worcester, also housed a specialized textile machinery manufacturer. In 1816 and 1829 Ira Draper acquired patents to loom temples, a critical part of the power loom, and, subsequently, his family members licensed them. Sometime in the 1840s, the Hopedale Company—part of the Christian socialist community of Hopedale—took over production of most Draper temples and also began turning out other parts for textile machines. Around the mid-1850s the socialist community went bankrupt, and George and Ebenezer Draper organized a business and entered into three partnerships. Warren Dutcher, a prolific inventor who held a patent on an improved temple loom, joined his W. W. Dutcher Company with the emerging Hopedale complex. Later he acquired twenty patents that dealt with temples and the machines for manufacturing them. Joseph Bancroft, George Draper's brother-in-law, partnered with the Drapers to form the Hopedale Machine Company. Finally, the third partnership brought together all of the principals (the Drapers, Bancroft, and Dutcher) in the Hopedale Furnace Company, a foundry that made castings for the other production units. By the late 1850s the Hopedale complex constituted a formidable industrial enterprise that manufactured sophisticated textile machinery. As the principal shareholders, the Drapers maintained control, and after 1860 the firm of E. D. and G. Draper Company became a leading producer of specialized textile machines, machine parts, and attachments.[10]

THE STATUS OF THE TEXTILE MACHINERY INDUSTRY IN 1860

The counties that housed most of the textile machinery industry in 1860 matched the leading centers in the early 1830s (see tables 6.13 and 6.4–8), confirming that the cotton textile boom of the 1820s set the contours of this industry's development for the next three decades (map 6.1). Boston's environs housed the nation's greatest concentration of textile machinery manufacturing. Although its total value of products exceeded the Providence nest's output, the latter remained close to the scale of Boston's environs based on the number of firms, employment, and capital invested. This testifies to the resilience of the machinist networks in Providence and vicinity, a legacy of Slater-trained mechanics before 1820. Together, the Boston and Providence nests accounted for roughly two-thirds of the nation's industry. Their powerful community of practice contributed innovations to the textile machinery industry and continually adapted to changes in technology and business organization.

Providence and Worcester Counties, the leaders of their respective nests of builders, shared the laurels as the foremost centers of textile machinery manufacturing in 1860; they had also been the largest production centers in the early 1830s. Bristol, Massachusetts, the second-ranked county in the Providence nest, housed the builders in Taunton and Fall River, both important manufacturers since the 1820s. Windham County, Connecticut, was a legacy of the Quinebaug Valley machinists who had roots from the Wilkinson family and Slater before 1820. Boston's environs also included the independent machinery firms that traced to Gay and Silver from North Chelmsford

TABLE 6.13
Textile Machinery by Nest of Builders and Other Sub-Areas Based on Counties, 1860

					Percentage of Nation		
Nest/County/Sub-Area	Number of Establishments	Number of Employees	Capital ($)	Value of Products ($)	Employment	Capital	Value of Products
Boston environs	54	1,687	837,600	2,124,995	35.1	33.6	43.3
Essex, Mass.	8	259	136,600	386,495	5.4	5.5	7.9
Middlesex, Mass.	3	160	81,000	203,000	3.3	3.3	4.1
Norfolk, Mass.	1	6	7,000	30,000	0.1	0.3	0.6
Plymouth, Mass.	4	97	124,600	129,600	2.0	5.0	2.6
Worcester, Mass.	25	899	424,100	1,192,590	18.7	17.0	24.3
Hillsborough, N.H.	12	241	63,700	164,470	5.0	2.6	3.4
Merrimack, N.H	1	25	600	18,840	0.5	0.0	0.4
Providence and vicinity	44	1,387	787,200	1,196,313	28.8	31.6	24.4
Providence, R.I.	27	939	597,200	802,571	19.5	24.0	16.4
Kent, R.I.	2	80	27,000	71,000	1.7	1.1	1.4
Newport, R.I.	5	160	36,500	145,426	3.3	1.5	3.0
Windham, Conn.	3	62	11,000	26,106	1.3	0.4	0.5
Bristol, Mass.	7	146	115,500	151,210	3.0	4.6	3.1
New York City environs	11	266	84,750	250,850	5.5	3.4	5.1
Passaic, N.J.	4	150	58,000	131,250	3.1	2.3	2.7
Columbia, N.Y.	1	75	6,000	90,000	1.6	0.2	1.8
Wayne, Pa.	6	41	20,750	29,600	0.9	0.8	0.6
Philadelphia	18	604	375,250	652,880	12.5	15.1	13.3
Four nests of builders	127	3,944	2,084,800	4,225,038	81.9	83.7	86.2
Other counties	21	234	80,700	170,767	4.9	3.2	3.5
Hampden, Mass.	8	128	29,700	77,067	2.7	1.2	1.6
Tolland, Conn.	7	25	12,000	16,700	0.5	0.5	0.3
Rensselaer, N.Y.	3	42	21,500	46,500	0.9	0.9	0.9
Oneida, N.Y.	3	39	17,500	30,500	0.8	0.7	0.6
Selected county total	148	4,178	2,165,500	4,395,805	86.8	86.9	89.7
Other counties in East	41	619	307,088	493,050	12.9	12.3	10.1
East total	189	4,797	2,472,588	4,888,855	99.7	99.2	99.7
Outside East	3	16	19,500	13,849	0.3	0.8	0.3
Nation	192	4,813	2,492,088	4,902,704	100	100	100

SOURCE: U.S. Bureau of the Census, *Manufactures of the United States in 1860, Eighth Census* (Washington, D.C.: Government Printing Office, 1865).

NOTE: The selected counties were chosen if they contained either 0.5 percent or more of the national value of textile machinery products ($24,514 or more) or 0.5 percent or more of the national employment in textile machinery (24 or more).

and the legacy machine shops of the large mill complexes of the Boston Associates in Essex (Lawrence) and Middlesex (Lowell) counties. Hillsborough County, New Hampshire, a mature center of cotton textile manufacturing and of machine shops (Manchester and Nashua), accounted for roughly 4 percent of the textile machinery industry.

Textile machinery manufacturing in New York City's environs had fallen behind the New England nests before 1820, and this relative status continued; it accounted for close to 5 percent of the industry in 1860 (see table 6.13). Much of the Hudson Valley portion had declined to insignificance, and Passaic County and its city of Paterson

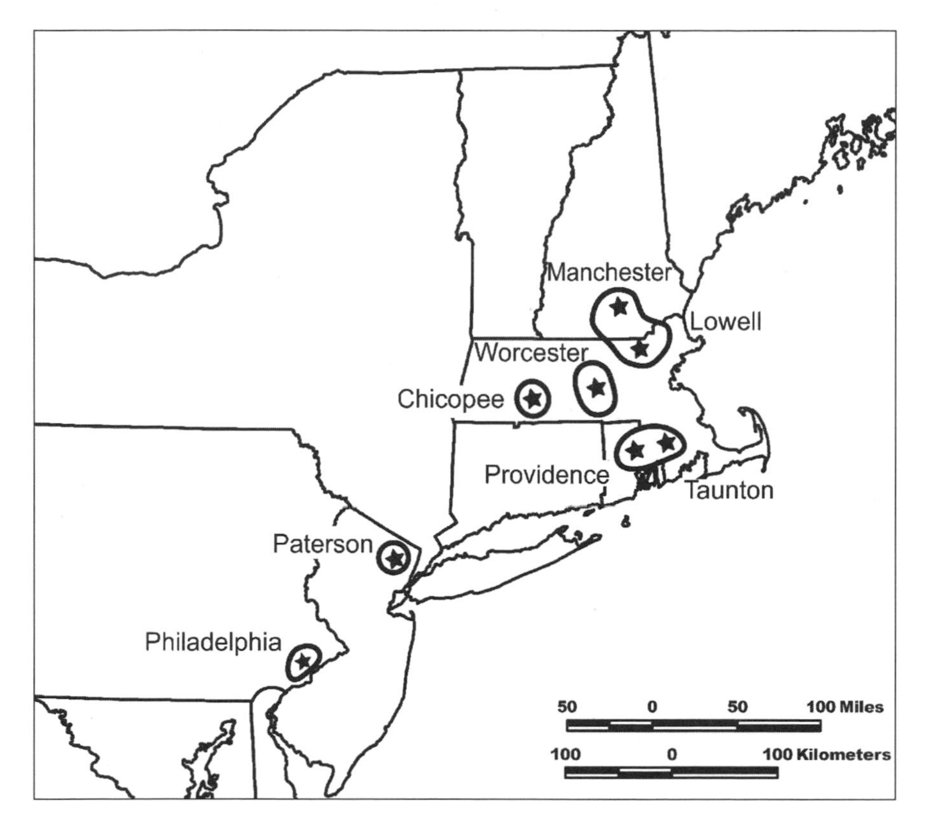

Map 6.1. Leading Clusters of Textile Machinery Firms in 1860

housed most of the production capacity for textile machinery. The city's importance had faded, however, as its firms moved into locomotive and other machinery manufacturing. The Philadelphia area nest of builders had mostly been reduced to those in Philadelphia County, and it ranked as the third largest center, accounting for about 13 percent of the nation's textile machinery production.

With more than 80 percent of the nation's textile machinery capacity, the four nests of builders exerted overwhelming control over the direction of the industry, as they had since the first few decades of the nineteenth century (see table 6.13). Nevertheless, a few other counties—includings the Connecticut Valley (Hampden and Tolland counties), the upper Hudson Valley (Rensselaer), and the Utica area (Oneida) in central New York—contributed to production, likewise operating in the early centers of cotton mills and machine shops that had emerged before 1820. Outside these counties and the four nests of builders, the remainder of the East's counties, collectively, housed about 12 percent of the textile machinery industry, but each of them contained only tiny amounts of it. No production of any consequence occurred outside the East because firms needed to be integrated into the major textile machinist networks in order to compete successfully for sales.[11]

The textile machinery industry, along with locomotive manufacturing and the iron foundry and steam engine sector, trained many of the machinists who developed metalworking machinery. Firms in these industries often built their own tools and sold them to other firms, but no distinctive set of firms specializing in metalworking equipment existed by 1860. Yet the forty-year period following 1820 was a seminal time for the development of a metalworking industry. Many of the leading firms that emerged subsequently had their origins in the machinist networks of iron foundries, steam engine firms, locomotive firms, and textile machinery firms. The arms industry would also make major contributions to metalworking machinery, and this development was rooted in the active participation of firearms machinists in networks before 1820.

CHAPTER SEVEN

The Cradles of the Metalworking Machinery Industry

> [At the Springfield Armory] we have the very singular and extraordinary operation going on, of manufacturing with the greatest care, and with the highest possible degree of scientific and mechanical skill, a vast system of machinery.
>
> *Jacob Abbott, "The Armory at Springfield"*

Antebellum machinists carried out much of their work with hammers, chisels, files and other hand tools, yet they made steady progress in substituting metalworking machinery for hand labor. These machines—broadly defined to include equipment for hammering, cutting (also drilling and boring), and grinding metal—reduced the drudgery of metalworking. As this equipment became more sophisticated and as power sources strengthened, machinists worked metal more precisely and faster than using hand labor. Machine tools, typically defined as equipment that cut metal, constitute a subcategory of metalworking machines, and many observers of the metal manufacturing industry focus attention on them. Nevertheless, before 1860 these distinctions remained unimportant because machinists worked with all types of metalworking equipment. They used the general term *machinists' tools* to describe a range of equipment, excluding trip hammers for pounding and shaping metal, which in some machine shops was central to the operation. The terminology of metalworking equipment and machine tools are used interchangeably here, and distinctions are noted when appropriate.

The emergence of independent machine-producing firms, especially those making machine tools, signifies the development of an advanced metalworking equipment industry. Yet few, if any, of these firms existed before 1880, which raises a puzzle. In the United States Census report on power and machinery employed in manufacturing in 1880, extraordinarily sophisticated equipment appeared in the lithographs. The two decades following 1860, however, witnessed extensive economic disruptions, including the Civil War and a severe recession from 1873 to 1879. The environment was not conducive to transforming metalworking machinery from a primitive sector to one of unusual sophistication. Therefore, one can assume that these machines were already highly refined by 1860 without being produced in specialized firms, which implies that antebellum machinists had achieved significant advances in metalworking machinery.[1]

Following 1820, mechanics in a wide range of industries contributed to developing metalworking equipment. They gradually added more machines to their shops and improved existing ones. Typically, they made their own equipment, and some firms sold metalworking machines as part of a diversified machinery business. Occasionally, improved or new equipment represented a quantum leap in metalworking capabilities, but incremental changes predominated. The use of this machinery in various industries multiplied the number of machinists and firms participating in transmitting technical skills through networks, which in turn contributed to the wider adoption of these machines and to innovations in them.

THE TALENT POOL

The skills required to build metalworking machinery matched those needed to make other types of equipment of equivalent sophistication. From the start steam engines ranked among the most complex machinery of the antebellum years; likewise, manufacturing cotton-spinning machines required substantial skills. By 1820 some machinists, especially those in the large metropolises, their satellites, and their prosperous inner agricultural hinterlands, had attained sufficient skills to build steam engines and textile machinery, and some mechanics in rich agricultural areas farther away also acquired these capabilities. During the economic upturn of the 1820s young boys from farms and villages left to work as apprentices in the expanding and new machine shops, enlarging the talent pool of machinists.

By the early 1830s many iron foundries, steam engine works, and machine shops could build textile machinery. Other shops thrived by building diverse machinery: most of them served statewide markets, some reached the New England market, and a few sold to larger territories (table 7.1). The large concentrations of diversified shops corresponded closely to the nests of textile machinery builders. The city of Boston, with 240 employed in thirty shops, housed one of the biggest talent pools in the East. Southwest of the city, Norfolk County had five sizable firms, including Shepherd

Leach's iron foundry in Foxborough, which mostly built machinery, and Hart's machine shop in Walpole. Each of them had 20 employees, ranking them among the larger machinery firms at that time. In Worcester County, home to one of the largest concentrations of machine shops producing textile machinery, eleven, mostly small shops, made diverse machinery (see tables 6.6 and 7.1).

In and around Providence a large number of shops manufactured various machines. Even assuming half or more of the 1,242 workers in Rhode Island's iron foundries and machine shops made textile machinery some of the time, probably as many as 300 or 400 built other machinery. Machine shops in Fall River, Massachusetts, and Norwich, Connecticut, also made diverse machinery. The 40 workers at Paul and Beggs in Paterson and the 110 at Jenks's works in Philadelphia, a firm that also produced textile machinery, provide a glimmer of the machine shops in these areas.

The Connecticut Valley did not rate as a nest of textile machinery builders because that sector did not remain important there; nonetheless, the valley became one of the leading machinery centers in the East. The American Hydraulic Company in Windsor, Vermont, on the border of the northern part of the valley, and a machine shop in Hartford, Connecticut, in the southern portion of the valley, ranked among the larger machinery manufacturers in the East. The scattering of machine shops in other counties outside the nests and the Connecticut Valley reflected the widespread distribution of diversified machinery firms by the early 1830s.

BUILDING METALWORKING MACHINERY

Most medium to large iron foundries, machinery firms, and machine shops made their own metalworking equipment as they needed it because a sizable talent pool of machinists possessed the appropriate skills. To remain competitive they had to stay abreast of innovations in this equipment. Machinist networks, principally, and other venues such as patent disputes and technical publications, secondarily, constituted the main avenues for acquiring this knowledge. Sales of metalworking machinery gradually became more important as channels for the diffusion of innovations. But, because firms sold most of this equipment to local or to subregional customers, knowledge gained through purchasing machines chiefly supplemented sources such as machinist networks. Some machinery sales reached regional and even multiregional markets, but these sales channels of knowledge acquisition, especially the multiregional, remained minor for much of the antebellum. Instead, machinist networks constituted the main mechanism for transferring knowledge across and within regions. Iron foundries, locomotive works, and textile machinery firms were the largest users of metalworking machinery, and by 1860 they were making substantial progress in employing more sophisticated equipment.

TABLE 7.1

Machinists' Firms Manufacturing Miscellaneous Machinery by Nests of Builders and Other Locations, 1832

City/County/State	Firm (date founded)	Number of Employees	Value of Products ($)	Products	Market Area
Boston and environs					
Boston	Thirty machinists' firms	240	195,000	printing presses, various machinery	Mass.
Norfolk County, Mass.					
Bellingham	Darling's Machine Shop	7	7,958	machinery	R.I.
Foxborough	Shepherd Leach Iron Foundry	20	24,600	machinery	Mass., R.I.
Sharon	G. H. Mann's Machine Shop (1831)	10	5,000	machinery	Mass.
Walpole	Hart's Machine Shop	20	20,000	machinery	Mass.
Wrentham	Slack's Machine Shop (recent)	8	8,000	machinery	New England
Plymouth County, Mass.					
East Bridgewater	Scott, Keith, and Co. (1825)	6	6,000	fire frames, pumps, machinery, castings	Mass.
Worcester County, Mass.					
Worcester	Washburn and Goddard	14	12,910	napping machines, wire	New England
Athol	Stillman Knowlton	2	2,320	turning lathes, misc. tools	Worcester and Franklin counties
Athol	James Young	3	2,400	shingle machines	United States
Brookfield	Unknown	17	14,100	shingle machines, machinery patterns, turning engines, lathes, other machinery	Mass.
Harvard	Benjamin J. Dwinnell	3	1,500	paper machinery	Worcester County
Lancaster	Josiah Fay	4	2,500	machinery	New England
Leicester	Two machine shops	6	4,500	machinery	Leicester, Mass.
Leominster	Three machine shops	6	4,500	machinery	New England
Hillsborough County, N.H.					
Hollis	Unknown	4	2,000	machines for coopering and working	N.H. and adjoining states

TABLE 7.1
continued

City/County/State	Firm (date founded)	Number of Employees	Value of Products ($)	Products	Market Area
Providence and vicinity					
Rhode Island	10 iron foundries, 30 machine shops	1,242	—	—	—
Fall River, Mass.	Brayton Slade and Co. (1826)	20	—	machinery	—
Norwich, Conn.	Unknown (1824)	—	10,000	machinery	—
New York City environs					
Paterson, N.J.	Paul and Beggs (1826)	40	—	machinery	Conn. to Ga., Mexico
Philadelphia	Jenks	110	—	machinery	—
Connecticut Valley					
Windsor, Vt.	American Hydraulic Co. (1827)	20	68,000	force pumps, hydraulic machinery, fire engines	United States, foreign
Windsor County, Vt.	R. Daniels and Co. (new)	25	—	machinery	local
Winchester, N.H.	Winchester Furnace Co.	—	9,000	machinery, holloware	—
Greenfield, Mass.	William Wilson (1824)	3	4,200	machinery, plows, castings	Mass.
Orange, Mass.	Unknown	—	1,100	machinery, plows	Mass.
Hartford, Conn.	Unknown	50	20,000	machinery	—
Other					
Gilmore, Maine	Unknown	2	1,000	machinery	Maine
Addison County, Vt.	A. Moses	8	6,000	machinery	local
Washington County, Vt.	Waterman and Co.	1	300	machinery	—
Great Barrington, Mass.	Unknown	28	16,000	machinery	Mass.
Columbia County, Pa.	George Mack (1831)	8	6,000	machinery, mill gearing, pattern work	local

SOURCE: Louis McLane, *Documents Relative to the Manufactures in the United States Collected and Transmitted to the House of Representatives, 1832, by the Secretary of the Treasury*, House Doc. 308, 22nd Cong., 1st sess. (Washington, D.C.: Duff Green, 1833).

NOTE: Iron foundries are included if their work was primarily machinery other than steam engines. Firms producing cotton or woolen machinery are excluded.

Iron Foundries

As specialists in shaping metal, iron foundries ranked among the earliest adopters of metalworking machinery, especially heavy equipment to form the large metal components such as those in steam engines. Machinists fitted a power supply to the industrial lathe, a metal-cutting machine to work large metal pieces. It shaped workpieces equivalent to the lathe's quality, its flexibility made it suitable for various cutting processes, its cost efficiency allowed firms to substitute it for some hand labor, and its low selling prices made it affordable to firms.

David Wilkinson's 1798 patent for a screw-cutting lathe constituted the most important domestic contribution before 1820, but major advances in designing industrial lathes originated in England, the first appearing there by 1820. Most, if not all, of the early industrial lathes in American foundries were imported from England or copied from English versions. The same acquisition process applied to other basic metal-cutting equipment such as boring machines. In 1820 the Columbian Iron Foundry of Robert McQueen in New York City boasted having five steam engines to power lathes, boring machines, and other equipment. At the same time, in nearby Morristown, New Jersey, Stephen Vail's Speedwell Iron Works owned three turning lathes, two trip hammers, one boring mill, and one engine for cutting large screws. These firms had little difficulty acquiring the lathes and boring machines because the business elite in New York City maintained ties to their English peers.

Over the next several decades the large foundries added to their repertoire of metalworking equipment, with the industrial planer constituting the most important new piece. As late as the mid-1840s, a leading steam engine firm such as Providence's Thurston, Gardner and Company sent a partner to Europe to purchase twelve thousand dollars' worth of machinery. Undoubtedly, most of it consisted of advanced machine tools that American sources did not make or which they produced of poorer quality. By the late 1840s and early 1850s major foundries added significantly to their supply of heavy machine tools. They built their own as well as purchased them from other firms in the East; a few of the foundries also sold machine tools.

The Providence foundry and steam engine works of Fairbanks, Bancroft and Company and its successor, Corliss, Nightingale and Company, ranked among the leaders in incorporating heavy machine tools into manufacturing processes and in building them for sale. Other prominent foundries—including West Point at Cold Spring, the Novelty Works in New York City, Watts and Belcher in Newark, the I. P. Morris and Company and the Southwark Foundry of Merrick and Sons in Philadelphia, and Betts, Pusey and Company in Wilmington—owned huge lathes, planers, slotting machines, and boring mills as well as smaller versions of them. Like the Corliss firm, they made their own equipment, and some of them sold heavy machine tools as part of their diversified machinery production.

In the early 1850s the committee that the British government sent to survey machinery use in the United States did not observe anything unusual about the metalworking equipment in the major foundries, although its members did not visit Providence's Corliss, Nightingale and Company. They criticized inferior machinery. Their failure to disparage American iron foundries, therefore, indirectly indicates that these firms were, if not fully up to British standards, not far behind the world's best-practice foundry work. Most domestic foundries employed similar types of metalworking machinery which they made themselves, except for the sophisticated machines purchased in Britain. This commonality in equipment testified both to the efficient exchanges of knowledge through the machinist networks that linked foundries locally and among cities and to the existence of an eastern community of practice of foundry workers and firms. Foundries contributed to the development of machine tools through the improvements they made and the dissemination of these innovations to their peers.[2]

Locomotive Works

After 1835 foundries, along with steam engine works, textile machinery firms, and machine shops, started to produce locomotives to meet growing demand for them. By the 1840s engines contained up to four thousand parts, and typical forge techniques of hammering or casting in a foundry sufficed for making many of them and chipping or filing to provide the finishing touches. Nevertheless, other components required additional metalworking, which raised new machining problems to resolve. As with general foundry work, locomotive production needed heavy metalworking equipment, and this spurred the search for better machine tools. Railroad repair shops used much of the same equipment, making these shops reservoirs of technical skills and diffusers of knowledge about this machinery.

Like other machinery producers, locomotive works made most of their own machine tools, which, at least up to the mid-1840s, consisted predominantly of general-purpose metalworking equipment such as lathes, planers, and slotters, which were common to foundries, machinery firms, and machine shops. Some engine works, such as Grant Locomotive and the Danforth, Cooke and Company in Paterson, took advantage of their extensive machine-making capacity and sold machine tools. Nonetheless, even these locomotive shops ordered selected, sophisticated machine tools from other shops, including those in Lowell and Worcester. Larger locomotive works that possessed ample capital imported machine tools from prominent British manufacturers such as Whitworth. When the Rogers Works in Paterson embarked on engine production in the mid-1830s, it imported several tools from England, including a planer reputed to be one of the largest in the East's locomotive shops. Even with the steady addition of metalworking machinery, much of the engine work of machinists consisted of hand labor with hammers, chisels, and files.

As railroads extended their tracks, they served more diverse territorial economies and terrain; thus, by the mid-1840s they demanded a wider variety of locomotives. During the next decade improvements in engine design to boost performance and to meet new, diverse technical demands resulted in a 50 percent increase in the number of engine parts to about six thousand. Following 1845, railroad investment surged and locomotive builders dramatically ratcheted up their production. All of these changes pressured them to enhance their metalworking capabilities. By the early 1850s the number, quality, and sophistication of machine tools increased significantly from a decade earlier. This transformation became apparent in the leading locomotive works, such as the Baldwin and the Norris firms in Philadelphia and, probably, Rogers in Paterson; it also extended to Taunton Locomotive and Mason's works in Massachusetts. Because these firms continued to make most of their machine tools, they must have contributed to innovations in them. Even when firms copied them, skilled machinists might devise incremental changes in design which improved a tool's performance. Yet the locomotive builders also increased their purchases of machine tools, indicating that some machine shops were placing a greater emphasis on selling tools.

The inventory of metalworking machinery at the Norris Locomotive Works in 1855 indicates how far a leading manufacturer had shifted its production techniques (table 7.2). At that time the firm employed six hundred to seven hundred workers and turned out fifty to one hundred locomotives annually. The writer of a magazine article on the Norris Works noted, with amazement, the large number and variety of metalworking equipment in the sprawling factory complex. The number of lathes, the standard tool in every machine shop for the previous three decades, exceeded every other type of tool, but the many specialized lathes stood out. This differentiation extended to a wide range of other machine tools, including various types of planers, cutting machines, bending and other shaping machines, drills, and boring machines. Steam power ran most of the metalworking equipment. The Norris Works contained two fifteen-horsepower engines, two twenty-five-horsepower engines, a sixty-five-horsepower engine, a seventy-horsepower engine, and a one hundred–horsepower engine. The amount of steam power made it one of the largest mechanized firms in the East.

The specialized, powered machine tools in locomotive works by the early 1850s epitomized the transformation that had been under way in machine tools since at least the mid-1840s. Norris's mechanics made most of this metalworking equipment, underscoring the efficiency of machinist networks in transmitting technical skills and innovations among the shops. Philadelphia machinists possessed superb network ties to the other leading machinist centers of New York City and its environs, Providence

TABLE 7.2
Metalworking Machinery of the Norris Locomotive Works, Philadelphia, 1855

- Iron Foundry (60 ft. by 72 ft.; L side 50 ft, by 25 ft.)
  - 2 cupola furnaces
  - numerous cranes
- Brass Foundry
- Boiler Shop (80 ft. by 100 ft.)
  - shears that chip iron plates ½ inch thick
  - heavy punches and drills to make holes for rivets
  - large rolls for bending plates
  - machines to bend flanges
- Tank Shop (30 ft. by 100 ft.)
  - similar equipment as in boiler shop
- Machine Shop (2 buildings, 3 stories each: 166 ft. by 153 ft. and 105 ft. by 45 ft.)
  - 2 large planing machines
  - 2 cylinder boring machines
  - 3 boring mills
  - 4 large wheel lathes
  - 13 large slide lathes
  - 17 slide lathes
  - 35 hand lathes
  - 4 large chuck lathes
  - 1 axle cutting machine
  - 1 large slotting machine
  - 3 slotting engines
  - 6 heavy drill presses
  - 10 drill presses
  - 1 large hoisting machine
  - 12 screw cutting engines
  - 1 perforator
  - 1 gear cutting engine
- Tender Shop (3 floors: 68 ft. by 100 ft.)
  - power tools of all kinds
- Blacksmith Shop (116 ft. by 153 ft.)
  - 48 smith forges
  - 3 large trip hammers
  - 2 furnaces
  - 1 bending machine
- Steam Hammer Shop (100 ft. by 80 ft.)
  - 2 Nasmyth Patent Steam Hammers (weight: 2,000 pounds, 4,500 pounds)
  - 2 heating furnaces
  - 9 smith forges
- Finishing Shop and Extension Building
  - 1 radial drill
  - 5 large slide lathes
  - 1 large wheel lathe
  - 1 large chucking lathe
  - 4 lathes
  - 1 compound planer
  - 1 compound planer with revolving tool box
  - 2 large planing machines
  - 32 planing machines
  - 1 wheel quartering machine
  - 1 wheel drawing machine
  - 4 drill presses
  - 1 shaping engine
  - 1 screw-cutting machine
  - 6 slotting engines
  - 1 trip hammer

SOURCE: "The Transportation of Passengers and Wares, a Visit to the Norris Locomotive Works," *United States Magazine* 2 (October 1855): 162–65.

and vicinity, and Boston and its environs. They also indirectly acquired technical skills and innovations from the Connecticut Valley shops through these other nests. Norris continued to import machine tools from England, but they generally consisted of highly specialized pieces such as two Nasmyth Patent Steam Hammers and a radial drill, a large slide lathe, and a compound planer with a revolving toolbox which Whitworth, the world's leading machine tool manufacturer, supplied. Increasingly, however, Norris, along with Baldwin, purchased machine tools from local firms, which were making this equipment more significant lines of their business.[3]

Textile Machine Shops

Early machinists who made textile machinery possessed the skills to build simple machine tools, such as the ubiquitous lathes, for their shops. If textile machine shops

TABLE 7.3
Metalworking Machinery of the Boston Manufacturing Company, 1817, and of the Saco Manufacturing Company, 1829

Boston Manufacturing Company, 1817
8 turning lathes
2 cutting engines
1 flutting engine
4 roller engines
1 polishing machine
Saco Manufacturing Company
12 large hand lathes
1 main-shaft hand lathe
12 polishing lathes
1 fluting engine
8 double roller engines
3 single roller engines
2 polishing engines
1 bevel gear engine
1 cylinder gear engine
5 spindle engines
1 filing engine
1 sloping engine
1 spiral gear engine
1 drilling engine

SOURCE: George S. Gibb, *The Saco-Lowell Shops: Textile Machinery Building in New England, 1813–1949* (Cambridge: Harvard University Press, 1950), 49, 108, tables 2, 5.

made small equipment for other markets, in addition to their machinery for textile mills, they did not require the heavy industrial lathes and, later, the large planers that foundries or locomotive works used. Textile machine shops might purchase a few machine tools. If they needed more of them, they copied ones they owned or built them following the patterns of tools in nearby shops and in firms where the machinists had previously worked. In 1814, when the Boston Manufacturing Company wanted equipment for its new machine shop in the first cotton mill at Waltham, it purchased a roller lathe, a fluting lathe, and a cutting engine from Shepard's foundry in Easton. Subsequently, the Waltham shop built most of its other metalworking equipment, and it probably copied some of the machines purchased from Shepard. By 1817 the shop owned at least sixteen machines, perhaps half of them powered by water (table 7.3).

The Waltham shop moved to Lowell in 1824, operating under the Locks and Canals Company to construct textile machinery for the mills. Like typical textile machine shops, it built most of its own machine tools. During the cotton mill boom of the 1820s the machine tool repertoire of major shops expanded as they increased their production of textile machinery. The equipment inventory of the Saco Manufacturing Company's shop in 1829 reveals the dramatic rise in the number of specialized machine tools, most of them power driven, compared to the equipment of the 1817 machine shop of the Boston Manufacturing Company (see table 7.3). The Saco shop, however, did not have water- or steam-powered lathes and therefore may not have been at the forefront of adopting machine tools.

In contrast, only five years later the Otis Pettee Machine Shop at Newton Upper Falls (near Boston) housed twelve powered lathes to carry out at least five different tasks, and Pettee's shop contained many other powered tools (table 7.4). Most, if not all, of them probably had been built no later than 1830–31 to complete an order for $150,000 worth of textile machinery from the Jackson Company in Nashua, New

TABLE 7.4
Metalworking Machinery of the Otis Pettee Machine Shop, 1834, and of the Saco Water Power Company, 1850

- Otis Pettee Machine Shop, 1834
  - 3 small trip hammers
  - 1 large engine lathe
  - 4 large double engine lathes
  - 1 large single engine lathe
  - 4 small engine lathes
  - 2 engine lathes for card cylinders
  - 40 hand lathes
  - 3 boring lathes
  - 1 machine to bore speeder cones
  - 1 punching machine
  - 1 machine to drill spindle rails
  - 2 machines to drill double speeder rails
  - 1 engine for cone gears
  - 4 spindle engines
  - 2 spiral engines
  - 1 bevel engine
  - 4 habbing engines
  - 1 flutting engine
  - 1 dividing engine
  - 1 machine to polish rollers
  - 1 tennon machine
- Saco Water Power Company, 1850
  - 9 large engine lathes
  - 9 large geared engine lathes
  - 54 engine lathes
  - 8 small lathes
  - 40 hand lathes
  - 4 drilling lathes
  - 1 chucking lathe
  - 4 large planing machines
  - 3 planing machines
  - 3 small planing machines
  - 3 slabbing machines
  - 1 vertical drill
  - 1 rail drilling machine
  - 1 machine for drilling hobs for card sheets
  - 1 machine for drilling whirls
  - 2 fluting machines
  - 1 splining machine
  - 3 spindle grinders
  - 1 machine for cutting flyers
  - 1 machine for turning pulleys
  - 3 machines for turning card cylinders
  - 1 machine for turning spiders for card cylinders
  - 1 machine for turning doffer cylinders
  - 1 squaring machine for large work
  - 1 machine for straightening shafting
  - 2 gear cutting machines
  - 1 set screw machine
  - 1 large bolt machine
  - 5 small bolt machines
  - 1 machine for squaring bolt heads

SOURCE: George S. Gibb, *The Saco-Lowell Shops: Textile Machinery Building in New England, 1813–1949* (Cambridge: Harvard University Press, 1950), app. 3, 635–37, tables A, B.

Hampshire. Pettee's equipment indicates that leading textile machine shops—also including the Locks and Canals shop at Lowell and top shops in Providence and vicinity—possessed numerous powered machine tools by the early 1830s, some of them consisting of heavy industrial tools such as Pettee's four lathes, each fourteen to sixteen feet long.

Textile machine shops increased their sales, albeit modestly, of machine tools, but up to the early 1830s they seldom listed them as separate lines of business (see tables 6.4–8). Between 1827 and 1832 the Nashua Manufacturing Company had sold lathes, along with its textile machinery. John Gage may have focused more on machine tools than on other machinery, however, in his new 1837 shop in Nashua. By the late 1830s Lowell's Locks and Canals shop sold small planing machines, but, like foundries and locomotive works, these sales remained minor. At that time the Matteawan Company expanded beyond its textile equipment line to machine tools and other machinery. During the decade of the 1830s textile machine shops added more sophisticated machine tools; like foundries and locomotive works, they sometimes purchased them from British suppliers. In 1839 the Locks and Canals shop imported a planer and a self-acting upright drilling and boring machine from England's Whitworth.

By the mid-1840s the sophistication of the machine tool repertoire of major textile machine shops had risen significantly, compared to the previous decade. Pettee's shop had added a trip hammer, three cutting engines, two screw machines, a filing machine, a punching machine, five planing machines, and a large number of lathes (see table 7.4). The planing machines indicate that Pettee probably engaged in heavy industrial machine work. These tools cut and scraped large metal pieces, taking the place of arduous hand labor with hammer and chisel. When William Burke, superintendent of the Amoskeag Manufacturing Company's shop, became head of the Lowell Machine Shop in 1845, William Bement accompanied Burke from Amoskeag. Bement spent the next six years designing and building an entire line of basic machine tools for sale, including a small chucking and reaming lathe, an upright drill, two models of engine lathes, two models of gear-cutting lathes, and a slotting and paring machine. These innovations positioned the shop as one of the leading suppliers of machine tools, comparable to the best of those being produced by Providence area shops.

Concurrent with the 1845 incorporation of the Lowell shop, William Mason equipped his state-of-the-art textile machine shop in Taunton with the latest tools. Many of them, such as the multiple-spindle drill press, performed single tasks, but lathes, planers, and drills represented the most common tools. Other equipment consisted of trip hammers, boring machines, punches, bolt and screw cutters, gear cutters, roller flutters, and bending and straightening machines. Mason's shop probably started selling light and heavy industrial machine tools shortly after beginning operations.

By the early 1850s textile machinery firms housed a wide array of special-purpose machine tools, many of them power driven, consisting of light and heavy industrial versions. The Saco shop's equipment, for example, differed dramatically from Pettee's 1834 shop, considered a premier facility at that time (see table 7.4). Saco's largest machine tools, such as a twenty-six-foot planer and a twenty-foot lathe, gave the shop the capacity to build heavy industrial machinery. Its advertisements prominently listed machine tools as a part of its diversified product line. The machine tool repertoire bore a striking resemblance to the equipment in the Norris Locomotive Works in 1855, suggesting that the East's greatest machine shops in locomotive works and textile machinery firms, as well as in foundries, possessed sophisticated metalworking capabilities (see table 7.2).

New large textile machinery firms, such as C. M. Marvel and Company (founded in Lowell in 1852), must have had a similar array of special-purpose machine tools because it also manufactured stationary steam engines, printing presses, and steam and gas fittings. In keeping with its technical capabilities, Marvel also sold machine tools. The ease with which these firms from Saco, Maine, to Philadelphia transformed their machine shops to house sophisticated, special-purpose machine tools testifies to the capacity of technical knowledge to move through machinist networks. Nonetheless, this equipment did not guarantee that firms would remain competitive in all machinery manufacturing. The Saco Water Power Company failed to stay abreast of technical changes and did not acquire licenses to produce the most advanced equipment, and its line of textile machinery became obsolete.[4]

THE FEDERAL ARMORIES ADVANCE THE "SYSTEM OF UNIFORMITY"

Iron foundries, locomotive works, and textile machine shops made decisions about adopting metalworking equipment based on its cost-effectiveness in producing quality, competitively priced products for the market. Federal armories did not have such market incentives, however, because their customer owned them. The government wanted to minimize the cost of firearms, but, in the absence of a competitive market, price signals to guide production decisions remained obscure. By 1820 most private manufacturers of firearms consisted of either firms under contract with the federal government's Ordnance Department or craft gunsmiths who produced high-quality, expensive weapons.

At this time the Ordnance Department did not resemble a typical slow-moving government bureaucracy. Following the congressional reorganization of the department in 1815, it had clearer, expanded authority to manage arms contracts, run the armories, and develop regulations to achieve uniformity in armaments. Its head, Col. Decius Wadsworth, had moved quickly and appointed Col. Roswell Lee as superintendent of the Springfield Armory and Lt. Col. George Bomford as deputy head of the Ordnance Department. These talented military engineers and decisive administrators aggressively advanced the system of uniformity in armaments. Lee continued as Springfield's superintendent until his death in 1833, and Bomford, as head of the department from 1821 to 1842, relentlessly pursued the goal of achieving uniformity; his successors followed suit.

Under Lee's direction by 1820 the Springfield Armory had made substantial progress in becoming an effective enterprise. In contrast, Harpers Ferry Armory remained plagued by weak management and labor problems, and its workers turned out poor-quality muskets. After 1820 the incorporation of increased numbers and new types of metalworking machines into the production process would make substantial contributions to achieving uniformity; these gains, however, did not come easily or quickly.[5]

The Springfield Armory
Stability in Metalworking

During the first two decades after 1820 machine tools played a minor role in the Springfield Armory's effort to achieve uniformity. Equipment accounted for only about 2 or 3 percent of its cumulative expenditures over that period. Lee devoted much of his effort to improving the organization and management of the plant. By 1821 the armory had adopted nearly all of the power-driven machine tools for manufacturing lock parts, the most complex limbs of firearms. Until the mid-1830s additions of metalworking machinery amounted to copying more of the same types of equipment, and armory machinists made most of them. The armory's gauging system remained stable, from the initial adoption of this system under Lee's stewardship around 1820 until the mid-1830s, when the armory incorporated more sophisticated gauging that drew on advances at other armories.

Between 1820 and 1840 total employment and the output of muskets at the Springfield Armory changed little overall (see fig. 3.1). During a few years bracketing 1820, some improvements that Lee implemented probably contributed to declining unit costs of muskets and to rising productivity of barrel welders and of workers overall, but these gains dissipated after 1830. John Robb, superintendent from 1833 to 1841, was a weaker administrator than Lee, and inadequate power supplies may have hindered efficient equipment operations. The cost per musket rose, overall productivity (muskets per worker) fell in a volatile manner, and barrel welding productivity stagnated (see figs. 3.2, 3.3, and 3.5). The cost reductions and productivity gains of the 1820s benefited armory workers in the form of soaring real wages. During the next decade their wages rose at a slower rate, and their pay fluctuated wildly during the economic contraction of 1837–43 (see fig. 3.4). For much of the 1820–40 period Springfield built muskets more cheaply and turned out greater numbers per worker than Harpers Ferry. This effort maintained the lead that Springfield had attained in the years immediately following Lee's appointment as superintendent.

Thomas Blanchard's gun-stocking machinery involved woodworking, not metalworking, and it constituted the chief innovation in machining at Springfield until the mid-1830s. He had worked on a barrel-turning lathe at Springfield in 1818, and Lee had sent Blanchard to Harpers Ferry to do similar work. Upon returning to his base in Millbury, the metalworking and machinery hub south of Worcester, Blanchard embarked on building a gun-stocking lathe. With help from Asa Waters, the local firearms contractor, and with assistance from other mechanics, Blanchard completed a working lathe by 1819. Information about his work percolated easily through machinist networks because Millbury mechanics possessed numerous ties to the machinist hubs of Worcester and Providence. Waters and Blanchard also had links to the Springfield Armory clearinghouse of firearms knowledge. Lee commissioned Blanchard to travel to Harpers Ferry to build a gun-stocking lathe, which he finished in 1819. The next year, following a patent dispute, he received a patent for a lathe to turn irregular shapes.

By 1821 Blanchard seems to have set up a gun-stocking machine at Lemuel Pomeroy's private armory in Pittsfield, Massachusetts, but Blanchard did not reach an agreement with Lee to establish mechanized gun stocking at Springfield until two years later. From 1823 to 1827 Blanchard operated as an "inside contractor" at Springfield, setting up his new lathe, as well as thirteen other machines, to perform specialized operations on the gunstock. He left the armory two years later and from that time on collected royalties for the use of his machinery. This equipment did not fully mechanize gun stocking—that was not achieved before 1860. But it raised gun stockers' productivity by about one-third by the late 1820s, thus contributing to the overall productivity gain of that decade (see fig. 3.3). Gun-stocking productivity did not improve much for another twenty years, as workers continued to engage in extensive handwork to finish gunstocks. Starting in the early 1840s, Cyrus Buckland—who became chief machinist at the Springfield Armory in 1842—began building second-generation gun-stocking machines, an effort that continued into the next decade. By the mid-1850s these machines raised productivity by about half, compared to a decade earlier, and by 1860 handwork on gunstocks consisted solely of smoothing, sanding, and oiling.

Upgrading Metalworking Capabilities

From the mid-1830s to the early 1840s the Springfield Armory upgraded its metalworking machinery, especially under the stewardship of Thomas Warner, master armorer from 1837 to 1841, and, secondarily, under Buckland. These machine tool innovators were among the few machinists employed at the armory who developed equipment while working there. William Smith, the master machinist at Simeon North's firearms factory in Middletown, moved to Springfield in 1837. He brought along many ideas for machine tools that had been implemented at North's factory, arguably the top private armory at that time. The Ordnance Department encouraged this transfer of technical skills through job mobility, and this know-how trading strengthened the community of practice of armory machinists. These top Springfield mechanics also shared technical knowledge when purchasing machine tools from other firms and building new equipment that drew on designs implemented elsewhere. For the most part, new machine tools installed at this time did not originate with the Springfield Armory.

New, specialized milling machines constituted the most important equipment the armory installed, including a machine to make lock plates with uniform thickness, a machine to mill the irregular shape of lock plates, a vertical spindle milling machine, and a set of five machines to finish bayonets. The armory also added powered equipment, including several drilling machines, at least five engine lathes, and several cutting engines—standard repertoire at major iron foundries since the early 1820s, at textile machine shops since the early 1830s, and at locomotive works since the late 1830s. At this time, however, these firms used few, if any, milling machines because they did not need to achieve the high degree of uniformity of small parts which the federal armories aimed for. By the early 1840s the armory had installed a staged production system, which persisted into the early twentieth century: it forged to near final shape, it milled or cut metal on lathes to eliminate excess amounts, it filed to final dimensions, and it employed an elaborate gauging system to check for uniformity.

The armory took much of the rest of the 1840s to master the use of the new equipment, and it did not achieve a high level of interchangeability until the end of the decade. In the mid-1840s it commenced a rapid multiplication of equipment which continued through the next decade. The number of tilt and drop hammers for forging rose from under twenty to over fifty, and the number of milling machines increased from about thirty to almost seventy; the count of drill presses and engine lathes also rose. Many of the machines carried out only one operation per component. From 1823 to 1850 the greater forging capacity of tilt and drop hammers and better metal-cutting capacity of engine lathes contributed to a decline in the consumption of grindstones (mostly used for barrel finishing) from fifteen pounds per barrel to just over two pounds. The lathes and the improved metal cutting of milling machines reduced the amount of metal that had to be filed off to make locks; thus, the number of files consumed per lock fell from one file to one-third of a file. In 1855 Buckland created an improved rifling machine for barrels which reduced that procedure to a trivial 1 percent of total work. The armory purchased some of the machines added after the mid-1840s, but, like other major metalworking firms, it continued to manufacture most of its own machine tools by adapting or copying machines it already owned or had newly purchased.

In the 1850s the Springfield Armory operated as a large industrial enterprise, employing about 250 workers, many of them skilled machinists, and turning out, annually, as many as twenty thousand high-quality muskets valued at about three hundred thousand dollars. Jacob Abbott, a New York City writer for *Harper's New Monthly Magazine,* visited the armory in 1852 and came away suitably impressed. Nonetheless, this scale of operations had changed little since the 1830s because the army, in the absence of wartime conditions, only needed a limited number of muskets and rifles (see fig. 3.1). Some private sector firms dwarfed the armory. The West Point Foundry at Cold Spring employed as many as five hundred workers and produced over a half-million dollars of machinery annually, and the Novelty Works in New York

TABLE 7.5
Metalworking Machinery of the Springfield Armory, 1851

20 tilt hammers	14 drill presses
1 rolling mill	5 screw cutting machines
1 cutting engine	1 bolt cutting machine
1 slitting engine	1 machine for tapping cone seats
16 turning engines	3 machines for drilling cones
14 lathes	2 machines for checking hammers
5 planing machines	2 hammer-straightening machines
57 milling machines	1 straightening machine
2 milling cutter grinding machines	2 barrel polishing machines
14 boring banks	1 machine for buffing bayonets
6 punch presses	

SOURCE: Michael S. Raber, Patrick M. Malone, Robert B. Gordon, and Carolyn C. Cooper, *Conservative Innovators and Military Small Arms: An Industrial History of the Springfield Armory*, 1794–1968, report to the U.S. Department of the Interior, National Park Service (South Glastonbury, Conn.: Raber Associates, 1989) 251, table 7.7.

City employed up to one thousand workers and turned out as much as one million dollars of machinery. Each of the big three locomotive works of Baldwin and of Norris in Philadelphia and of Rogers in Paterson was over twice as large as the Springfield Armory, measured by employment and sales. Likewise, some of the greatest textile machine shops, such as the Lowell Machine Shop and Mason and Company, operated at about twice the scale of the armory.

Along with Harpers Ferry and some of the private armories, Springfield ranked as one of the foremost machine shops in its use of milling machines for precision metal cutting of small components. Excluding this equipment, however, the armory did not possess exceptional metalworking capacity compared to leading iron foundries, locomotive works, and textile machine shops. The armory's range of equipment by 1851, when its machining operations had stabilized for the rest of the decade (other than duplication of machines), provides a point of comparison (table 7.5). Except for the millers and a few machines that only firearms producers used, the equipment mirrored machine shops of similar scale, with the qualification that the limited product line, namely firearms, required a narrower range of machines than did the diversified shops.

In contrast, the 1850 machine shop of the Saco Water Power Company, probably less sophisticated than either Lowell's, Amoskeag's, or Mason's shop, possessed far more powered engine lathes and planing machines than the Springfield Armory, and much of this equipment was highly specialized by size and type (see table 7.4). Moreover, Saco had a greater variety of other specialized machines because it built diverse machinery. Similarly, in 1855 Philadelphia's Norris Locomotive Works, along with other large locomotive works (Baldwin and Rogers), employed greater numbers of powered lathes and planing machines, many of them specialized by size and operation, and it used a good deal of other specialized equipment (see table 7.2). Although locomotive works purchased many components from outside suppliers, these enterprises had an incredible variety of parts to forge, machine, and file, which required them to operate a wide range of equipment.

A Mixed Record

During the several years bracketing 1840 the restructuring of metalworking and gun-stocking capabilities of the Springfield Armory advanced the system of uniformity, as witnessed by the high degree of interchangeability attained by the end of the decade. Yet the overall record of accomplishment for the twenty years following 1840 remained mixed. The federal government had no intention of enlarging the armory beyond a capacity to supply firearms, along with Harpers Ferry and the private armories, for a small peacetime army. Musket output grew slowly, employment changed little, and episodic sharp declines in each area occurred during transitions to new firearms models (see fig. 3.1). These model changeovers, sometimes accompanied by significant changes in machinery, contributed to a sharp rise in the cost per musket and a plummet in productivity as measured in muskets per worker. It took much of the 1840s to reduce the cost per musket to that prevailing in the late 1820s, and then costs climbed in the 1850s to some of the highest levels of the antebellum years, seemingly wiping out any benefits of the new machine tools (see fig. 3.2).

Similarly, worker productivity rose during the 1840s, but it did not exceed the prior peak around 1830 until after 1848 (see fig. 3.3). A firearms model changeover commenced after the brief productivity peak of the early 1850s, and even the recovery of production volume between 1858 and 1860 did not bring with it a return to high productivity. Instead, productivity matched the level that had existed during the early 1820s, when primitive machine tools (by 1850s standards) had been installed. The armory fared much better at increasing the productivity of powered hammer welding of barrels, from just over two thousand barrels annually per welder around 1840 to about four thousand within four years (see fig. 3.5). After that it made few sustained productivity gains until around 1858, when it finally achieved success in welding by barrel rolling, following twenty-five years of occasionally failed efforts. The armory adopted British barrel-rolling technology and imported William Onions, an expert barrel welder, to introduce the technique. Productivity quickly doubled, and quality improved, and by 1859 the armory welded all of its barrels by rolling.

Could excessively paid laborers have caused the failure of the Springfield Armory to lower the costs consistently of producing muskets with new machinery technology? Following the fluctuating real wages during the economic contraction of 1837–43, wages increased over the next two decades at a rate comparable to the rapid gains of the 1820s, and this matched the rate of real wage growth of unskilled workers in the nation (see fig. 3.4). Firearms workers ranked among the most highly skilled and well-paid industrial employees, and their wages approximated those of machinists in other manufacturing sectors who did comparable work. The Ordnance Department explicitly aimed to maintain Springfield's wages close to, but somewhat higher than, workers in private armories, in order to retain a stable, skilled workforce. The department achieved that goal, but some private armories and machine shops occasionally paid higher wages. Even though some observers asserted that armory workers achieved higher wages because they used their status as federal employees to mobilize political support, their wages did not depart much from skilled machinists elsewhere in the Connecticut Valley. These workers stayed in short supply relative to demand for them.

Under prodding from the Ordnance Department, the attention armory officials paid to achieving uniformity in firearms production probably hindered them from lowering costs much. In order to make parts as close to being fully interchangeable as possible, the armory made extensive use of gauges to check components that had been carefully filed after using machine tools, resulting in high costs. Even after installing the new machine tools around 1840 and multiplying the number of copies of equipment, the share of time spent filing firearms' parts in the 1850s had declined little from several decades earlier. The armory achieved labor-saving efficiency gains during the 1840s, but these gains were not motivated by problems of labor scarcity—that is, installing machinery to compensate for the rise of wages relative to the cost of existing technology. For the 1820–60 period productivity gains probably came from the increasing quality of the labor force and its work, not from labor-saving characteristics of machine tool technology. The new machine tools and the greater amounts of all types of equipment allowed workers to improve musket quality, which permitted the armory to achieve a high level of interchangeability by the late 1840s. Highly efficient machine tools that sharply reduced labor costs did not appear in large numbers until after 1860.[6]

Harpers Ferry Armory

The production and employment levels at Harpers Ferry closely matched Springfield's for most of the forty years after 1820 (see fig. 3.1). Nonetheless, for much of the time the cost per musket at Harpers Ferry remained higher and its productivity stayed lower, thus continuing the laggard status that had characterized it before 1820 (see figs. 3.2 and 3.3). During the years leading up to 1829, Springfield's Superintendent Lee sent as many as twenty-five mechanics to Harpers Ferry to help Superintendent Stubblefield improve musket manufacturing, whereas the latter sent only three mechanics to Springfield. This worker mobility epitomized the direction of know-how trading between the armories. Stubblefield failed to implement improvements, and, when his obstinacy became too great, Bomford, chief of the Ordnance Department, ordered Lee to replace Stubblefield temporarily in 1827 and to institute changes. During Lee's brief time at Harpers Ferry he installed Blanchard's full line of fourteen gun-stocking machines, which had been completed at Springfield four years earlier. But he made little other progress because the entrenched workers and powerful local residents resisted efforts to improve the operation of the armory. Bomford finally succeeded in forcing Stubblefield's resignation in 1829.

Hall's Rifle Works

The Ordnance Department took another approach to improving the manufacturing of firearms at Harpers Ferry. In 1819 Bomford appointed John Hall as director of the Rifle Works. It would operate as a separate branch of the armory, and Hall would act as a private government contractor to make the U.S. Model 1819 rifle that he designed; this approach continued until Hall's death in 1841. The department viewed Hall, a prominent machinist, as a vehicle to advance the system of uniformity through the development of methods, tools, and machinery to make firearms. He completed his equipment in 1823 and finished his first contract for one thousand rifles the following year. The quality of Hall's rifles pleased Bomford and Secretary of War John Calhoun, and they awarded him another contract for one thousand rifles.

An ironic political twist advanced Bomford's goals and Hall's recognition as a leading firearms manufacturer. Stubblefield and his friends hoped to discredit Hall's success, which highlighted the failures at the Harpers Ferry musket works. They persuaded supporters in the United States Congress to pass a resolution ordering a test of Hall's rifles and an examination of his tools and machinery. Extensive military tests in 1826 and the Carrington Committee's evaluation of Hall's methods, tools, and machinery in its report of early 1827 instead vindicated Hall's claims that he produced superior firearms. His rifles achieved greater uniformity than those both from the Springfield Armory and from prominent private armories in Connecticut, such as Whitney's in New Haven and North's in Middletown. Thereafter, the Ordnance Department wanted Hall to focus on improving rifle manufacturing through the use of mechanized methods.

Nevertheless, over the years Hall's rifles cost from sixteen to twenty-one dollars each, about one-fifth more than muskets from the inefficient Harpers Ferry Armory and as much as one-third more than the Springfield Armory's muskets (see fig. 3.2). His achievement of a high degree of uniformity required the use of expensive machinery whose limited cutting capacities left large amounts of metal to be removed by filing to gauge. The approach that Hall had developed by the mid-1820s therefore remained too expensive for most private armories to follow closely, and the Springfield Armory would wait until around 1840 to attempt Hall's extent of mechanization. Bomford took advantage of the demand for Hall's Model 1819 rifles and encouraged him to work with North, who was given a contract in 1828 to produce Hall's rifles. Hall exchanged knowledge about rifle manufacturing and about technology with North, and over the next few years he upgraded his machinery and acquired gauges. By 1834 rifles produced by both manufacturers possessed components that interchanged with one another.

Hall's contributions to machine tool development primarily consisted of improving existing equipment and building a more elaborate line of machinery than his peers. His Rifle Works built most of his machinery, though he relied on foundries for castings. Based on the Carrington Committee's report, much of his machine tool innovations must have been completed by 1827. Like most talented machinists, Hall obtained ideas by know-how trading through networks of mechanics and by copying machine tools, and, like them, he made incremental, yet important, improvements and adaptations. He did not follow Blanchard's gun-stocking system of fourteen machines; instead, Hall developed five machines, and his equipment left substantial handwork for finishing the gunstock. His elaborate drop hammer forging system, which reduced dramatically, if not eliminated, much hand hammer work, preceded the forging system of the Springfield Armory by as much as fifteen years. He improved on well-known drilling and milling machines for hollow milling and obtained patents for metal-cutting machines. The metal cutters, all of them heavy duty and water powered, apparently consisted of various types of milling machines for straight cutting, curved cutting, and lever cutting.

Hall built on the ideas about milling machines which circulated among machinists in New England and perhaps elsewhere. Around 1816 North had developed a milling machine, and he visited Harpers Ferry the same year to share his approach to achieving uniformity. When Hall arrived at Harpers Ferry in 1818, he may have picked up ideas from some of the skilled machinists who had met North. He also had visited the Springfield Armory in 1818 and might have heard about North's innovative milling machine because Superintendent Lee and his top machinists maintained close contact with North and the other Middletown armory machinists.

During the 1820s neither ignorance nor stubbornness caused Hall's private sector peers to fail to increase their use of milling machines; machinist networks continually transmitted Hall's advances to New England. Cost did not rate high for Hall because the Ordnance Department subsidized his machine tool work, whereas the private sector owners calculated the high cost of building milling machines and compared it with their limited metal-cutting capabilities. To them cutting with low-cost lathes and extensive filing rated as better methods than heavy use of milling machines. In keeping with the goals of having a uniform system, Hall devoted great effort and investment to substituting machines with work-locating and work-holding fixtures for filing jigs and other hand tools, and he completed this work by 1828. The proper positioning of the workpiece relative to the cutting tool would loom large in subsequent machine tool development. His elaborate gauging system for checking parts exceeded that of both federal armories. Hall's contributions to machine tools ended by the mid-1830s, as he struggled with illness and engaged in various disputes over contracts until his death in 1841. Nonetheless, his advances became embedded in the networks of machinists, which became part of armory practice even after he became less active.

Mediocrity at Harpers Ferry

Between Stubblefield's resignation in 1829 and Maj. Henry Craig's appointment in 1841, various civilian superintendents of Harpers Ferry Armory engaged in practices similar to Stubblefield's. The armory remained a mediocre operation and an embarrassment to the military. Although Craig instituted some labor reforms during his brief tenure from 1841 to 1844, political and financial problems prevented renovation of the dilapidated physical plant. Under the leadership of Benjamin Moor, a talented machinist and the master armorer since 1830, the armory's machine tools were upgraded, but the inept superintendents undoubtedly hindered his efforts. Finally, around 1838 Moor started to make progress adopting Hall's gauging system and replacing obsolete equipment. During the 1840s, coincident with the military superintendents' tenure at Harpers Ferry, Moor and his deputies made trips to New England machine shops and armories. They borrowed patterns and drawings and built many of the new machines and purchased specialized machine tools, especially from the Springfield Armory and from firms in New England; Middle Atlantic firms also supplied some of them. Under Moor's successor the upgrade of the armory's machine tools continued until the mid-1850s.

Starting in 1844 and continuing for a decade, the armory's physical plant underwent a major renovation and modernization. It established new shops for the special tasks of boring, forging, and machining, and it added a foundry and rolling mill. Like the experience at Springfield, greater numbers of machine tools and the adoption of more sophisticated equipment allowed workers to turn out higher-quality muskets. Yet, like Springfield, these improvements did not lower the cost per musket below the level of the 1820s, when Harpers Ferry suffered from poor management and labor practices. It took until the late 1840s for productivity of workers, measured in the cost per musket and in the number of muskets made per worker, to reach the level of the 1820s (see figs. 3.2 and 3.3). Finally, during 1850–55 the armory reached its highest productivity ever. In 1854, when Henry Clowe, the new civilian superintendent, took over, he returned to the same practices Stubblefield had followed more than twenty-five years earlier, and the armory quickly deteriorated as a firearms manufacturer.[7]

The Federal Armories' Contributions

During the mid-nineteenth century the status of the federal armories as metalworking hubs in firearms networks waxed and waned. Although the Springfield Armory did not always operate as a leading machine tool pivot, it exhibited remarkable technical resilience over the forty years following 1820. The start of an upgrade of its metalworking technology around 1840 vaulted it to a prominent position as a user of machine tools by the early 1850s. In contrast, Hall's Rifle Works, which operated as a separate unit of the Harpers Ferry Armory, remained a leading machine tool axis for about fifteen years following 1820. After the mid-1830s its influence quickly waned, as Hall faced personal difficulties. The musket works of the Harpers Ferry Armory fared poorly as a machine tool center for much of the four decades after 1820, except for a fifteen-year period following 1840, and even its improvement resulted from borrowing New England firms' technology. Although the federal armories' success as metalworking hubs varied over time, they joined with private armories to play pivotal roles in machine tool networks. These arms makers, as well as firms in other industries and the machinists in all of these manufacturing establishments, operated within a sophisticated networked community that produced extraordinary advances in machine tools.

CHAPTER EIGHT

Machine Tool Networks

> The Lathes, Planers, Drills, Borers, and the machinery for working metals generally, made in Philadelphia, are wonderful specimens of workmanship, and celebrated not only throughout the United States, but in portions of Europe.
>
> *Edwin T. Freedley, Philadelphia and Its Manufactures*

A wide range of metal-fabricating industries, including iron foundries, locomotive works, textile machinery firms, general machine shops, and firearms manufacturers, both public and private, built machine tools within their shops for their own use and, occasionally, for sale. This practice provided diverse channels for mechanic mobility and for the transmission of technical innovations among firms. Pivotal hubs of firms and individuals constituted the organizational framework for machine tool networks whose structures and processes mirrored those of other mechanic networks.

THE FEDERAL GOVERNMENT

The Ordnance Department

The pivotal hub status of the Ordnance Department in firearms networks solidified during the first few years after the congressional reorganization of the department in 1815. Until his death in 1848, Chief of Ordnance George Bomford provided administrative leadership that reinforced the department's hub position, and his successors maintained his policies. The chief of ordnance exerted strong control over the federal armories when military officers served as superintendents, but that control weakened under civilian administration. As the lead agency in achieving the system of uniformity for the U.S. Army, the department possessed the administrative means to be a knowledge hub about technological changes in machine tools. Bomford maintained a lively correspondence with armory superintendents—encouraging, prodding, and sometimes scolding them—on the subject of improving firearms quality. The contract system that put him in contact with private armories gave him the opportunity to stay abreast of changes in metalworking equipment.

By inspecting firearms, the department directly monitored machine tool developments at the federal and private armories. Springfield's superintendent, Roswell Lee, led this effort until 1831, monitoring the New England private armories, the key contractors. After that date the department centralized contract inspection under the chief inspector of contract arms, who sent assistants to visit the armories, and they provided reports on machine tools being used by contractors. In 1832 Bomford established the position of "Inspector of Armories, Arsenals and Depots," which gave the department direct access to all operations of the federal armories and supplied information on their machine tool practices. Many of the military officers working for the department possessed engineering training from their educations at West Point, giving them an understanding of technical knowledge.

This administrative organization positioned the Ordnance Department as a hub of know-how trading and ranked it among the leaders in shaping the community of practice of machine tool building. It collected information from federal and private armories, digested it, and recirculated it back to the armories. The deputies of the chief inspector of Contract Arms and of the inspector of Armories, Arsenals and Depots could directly transmit technical knowledge about machine tools on their visits to federal and private armories. This on-site transmission through the mechanism of the artifact-activity couple gave each armory access to machine tool developments at every other armory; any failure to adopt a machine tool was therefore not the result of inadequate knowledge. Instead, armories probably did not adopt some machine tools because labor-cost savings remained too low to compensate for the high cost of the equipment.

The contract system managed by the department had its greatest consequences for machine tool networks in New England because the Springfield Armory possessed far more linkages with private armories than the Harpers Ferry Armory did. The underpinnings of Springfield's knowledge exchanges had been established between 1815 and 1820, and the interactions under the contract system continued for two more decades. Thereafter, as the relative importance of contracts declined and highly competitive private firms emerged, the contract system atrophied.

During the 1820–40 period private armories could send machinists to Springfield to observe operations, collect patterns for equipment, and borrow machine tools to take back to their firms. The armory lent the private firms skilled machinists, and the borrowers compensated them at Springfield's wage levels; occasionally, machinists stayed permanently or moved on to other firms. Machinists from Springfield also visited private armories and gathered information about machine tools. The Ordnance Department conveyed an implied threat that, if private armories did not participate in this interchange, they would not receive contracts in the future, and any innovations should be shared without payment of royalties. These exchanges of machinists among the federal and private armories sustained the operation of the artifact-activity couple—the transfer of technical skills and innovations during the on-site confrontation of machinists with equipment in a learning environment with other machinists. The types of linkages which existed among armories had precedence among other metalworking firms, but the firearms networks also possessed elements of management oversight from the Ordnance Department and from the Springfield Armory whose goal was to achieve a system of uniformity. This structured approach went beyond the more freewheeling networks in other industries.

The department's management role also provided indirect, yet equally powerful, means to diffuse technical innovations. The political effort of Superintendent James Stubblefield at Harpers Ferry Armory and of his supporters to discredit John Hall and his Rifle Works convinced Congress to pass a resolution in 1826 to evaluate Hall's methods, tools, and machinery. Maj. William Wade, Bomford's assistant, consulted with Lee at Springfield, but he refused to serve as an evaluator. On his advice James Carrington (chairman), Luther Sage, and Col. James Bell were chosen as examiners, the first two as key members.

Carrington, one of the most widely known firearms machinists in the nation, had served as armorer at Eli Whitney's firearms works in New Haven from 1799 to 1825. After Whitney's death Carrington established his own foundry, coffee mill, and razor strap factory in Wallingford, near New Haven. His extended service under Whitney, who had had strong connections to the Springfield Armory and the Ordnance Department, gave Carrington wide access to knowledge about metalworking machinery. He also had many contacts with the nearby Middletown armories of Simeon North, Nathan Starr, and Robert Johnson. Luther Sage, an assistant armorer under Lee at Springfield from 1815 to 1823, had served much of the time as an inspector of contract arms. This gave him direct contact with all of the major private armories in New England, including Whitney's, the Middletown cluster, and Asa Waters's in Millbury. After 1823 Sage worked as an armorer and inspector of contract arms for the Frankford Armory in Philadelphia, providing him with ties both to firearms makers, such as Marine Wickham and Henry Derringer, and to machinery and machine tool builders in the city's highly networked community of practice.

The Carrington Committee's flattering evaluation of Hall's Rifle Works made special note of his innovative use of machine tools to achieve a high degree of interchangeability. The key metalworking equipment consisted of powered drop hammer forges for getting metal components close to final dimensionality as well as many types of milling machines to remove excess metal prior to filing. The report was significant beyond simply vindicating Hall's operations, rebutting his critics, and documenting systems for use by the Ordnance Department. Armed with the findings of his two seasoned inspectors, Carrington and Sage, Chief of Ordnance Bomford called for achieving uniformity among machinists and shops, and improved and innovative machine tools, he argued, would help that effort.

Bomford awarded a contract to North in 1828 and appointed Hall to work with North over the next few years to achieve interchangeability between their rifles. This collaboration became a vehicle to insert Hall's machine tool innovations into New England's firearms networks. After Carrington returned to Wallingford, he became another source of knowledge about Hall's innovations, and he could monitor Hall's subsequent work with North in Middletown. Springfield's Lee kept appraised of Hall's innovations because Lee had served as the temporary superintendent at Harpers Ferry Armory in 1827. Furthermore, he recognized the significance of the machine tools, given that respected machinists whom he knew, and had recommended for the Carrington Committee, had written the report.

Even though these machine tools were not adopted extensively at that time, technical knowledge about their design had been transferred from Hall at Harpers Ferry, where he was peripheral to the New England machinist networks, to key hubs of that region's networks, especially at Middletown and Springfield. The Ordnance Department must have encouraged leading New England firearms manufacturers to gain this technical knowledge directly; between 1821 and 1835 North (Middletown), Whitney (New Haven), Lemuel Pomeroy (Pittsfield), and Waters (Millbury) all visited Hall's Rifle Works. The department retained a full set of specifications for Hall's machinery which it could disseminate without considering royalties. Although the department served as a pivotal hub of technical knowledge about machine tools, its officers could not directly adopt equipment or improve technical skills, but the federal armories could do both. They played quite different roles within the networks which mirrored somewhat the roles they had played before 1820.

The Federal Armories

The Springfield Armory

Superintendent Lee maintained Springfield as a clearinghouse for technical knowledge about metalworking until his death in 1833, but his efforts did not guarantee it would continue as a network hub. Through the 1820s the reports of Springfield's inspectors of contract arms kept Lee's machinists informed about machine tool innovations at private armories. Lee remained in charge of these inspections until 1831, when the Ordnance Department centralized that task under its direct control. For two decades the contract system provided a vehicle to encourage the sharing of technical knowledge about machine tools, and the exchange of machinists between Springfield and the private armories facilitated the direct, hands-on transfer of machine-building skills through the mechanism of the artifact-activity couple.

The network links between Springfield and the private armories changed between the late 1830s and the mid-1840s. Older, private firms could not satisfy the demands for interchangeability from the Ordnance Department, especially the need to invest more capital in machine tools, as both federal armories were doing. Many of these firms, including North and Starr in Middletown, Waters in Millbury, and Pomeroy in Pittsfield, exited firearms manufacturing. New firms such as the Ames Manufacturing Company in Chicopee, which Boston capitalists funded in the mid-1830s, and Samuel Colt's factory in Paterson, which wealthy New York City residents funded around 1840, added many machine tools. With the demand for mechanics consistently outstripping supply, capable machinists easily relocated to new firms or entered others that needed their skills. The networks of the Springfield Armory changed, but technical knowledge continued coursing through them. Nonetheless, the rise of the new, well-capitalized firms that hired large numbers of top mechanics interjected more machinist hubs into firearms networks, diminishing somewhat the centrality of Springfield.

From 1830 to 1860 even a large industrial enterprise such as the Springfield Armory could not build all of its equipment, so it reached out to various suppliers for machinery, steam engines, and machine tools (table 8.1). General machinery that could be produced by most large foundries and machinery firms came from nearby suppliers in Springfield and Holyoke. The armory looked more widely for steam engines, tapping firms in the leading steam engine centers of Boston, Providence, New York City, and Pittsburgh. Springfield built the bulk of its equipment; nonetheless, like most major industrial firms and even machine shops, it purchased some specialized machine tools. In some cases patents protected this equipment, or they required too much time and money to build, especially if the armory's machinists did not have experience making them. The suppliers represented some of the leading firms of that period, including Robbins and Lawrence, which had factories in Hartford, Connecticut, and Windsor ,Vermont; the Gage, Warner and Whitney firm of Nashua, New Hampshire; Fairbanks, Bancroft and Company of Providence; and Merrick and Towne of Philadelphia. Through its purchases the armory received technical knowledge and in turn disseminated it, enhancing the armory's hub position.

TABLE 8.1
Supplier Networks of the Springfield Armory, 1830–1860

Equipment/Firm	Location
Machinery	
American Machine Works	Springfield, Mass.
Bemis and Co.	Springfield, Mass.
Springfield Tool Co.	Springfield, Mass.
Hadley Falls Co.	Holyoke, Mass.
Steam Engines	
Ethan Earle	Boston, Mass.
Lazell and Perking Co.	South Bridgewater, Mass.
Babbitt's Fan Blower	Providence, R.I.
Many and Ward	Albany, N.Y.
Clark's Patent Steam and Fire Regulator Co.	New York, N.Y.
William Wade	Pittsburgh, Pa.
Machine Tools	
White and West	Windsor Locks, Conn.
Weaver	Windham, Conn.
Robbins and Lawrence	Hartford, Conn.; Windsor, Vt.
Gage, Warner and Whitney	Nashua, N.H.
Fairbanks, Bancroft and Co.	Providence, R.I.
New York Screw Co.	New York
Stevens Brothers and Co.	New York
Gage and Campbell	Waterford, N.Y.
Merrick and Towne	Philadelphia
Philos B. Tyler	Philadelphia

SOURCE: Felicia J. Deyrup, "Arms Makers of the Connecticut Valley: A Regional Study of the Economic Development of the Small Arms Industry, 1798–1870," *Smith College Studies in History* 33 (1948): 147.

The Harpers Ferry Armory

The musket works of the Harpers Ferry Armory remained a backwater of technical innovation in machine tools until around 1840 and therefore did not operate as a hub in firearms networks before then. In contrast, Hall's Rifle Works at Harpers Ferry, a hybrid public/private unit, served as an outlier to New England firearms networks rather than as a hub in its own right. Hall collected machine tool ideas from that region and copied, modified, and improved on them; his firm's innovations were then disseminated back to New England. The Ordnance Department structured a variety of the channels for transferring Hall's machine tool innovations, including the Carrington Committee's visit in late 1826 and its report a few months later, Lee's service as temporary superintendent of Harpers Ferry Armory in 1827, Hall's work with North from 1828 to 1834 on the interchangeability of its rifles, and the visits of leading New England firearms manufacturers to Hall's Rifle Works over the years from 1821 to 1835.

The mobility of Hall's machinists provided an alternative channel of know-how trading of his machine tool innovations with New England mechanics. His employees acquired important technical ideas by working with him, but they had few job opportunities nearby. Philadelphia, about 150 miles northeast of Harpers Ferry, was the nearest large source of machinist jobs. Hall occasionally lost machinists to New England, and some of them entered the Springfield Armory. Nathaniel French, one of the nation's leading firearms machinists during the two decades after 1820, started work in Hall's Rifle Works in 1821, and his extensive experience earned him the third highest pay among the employees. After working with Hall during his seminal period of machine tool innovations, French moved to North's firearms factory as a master machinist in 1827. He returned to Hall's works two years later; in 1831 he accepted a permanent job at the Springfield Armory. French's job mobility—which brought him into contact with the prominent machine tool innovation center in Middletown and with the information clearinghouse of the Springfield Armory—made him a leading transmitter among machinist hubs of Hall's technical skills and innovations in machine tools.

The Harpers Ferry Armory undertook a major upgrade of its machine tool repertoire in the 1840s, when master armorer Benjamin Moor and his deputies visited many of the leading machine-building centers in New England, including Boston, Lowell, North Chelmsford, and Lawrence in eastern Massachusetts; Keene and Nashua in southern New Hampshire; Chicopee, Springfield, and Middletown in the Connecticut Valley; and Providence. They drew from various industries—iron foundries, textile machinery firms, and firearms factories—for their machine tool ideas. The Springfield Armory, however, probably contributed most of the key ideas. Its top machinists, led by Thomas Warner and Cyrus Buckland, were in the midst of a substantial upgrade of the armory's metalworking equipment. Although the Harpers Ferry machinists built many of their new machine tools, they followed patterns and drawings obtained on their travels. They purchased specialized machine tools mainly from firms in New England, especially the Ames Manufacturing Company of Chicopee and the American Machine Works of Springfield, both closely allied with the Springfield Armory. These purchases continued up to the mid-1850s, after which Harpers Ferry regressed back to a second-rate armory. It never served as more than a minor hub of firearms networks, and most of its machine tools consisted of adaptations of innovations generated elsewhere.[1]

THE PRIVATE FIREARMS MANUFACTURERS

Starting around the mid-1830s, new, private manufacturers transformed firearms networks, as some of them became hubs in machine tool networks. They hired top machinists, trained future leaders, and their mechanics departed for other firms. Their contributions to machine tools had greater consequences for industrialization than their firearms production. They joined with the iron foundries, locomotive shops, textile machine shops, and other mechanic shops in molding the community of practice of machine tool building.

The Ames Manufacturing Company

In 1829 Nathan and James Ames transferred their edge tool and cutlery business from the Lowell satellite of Chelmsford to Chicopee, and they continued the same lines of work in a small shop with nine workers. Edmund Dwight, whose pedigree included membership in the Boston Associates, partnership in the merchant house of J. K. Mills and Company, investor in the Chicopee Manufacturing Company, and member of the prominent Dwight merchant family of Springfield, had suggested this move. The twenty-six-year-old Nathan and nineteen-year-old James must have impressed Dwight. He offered rent-free space for about four years in one of his cotton mills, suggesting that he contemplated a bigger plan.

The Ames brothers soon carried on a diverse business, including mechanical repair work for local cotton and paper mills, making spindles for the Chicopee Manufacturing Company, supplying swords to the federal government, producing tools and machinery for nearby paper mills, and making tools and cutlery for sale through merchants. The sword contracts probably resulted from the political contacts that Mills and his fellow Associates maintained with Congress and the federal bureaucracy. The Ames brothers sold their various metal products locally, and they distributed the rest through merchant wholesalers, including Dwight, to markets from Hartford to New York and Philadelphia. This market area, coupled with government military contracts, was considerable for a small metalworking shop; its scale had been reached within three years of starting production in Chicopee.

Dwight's grander industrial scheme became apparent with the incorporation of the Ames Manufacturing Company in 1834. It started with thirty thousand dollars of capital, which went to seventy-five thousand dollars in 1841, when Ames purchased the property of the failed Chicopee Falls Company, including its machine shop. After the firm took over the textile machine shop of the Springfield Canal Company four years later, the capitalization reached two hundred thousand dollars. Dwight and Mills partnered with the Ames brothers to underwrite the firm's capital, and by the late 1840s the brothers held slightly over 20 percent of the stock, while Boston investors controlled most of the remainder.

At the time of the 1834 incorporation of the company, it employed thirty-five workers, barely one-ninth the size of the Locks and Canals machine shop at Lowell. During the rest of the 1830s profitable government contracts continued, boosting Ames's reputation as an ordnance manufacturer. By the end of that decade the contracts, which then included swords, firearms, and brass cannons, generated the largest share of the company's profits. Agricultural and small trade tools became less important, whereas machinery, machine parts, and cutlery continued as significant lines. By 1845, with the inclusion of the textile machine shop of the Springfield Canal Company, the Ames Manufacturing Company had become one of the largest machine shops in the East, with 130 employees and annual sales of over $130,000. Its product line included swords, firearms, cannons, cutlery, machinery, and bronze and brass castings; in other words, it operated as a full-line machine shop with its own foundry.

By 1848, with the economic upturn well under way, Ames quickly expanded to three hundred workers concentrating production mainly on textile machinery. This focus did not last, however, because too many textile machine shops, including Lowell, Amoskeag, Mason, and Whitin, had expanded their textile equipment production, thus most of them also had to build other machinery. By the 1850s Ames had returned to a diversified strategy, producing textile machinery, machine tools, castings, swords, firearms, plated ware, shafting, and general factory equipment.

Machine Tool Manufacturing

The mechanics of the Ames Manufacturing Company did not make innovative contributions to machinery design; the firm's hub position in machinist networks rested, instead, on other sources. The company possessed extraordinary network ties that allowed it to produce machine tools incorporating the latest improvements which it sold, along with a diversified line of machinery, to firms throughout the East. These sales and the mobility of mechanics who came to work at Ames and left to work for other firms positioned it as a hub within machinist networks. From the beginning the Ames brothers maintained strong ties to the federal government because their shareholders had the political heft to ensure that the Ordnance Department considered them a serious supplier of ordnance. This political backing also won them a warm reception by the nearby Springfield Armory. As firearms contractors, the Ames brothers and their machinists could visit the armory to examine machine tools and borrow patterns. By the late 1830s they had built a superb, sizable machine shop.

As early as 1835, Ames supplied machine tools to prominent Philadelphia machinery firms, including Merrick, Agnew and Tyler, which purchased a milling engine and a screw-turning lathe, and Coleman, Sellers and Sons, which bought an engine lathe. These sales testified to Ames's pivotal position in machinist networks, and, by selling to top Philadelphia metalworking firms, it acquired contacts in that hub of machinery manufacturing. The company benefited from the large-scale upgrade of machine tools at the nearby Springfield Armory led by Warner, Buckland, and other top machinists, who brought together innovations in machine tools that had been developed elsewhere in the East.

The company's skills in building machine tools grew, and its reputation with the

Ordnance Department grew more solid as it supplied equipment to the federal armories at Springfield and Harpers Ferry. From 1845 to 1854 Ames ranked as one of the largest suppliers to Harpers Ferry and furnished ten gun-stocking machines designed by Buckland and at least twenty-six plain milling machines. When Ames hired Jacob MacFarland, who had trained at the Springfield Armory in the late 1830s, the firm strengthened its bonds with the federal establishment. MacFarland probably learned about the adaptations to milling machines being developed at the armory by Warner, French, and others. From 1845 to 1858 MacFarland served as foreman of the Ames machine shop, and his name became attached to plain milling machines—the "M'Farland" millers. The extensive know-how trading between Ames and the armory epitomized a vibrant local community of practice of machine tool builders.

In the 1850s astute British visitors testified to the prominence of the Ames company. George Wallis, headmaster of the Birmingham School of Art and Design and a distinguished industrial design educator, visited the factory in 1853. He noted that "this extensive manufactory is curious from the great variety of articles and the multiplicity of operations carried on." The following year the British Committee on Machinery, which included two military officers and an industrial engineer, arrived at Ames. They inspected gun factories and purchased equipment for Britain's new Enfield Armory; the committee described Ames as "an extensive foundry and machine works." Their visit came at the recommendation of Colonel Ripley, superintendent of the Springfield Armory, who sent them to Chicopee to consult with James Ames about purchasing gun-stocking machinery for Enfield.

Subsequent interactions of the committee with the Ames company revealed its pivotal position in firearms networks. The committee discovered that Ames supplied machine tools and machinery to the Springfield Armory. This close linkage encouraged James Ames to submit a contract proposal to the committee to build gun-stocking machinery based on patterns developed by Buckland, the armory's top machinist. Ames would build the equipment only if Buckland agreed to serve as a consultant, for a fee of $1,000, on the design of the 15 different machines that would be adapted for the British Enfield rifle musket. Superintendent Ripley agreed to this arrangement, and, ultimately, Ames received contracts for $46,845. This included gun-stocking machinery ($38,220); 2 milling machines for edging the lock plate of firearms ($2,400); a total of 114 gauges, tools, and patterns ($5,600); and miscellaneous equipment and fees ($625). The willingness of the British committee to give such a large, prestigious order to Ames to supply important machine tools and machinery for Britain's new state-of-the-art armory enhanced the firm's reputation as a hub of machine tool networks.[2]

Robbins and Lawrence

The firm of Robbins and Lawrence in Windsor, Vermont, however, surpassed the Ames Manufacturing Company as a supplier of metalworking machines to Britain's Enfield Armory. The Windsor firm's entire order of $41,245, including packing and delivery in Boston, consisted of machine tools, whereas Ames supplied less than $3,000 worth. The British committee had sound reasons for choosing Robbins and Lawrence for such a significant order. Its roots traced to 1843, when Richard Lawrence and Nicanor Kendall started a small gun shop in Windsor. The following year the businessman Samuel Robbins joined them in bidding for a government contract, which they received in 1845. They fulfilled this contract for ten thousand rifles and obtained a second contract for fifteen thousand rifles in 1849. Shortly afterward, Kendall sold out to his partners, and the firm became Robbins and Lawrence.

The Talented Trio

The firm's machine tool expertise rested especially on the talented mechanics Robert Lawrence, Henry Stone, and Frederick Howe. Without dismissing the significance of Lawrence and Stone, Howe's arrival in 1847 probably constituted the seminal point in vaulting Robbins and Lawrence into the ranks of leading machine tool builders. In 1838 the sixteen-year-old Howe began an apprenticeship at the Gay and Silver machine shop in North Chelmsford, which gave him experience building milling machines, turret lathes, and other metalworking equipment at one of the top machine shops in New England. Through the contacts of Gay and Silver, Howe must have participated in the networks of the many leading mechanics at the other machine shops in the Lowell–North Chelmsford area, including the Lowell Machine Shop. During his nine years with Gay and Silver he acquired superb skills as a machinist and draftsman, before becoming a draftsman at Robbins and Lawrence (map 8.1). Within a year the firm promoted the twenty-six-year-old to be superintendent of the machine shop, an exemplar of talented, young machinists riding a rapid upward career trajectory to senior levels.

Around 1848 Howe had designed an improved milling machine, undoubtedly drawing on ideas from the Lowell–North Chelmsford area and from the Connecticut Valley machinist networks, which included the Springfield Armory and the Ames Manufacturing Company. Over the next few years he improved it, and around 1850–52 he designed a version of the universal milling machine with mechanisms that allowed the workpiece to be moved to different positions for the cutting process.

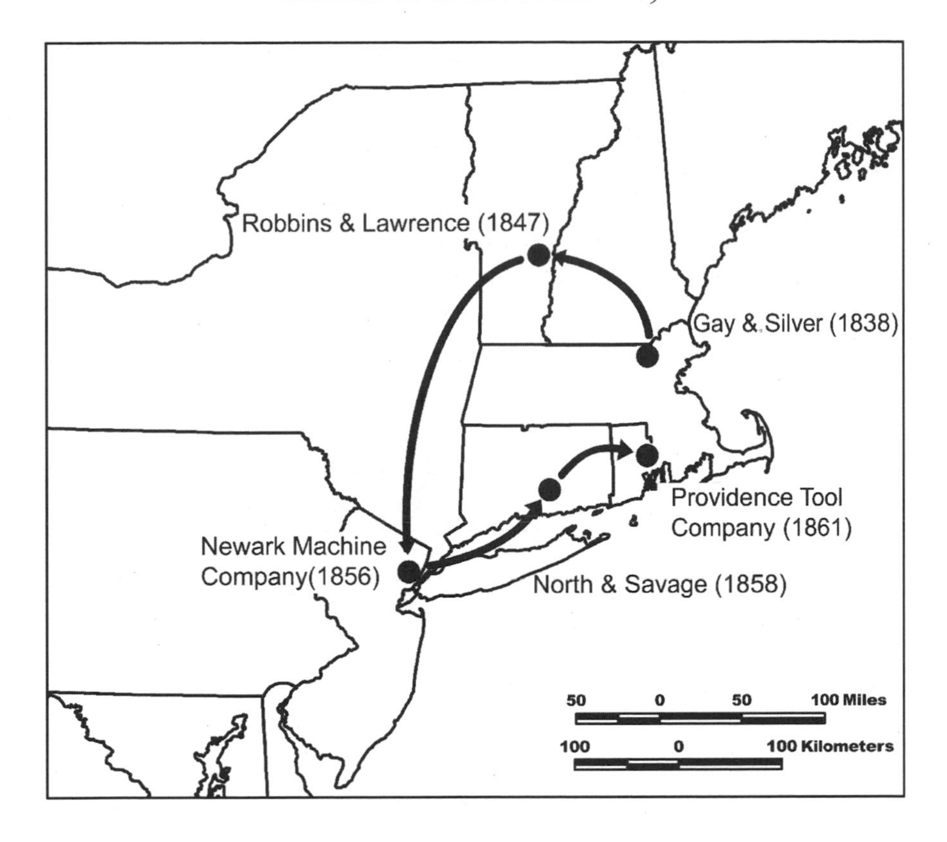

Map 8.1. Career Mobility of Frederick Howe, 1838–1861

Attracting Talented Machinists

Lawrence, Stone, and Howe constituted a dynamic team of mechanics, making it difficult to assign improved machine tool designs and innovations to only one of them. From the late 1840s to the mid-1850s Robbins and Lawrence acquired firearms contracts from private firms, and it produced and sold many types of machine tools, such as milling and profiling machines, barrel rifling machines, drilling machines, and turret and slide lathes. Its capacity to build this sophisticated equipment rested, in part, on its three leading machinists, yet it also had a deep pool of skilled mechanics. When the firm received its first contract in 1845, it distributed print advertisements in the Windsor region and used its machinist networks in New England to spread the news about its sizable factory expansion by word of mouth. Employment soared from 25 workers to over 150 as machinists flocked to Windsor from the surrounding region and from leading firms such as the Whitney armory near New Haven, Silver and Gay in North Chelmsford, the Amoskeag Machine Shop in Manchester, the Springfield Armory, and even as far away as the Harpers Ferry Armory.

These mechanics vaulted Robbins and Lawrence into a hub position in machinist networks, and, as the firm began to build machine tools for sale, it strengthened this rank because its job-hopping machinists carried the technical advances to other firms where local mechanics could reproduce them. In 1852 it received a contract from the Sharps Rifle Manufacturing Company of Hartford to make five thousand Sharps carbines at the Windsor plant. Shortly after signing this contract, Robbins and Lawrence entered into another agreement with Sharps to manufacture fifteen thousand rifles and carbines at a new factory to be built in Hartford. Sharps advanced them forty thousand dollars to build a state-of-the-art factory, and in 1853 Lawrence moved to Hartford to oversee the construction of the plant and the installation of machine tools. Howe and Stone, with Lawrence's guidance, supervised the construction of most, if not all, of this equipment at the Windsor plant.

The building of such a sophisticated factory in Hartford equipped with the latest machine tools attracted ambitious machinists such as Joseph Alvord, who had accumulated skills and experience moving among the leading machine shops of the era. He apprenticed at the Ames Manufacturing Company, and then the nineteen-year-old moved to the nearby Springfield Armory, where he worked for about eight years. In 1851 Alvord became a contractor at the Hartford plant of Robbins and Lawrence, and six years later William Perry, superintendent of the Wheeler and Wilson Manufacturing Company and a former trainee at Colt's firearms factory in Hartford, hired Alvord. Wheeler and Wilson had recently moved from Watertown, Connecticut, to Bridgeport, soon expanding its manufacture of sewing machines. Perry also hired Alvord's friend and fellow contractor, James Wilson, who had worked with Alvord at Robbins and Lawrence in Hartford. These machinists and others created a sewing machine factory modeled in part on Colt's armory practice of drop forging, machining, and a gauging system with a rational jig and fixture system to realize uniformity of parts. Robbins and Lawrence, therefore, mutated into a dual hub in machinist networks, and its Hartford branch became integrated into the Connecticut Valley's networks, which transmitted many machine tool advances among firearms and other industrial sectors.

In 1851 Robbins and Lawrence had gained notoriety at the Crystal Palace Exhibition in London for the rifles it displayed in the American exhibit. Observers admired them for the machining of the locks and the interchangeability of the components, especially the lock mechanism. When the British Committee on Machinery arrived in the United States three years later to examine firearms manufacturing and to purchase equipment for the new Enfield Armory, they wanted to visit the Robbins and Lawrence operations. They first went to the new factory the firm had recently built in Hartford, although all of the machinery had not been installed yet. Lawrence gave them a tour of the works, and the committee later reported that they were "so struck with the beauty and efficiency of the machines." This led to a tender by Robbins and Lawrence, which the committee accepted after revisions, to build machine tools for the Enfield Armory. The committee then visited the Windsor factory and must have been equally impressed.

Ordnance officials in England granted the committee authority to award a contract, which turned into the order for $41,245 worth of machine tools from Robbins and Lawrence. It consisted of numerous milling machines for all types of specialized operations on different parts, including the lock plate, hammer, tumbler, main spring, lock screws, and trigger plate. Robbins and Lawrence also supplied many drilling machines and some clipping machines, and they made one twelve-foot lathe and six six-foot lathes. In addition, they built two machines for rifling barrels and six universal milling machines at $850 each, among the most expensive machines. They were probably based on Howe's 1852 design of a miller in which work could be adjusted to different positions. As superintendent, Howe oversaw much of the design and building of these machine tools. This large, prestigious order ratified the firm's hub position in machinist networks.

Robbins and Lawrence ran into financial difficulties, however, by overextending themselves. In 1855 the firm took on a contract, based on the Enfield rifle design, for twenty-five thousand rifles to be supplied to the British government, because the Enfield Armory had not yet started production. They failed to meet the contract provisions, and the firm went bankrupt two years later. The Windsor business of Robbins and Lawrence, therefore, operated as a premier hub of machinist networks for only about a dozen years. Nevertheless, the firm's demise underscores the principle that technical knowledge and skills were embedded in networks; they were not solely the property of individuals or firms. Once their innovative machine tools had been created and sold, they embodied technical knowledge that was transmitted to other locations in the East where local machinists could copy them and make further adaptations. The leading mechanics went on to greater fame and operated as hubs in machinist networks. Lawrence continued as the master armorer of the Sharps Rifle Works in Hartford, and machinists from the Robbins and Lawrence factories in Windsor either stayed to work at other machinery firms or moved elsewhere. Some mechanics at the Hartford branch remained with the facility, which became the Sharps Rifle Works, while others moved to top firms such as the Colt Firearms Factory or the Wheeler and Wilson Manufacturing Company, maker of sewing machines.

Frederick Howe's Career—Continued

In 1856 Howe left Robbins and Lawrence and took a position at the Newark Machine Company, a manufacturer of machine tools and gun machinery; here he continued work on his universal milling machine (see map 8.1). This placed Howe in the machinist networks of New York City and its environs, which also included the nearby Paterson locomotive and machine shops. Within two years he moved to Middletown to work at North and Savage, manufacturers of firearms and gun machinery, which strengthened his network ties to the big nest of machinists in the Hartford region, including that of his coworker Lawrence (now at the Sharps Rifle Works) and the large number of machinists at the Colt factory, headed by the renown Elisha Root. Finally, in 1861 Howe moved to Providence, one of the most significant nests of machine tool builders. He was appointed superintendent of the Providence Tool Company at the princely salary of five thousand dollars. This thirty-nine-year-old had been one of the East's leading machinists for over a decade, and his networks reached to most of the top nests of machine builders. Nonetheless, he became an even more influential hub when he supervised construction of a sewing machine factory, among the most technologically advanced users of machine tools at that time. In 1868 Howe moved to Providence's Brown and Sharpe, taking the position of superintendent and then president, with an equity stake of almost 20 percent. The company became one of the nation's greatest machine tool firms, and Howe continued to develop innovations in that equipment.[3]

Colt's Hartford Armory

The legacy of Samuel Colt's firearms company continued into the twentieth century, whereas Robbins and Lawrence experienced a short life span. Yet, like that firm, the greatest impact of Colt's Hartford Armory on machine tool networks occurred within a brief period of about sixteen years, from 1849, with the hiring of Elisha Root as superintendent, until 1865, when he died. Colt's early efforts in the late 1830s and early 1840s to manufacture his patent revolving pistol in Paterson failed, in part, because he employed poor production methods. In 1847 he received a government contract for one thousand pistols and subcontracted with Eli Whitney Jr., head of the Whitney armory. This firm remained embedded in the networks of firearms machinists, especially those linking armories in the Connecticut Valley. Colt contracted with Thomas Warner, former master machinist at the Springfield Armory, to supervise operations and the construction of machine tools for making the pistols. Warner had been a leader in the upgrade of machine tools at Springfield and could thus ensure that Colt's equipment incorporated the latest technological advances. In 1848 Colt relocated his equipment to Hartford and converted a vacant textile mill to his new armory. Here he produced the revolvers and made machine tools for the new London armory and for a new factory in Hartford.

Elisha Root

Colt recognized that he needed an accomplished manager/machinist as his superintendent if he wanted to be profitable in manufacturing large quantities of his complex patent revolvers, whose revolving chamber required precision operation. Colt tried to hire Albert Eames, a former machinist at the Springfield Armory who worked at the New Orleans mint, but Eames declined the job. Then Colt made a determined effort to lure Elisha Root, superintendent of ax manufacturing at the Collins and Company factory in nearby Collinsville, arguably, the greatest ax factory in the nation. Colt had probably turned to Eames first because Root, one of the most sought-after machinists of that era, had previously turned down the prestigious job of master armorer at the Springfield Armory and had declined offers from two other manufacturing firms in Massachusetts. In 1849, when Colt made the offer to Root, he had completed his integrated system of ax manufacturing, a project that seemed to intrigue him. He received a salary of about three thousand dollars, and, even more significant, he became a stockholder. This lucrative financial package made him one of the most highly remunerated factory superintendents in the East.

Root brought unusual talent and experience to Colt's company. In 1823 the fifteen-year-old became an apprentice at a machine shop in Ware, Massachusetts. Over the next nine years he followed the path of many young machinists who wanted to acquire technical skills. He moved among machine shops, working in Stafford, Connecticut, and in Cabotsville and Chicopee Falls in Massachusetts. In 1832 the twenty-four-year-old Root became a journeyman at the Collins and Company ax factory, about fifteen miles northwest of Hartford. Within several months President Samuel Collins appointed him foreman of the repair shop, which also built machinery. At that time the firm operated as a skilled craft shop whose workers employed traditional ax-making techniques, but this quickly changed. Besides Root, Collins had assembled five other innovative machinists—David Hinman, Erastus Shaw, Benjamin Smith, Isaac Kellog, and someone named Burke. They constituted a local community of practice, and their know-how trading while they worked on equipment epitomized the mechanism of the artifact-activity couple.

These six talented machinists contributed many innovations to the design of metalworking equipment. During 1832–33 Hinman and Smith received patents for die-forging machines, which had two dies with the shape of a forging cut into them. The dies were brought together with force, and the hot metal pressed between them took on the shape that had been cut into the dies. The principle of die forging had been developed already, but the mechanism to work the machines effectively had not been perfected sufficiently to replace traditional trip-hammer forging, which consisted of pounding a metal object into shape by repeated blows. Smith also received a patent for milling machines that worked on axes. Although the Collins factory did not use the milling machine, the die-forging machines and others developed by these innovative machinists were aimed at mechanizing ax manufacturing.

In 1836 the twenty-eight-year-old Root took out his first among the numerous patents he would acquire over the next two decades, and many of his innovations were never patented. His first patent foreshadowed a theme in much of his subsequent efforts; the die-forging machine combined several mechanical processes in sequence. Hinman and Root shifted from designing machines that incorporated craft skills to creating systems of production line equipment for manufacturing large quantities of axes. Collins appointed Root superintendent of the firm in 1845, and within two years he had equipped a large, modern, three-story building with metalworking equipment and ovens. The integrated production system generated high-volume, low-cost output. It replicated, to some extent, what cotton mills and Blanchard's gun-stocking line of sixteen machines had achieved more than two decades earlier. Yet Root shifted production far beyond these antecedents. Drawing on his conceptual and design skills, he built machine tools and other equipment that carried out multiple steps on each machine, thus eliminating the time spent moving a part from one machine to another and setting it up. In 1849 Root was ready for the next challenge—the job of superintendent at Colt's factory and the opportunity to produce his patent revolver.

Machinists' Roots

Root's work at the Collins factory had placed him in a pivotal position among machinist networks. He had designed and supervised the establishment of one of the nation's most modern factories, and he had associated with top-notch machinists. Each of them had their own networks, and some, including Root, also participated in the patent networks that reached across technical journals and court cases. When Colt hired Root, therefore, a large number of talented machinists, especially in New England, recognized an opportunity to enhance their careers and make a lucrative income from a profit-oriented manufacturing firm. They brought the latest ideas for machine tool design, and, working together under a brilliant superintendent, they made substantial contributions to machine tool technology. Clearly, machinist networks were also sophisticated job recruitment systems, and, when one of the greatest manager/machinists took a leadership position, the attraction of so many talented workers created a new network hub.

Quite a few of the mechanics ranked among the nation's premier machinists of the mid-nineteenth century, and several, including Francis Pratt, Amos Whitney, George

Fairfield, Charles Billings, and Christopher Spencer, went on to greater distinction later. Pratt grew up in the Lowell–North Chelmsford area and apprenticed there as a machinist, which put him in contact with the top machinists who worked in that nest of textile machine shops and machinery firms which included the Lowell Machine Shop and Ira Gay's firm. In 1848 the twenty-one-year-old Pratt moved to the Philadelphia satellite of Gloucester, New Jersey, where he worked as a journeyman and then as a contractor in the Gloucester Machine Works. Thus he became established in the Philadelphia area nest of machinists. In 1852 he took a job at the Colt armory, and, two years later, the nearby Phoenix Iron Works, owned by Levi Lincoln, offered the twenty-seven-year-old the position of superintendent. Like other top machinists, Pratt had vaulted to the top of his profession at an early age, and he possessed bridges to some of the leading local networks of machinists in the East.

Whitney moved from his Maine home to Lawrence, where he apprenticed at the famous Essex Machine Shop, builders of cotton textile machinery, locomotives, and machine tools. Like Pratt, Whitney developed close ties to the Lowell–North Chelmsford nest of machine shops. In 1850 he and his father took positions at the Colt armory, and four years later Whitney shifted to the nearby Phoenix Iron Works, where he became a contractor while Pratt served as superintendent. Whitney's move had been instigated by Pratt, who knew him from their time spent working together at Colt. In 1855, following a design by Pratt and with help from Root, the Phoenix Works completed the famous Lincoln miller for the Colt armory. It improved on Howe's 1848 design, which Lawrence must have brought to Hartford when Robbins and Lawrence opened their branch factory. This machine exemplifies the capacity of machinist networks to transmit technical expertise among widely separated machine tool centers. A significant advance could then emerge within the Hartford hothouse of innovation, where some of the greatest mechanics worked together—the mechanism of the artifact-activity couple writ large.

Fairfield and Billings had worked at the Robbins and Lawrence factory in Windsor. In 1856, when the firm's bankruptcy became evident, Fairfield moved to the American Machine Works in Springfield, Massachusetts; networks linked these machinery centers, so it was relatively straightforward for him to make this job change. The ties had been manifested two years earlier when the British Committee on Machinery, which canvassed American firearms firms, had visited the Springfield Armory, the nearby Ames Manufacturing Company, and the Robbins and Lawrence branch factory in Hartford and its main factory in Windsor. Their itinerary evolved that way because managers and top machinists at each place knew peers in the other places. In 1857 Fairfield joined his friend Billings at the Colt armory. Billings's move to Colt had been a simple step because Windsor and Hartford machinists maintained close bonds through the branch factory that Robbins and Lawrence had in Hartford.

Spencer apprenticed in the machine shop of the Cheney silk mills in Manchester, near Hartford, and he continued there as a journeyman machinist until 1853. His next move, to Rochester, New York, to work in a machine shop of a locomotive works, exposed him to techniques for building machine tools to make engines. Returning to Hartford, he took a position in the Colt armory, where he met Billings, and he also spent time working again at the Cheney silk mills. In 1860 Spencer settled down, embarking on a career inventing firearms, finally partnering with Billings in 1869 to manufacture drop forgings.

The machinists who came together around Root founded some of the most famous machine tool firms of the post–Civil war period, including Pratt and Whitney, Hartford Machine Screw Company, and Billings and Spencer. Their coming together, along with their talented peers, at Colt constituted one of the most extraordinary local communities of practice of the mid-nineteenth century.

The State-of-the-Art Factory

Constructed in 1854–55 and capitalized at one and a quarter million dollars, the Colt Patent Fire-arms Manufacturing Company in Hartford constituted the largest modern private arms plant in the world. Although Colt credited himself with the overall strategic vision for producing revolvers by machinery, Superintendent Root, the brilliant production system manager and machinist, deserves accolades for creating the state-of-the-art factory. Its fame swept through machinist networks, and it continued to attract leading machinists through the late 1850s. They sought the opportunity to work with the many talented machinists who continued to develop innovations in machine tools while learning directly how the integrated production system worked—the mechanism of the artifact-activity couple in operation.

These machinists gained incalculable benefits working with the renown Root, who continued to innovate machine tools even while serving as superintendent, further testimony to his brilliance. Among his many innovations, often designed in cooperation with other talented machinists, were the compound crank drop for die forging developed around 1850 (patented in 1858); the double-turret machine, which performed multiple operations on metal components (designed around 1852); the compound screw drop hammer (patented in 1853); an improved machine for boring cylinders of revolvers (patented in 1854); and an improved slide lathe for turning the taper and a compound rifling machine (both patented in 1855).

A writer for the *United States Magazine* visiting Colt's armory in 1857 provided a vivid description of its internal workings. It housed almost six hundred workers who turned out roughly 250 firearms daily. One of the largest industrial steam engines in existence, rated at 250 horsepower, drove the extensive inventory of machine tools. Befitting Root's experience with, and innovations in, drop hammer forging, a pistol went through thirty-two distinct forging operations. The second floor of the front parallel building, which contained almost four hundred machine tools, awed the writer most of all. Individual or sets of machine tools performed a bewildering array of specialized tasks, such as chambering cylinders, turning and shaping them, boring barrels, milling and drilling lock frames, boring and screw-cutting nipples, and making screws. Many of these tasks were subdivided into separate steps, often with their own machine tools. The lock frame, for example, went through thirty-three distinct operations, many carried out by different machines.

Colt had decided to branch into making machine tools for sale to producers of firearms. The first floor of the front parallel building housed a fully equipped machine shop, where about sixty workers turned out machine tools. Sales of the tools probably did not amount to much because the machinists had plenty to do equipping the factory and adding new, better machine tools. Hartford firms, such as the Phoenix Iron Works (owned by Lincoln), and other Connecticut Valley sources, such as the Ames Manufacturing Company and the Springfield Armory, also supplied machine tools to Colt. Under Root's supervision his machinists probably made multiple copies, subject to license agreements, of some of this equipment.

In spite of this factory's sophistication, the components of the firearms were not perfectly interchangeable by 1860. This profit-oriented business, which extensively used machine tools and gauges (to check for uniformity), aimed at achieving high productivity and low cost. Near-perfect interchangeability cost too much; only the federal armories could afford it. The hub position of the Colt firm in machine tool networks probably peaked by the late 1850s. Its leading machinists recognized that, once the factory had reached full development for firearms manufacturing, relative to the existing technology, it was time to move on to new challenges in other firearms firms or in other industries. Companies sought these gifted machinists because they had acquired experience building machine tools and incorporating them into production systems in one of the nation's most advanced factories.

The exodus from the Colt armory did not mean its end as a hub of machine tool networks, and the mobility of the top machinists such as Pratt, Whitney, Fairfield, Billings, and Spencer created additional firms as hubs. The concentration of many of these firms in the Hartford area added that city to the existing pivots of machine tool networks, including the Springfield, Providence, and Lowell–North Chelmsford areas. Under Root's leadership the Colt armory constituted a remarkable community of practice of integrated production systems and machine tools, and, through the Colt alumni and their networks, over subsequent decades that community permeated other industrial sectors.[4]

The Hubs of Private Firearms Networks

The active engagement of mechanics in advancing the sophistication of machine tools and in incorporating them into firearms manufacturing caused firearms and machine tool networks to overlap significantly by 1860. These networks had evolved on a base of private arms networks that concentrated in or near the Connecticut and Blackstone valleys as early as 1820. The leading firms were embedded in networks of machine shops that populated these prosperous farming areas, and the private firearms networks emerged before the Springfield Armory's ascendance as a clearinghouse for technical knowledge around 1815. Over the next four decades Springfield and the private arms producers built a highly networked community of manufacturers and of skilled machinists who moved among firearms firms, machine shops, and machinery firms in the region.

By 1860 few firms outside these networks broke into the top ranks of firearms manufacturing (table 8.2). Only twenty-four firms possessed either a capitalization or a product value greater than ten thousand dollars, and, of these, two-thirds manufactured firearms and one-third assembled arms. They concentrated in New England, which housed over 80 percent of firearms manufacturing (measured by employees, capital, or sales), and in Philadelphia and its satellite of Lancaster, a craft gunsmith center prior to 1820, which contained over 70 percent of firearms assembling. Yet major firearms producers clustered even more than this. Hartford, home of the Colt and Sharps firms, dominated manufacturing, New Haven contained the second largest concentration, and the Springfield-Chicopee area followed in third place. Worcester, at the northern end of the Blackstone Valley and a leading machinery center, also housed a major firm.

The Remington firm in Ilion, New York, a future giant in the firearms industry, seemingly stands as an exception to the highly concentrated hubs in New England (see table 8.2). Yet its modest success by 1860 and its transformation to large-scale production during the Civil War rested on network ties to the machinist hubs in the Connecticut Valley. Until the early 1840s Remington functioned as a small gun-making concern. In order to meet a government arms contract that Eliphalet Remington had bought, he took a trip in 1844 to New England to purchase machinery for firearms manufacturing. Among several firms he visited, the Ames Manufacturing Company in Chicopee ranked as the most important. It supplied him with a substantial amount of machine tools and other machinery for firearms production. He also hired William Jenks, the inventor of a carbine. As production at Ilion increased, Remington realized he needed a sophisticated superintendent of manufacturing. Again, he headed to the Connecticut Valley and hired Fordyce Beals, a machinist and inventor of a revolver,

TABLE 8.2
The Hubs of Firearms Networks, 1860

Hub/Firm	No. of Employees	Capital ($)	Sales ($)
Manufacturers			
Springfield-Chicopee, Mass.			
Massachusetts Arms, Chicopee	75	70,000	112,000
Smith and Wessen, Springfield	57	40,000	75,600
Springfield Arms, Springfield	24	42,000	27,000
Hartford, Conn.			
Colt	369	1,250,000	1,050,000
Sharps Rifle	300	660,000	325,000
New Haven, Conn.			
Bliss	16	8,000	15,000
New Haven Arms Company	138	50,000	25,000
Stafford	9	6,000	18,000
Whitney	70	50,000	50,000
New London, Conn.			
Bacon	40	20,000	50,000
Worcester, Mass.			
Allen and Wheelcock	130	50,000	100,000
New York City Area			
Marston, New York City	60	10,000	35,000
Manhattan, Essex County, N.J.	6	32,000	50,000
Ilion, N.Y.			
Remington	75	15,000	60,000
Philadelphia			
C. Sharps and Company	87	31,987	82,500
Bolton, Pa.			
Henry	6	14,000	900
Total	1,462	2,348,987	2,076,000
Assemblers			
Philadelphia			
Geisinger	15	2,000	30,000
Justice and Steinman	10	10,000	25,000
Krider	27	10,000	25,000
Lamb and Clarke	6	500	10,000
Tryon	25	25,000	25,000
Lancaster County, Pa.			
Leman	62	30,000	40,000
Erie County, Pa.			
Brown and Tetley	25	21,000	23,000
Erie County, N.Y.			
Smith, Patrick	4	13,000	6,500
Total	174	111,500	184,500

SOURCE: Robert A. Howard, "Interchangeable Parts Reexamined: The Private Sector of the American Arms Industry on the Eve of the Civil War," *Technology and Culture* 19 (October 1978): 638–39, table 1.

who worked at the Whitney armory. Remington thus forged network ties to major firms and machinists working on improving firearms manufacturing and machine tools.

About 1856 the Remington firm again enlarged its plant facilities and added a new foundry, but four years later it remained small. During the Civil War, however, the firm acquired big government contracts and changed dramatically. It installed large steam engines to power the substantial increase in machine tools and machinery for producing firearms and other armaments, and it enlarged its buildings and erected new ones. In 1862, in typical fashion, Remington drew on its network ties to the Connecticut Valley and hired Charles Billings, who had apprenticed at Robbins and Lawrence in Windsor and trained under Howe and then worked as a machinist at Colt's armory. At that job he had absorbed the technology of drop hammer forging and the integrated machine tool manufacturing system that Root had designed. Billings supervised the installation of drop hammer forging at Remington and stayed there until the war ended, thus transferring this technology through the mechanism of the artifact-activity couple. He also must have passed to Remington the innovations in machine tools which had been developed (or copied) by Colt's machinists and shared the strategy of using specialized machine tools for each process.

During this period the scale of the Remington factory complex approached Colt's state-of-the-art factory, employing about six hundred to seven hundred workers. Remington's physical separation from the firearms and machine tool hubs in New England obscured its strong network ties, which allowed it to draw on the talent and technology of leading firms and machinists. Its acquisition of the architectural knowledge of successful firearms manufacturing set the stage for its emergence as one of the nation's most important arms producers.

Only the Colt and Sharps firms in Hartford possessed sufficient capital to operate large, mechanized firearms factories filled with numerous machine tools (see table 8.2). The Massachusetts Arms in Chicopee, the New Haven Arms and the Whitney factories in New Haven, and the Allen and Wheelcock firm in Worcester invested significantly in machine tools, but the Hartford behemoths dwarfed them. Although the remainder of the manufacturers used machine tools, their smaller investment suggests that most of them did not run integrated, sequential production lines. In Philadelphia, for example, C. Sharps and Company operated powered machinery and machine tools, most of which they built in-house. Their simple production processes bore little resemblance, however, to the sophistication of Colt's armory.

By 1860, therefore, firearms manufacturing utilizing large, integrated production lines and machine tools remained in its infancy. The situation quickly changed under government pressures arising from Civil War demands, as arms firms such as Remington expanded dramatically and metalworking firms such as the Providence Tool Works added large firearms production facilities. The rapid addition of sophisticated machine tools and expanded production could occur because machinist networks facilitated the quick assembly of talent to build machine tools for arms manufacturing.[5]

THE WIDER CIRCLE OF MACHINE TOOL HUBS

Many iron foundries, locomotive works, textile machinery firms, and general machine shops built machine tools for their own use, and some also sold them to other companies. Several agglomerations of firms in these sectors, including Worcester, Providence, and Philadelphia, constituted hubs in machine tool networks, and a few other places also contributed to the emerging machine tool industry.

Worcester

Since the 1820s Worcester and its surrounding county, of the same name, had been a major iron foundry, machine shop, and textile machinery center. Its firms and machinists were embedded in networks that encompassed Boston and its environs, Providence and vicinity, and the Connecticut Valley. These networks gave Worcester's machinists access to the latest technical knowledge regarding machine tool advances. Some of the area's machine shops must have made machine tools, especially engine lathes, for their own use prior to 1840. Most of this equipment was hand- or horse-powered. During the 1830s and early 1840s iron foundries, locomotive works, textile machine shops, and firearms factories increased their deployment of water- and steam-powered machine tools. Over the next two decades Worcester's many skilled machinists formed shifting partnerships to build machine tools, yet most of them also built other equipment because markets remained too small to support specialization.

Between William Wheeler's first partnership in 1825 and the early 1830s, he acquired a checkered history in the iron foundry trade, moving among various towns, starting and exiting businesses. After reconstituting his foundry work in Worcester, he built a diversified business during the rest of the decade, making, for example, heavy gearing, plow castings, and boiler doors. Around 1840 he designed his first machine tool, a boring machine, and over the next two decades he introduced innovations in other heavy metalworking equipment. While maintaining a diversified foundry business, the William A. Wheeler Iron Foundry focused on selling the heavy machine tools—boring machines and trip and drop hammers—for which its founder contributed innovations.

Samuel Flagg, another early machine tool builder, moved from nearby West Boylston to Worcester and opened a machine shop around 1840. One of the first machine tools he sold was a turning lathe for Wheeler's foundry. Flagg concentrated on building hand and engine lathes, but a sales slowdown during the economic contraction of the early 1840s led Thomson, Skinner and Company to take over Flagg's business in 1845. Within two years he started anew in a partnership with three other machinists, including his former apprentice Lucius Pond. They probably worked mainly on engine lathes, among the most widely purchased machine tools at this time. Eventually, Flagg left the firm, and sometime before 1860 Pond bought out his partners, setting the basis for the Pond Machine Tool Company, which specialized in heavy engine lathes and became one of the predecessors of Niles-Bement-Pond, among the nation's most prominent machine tool companies of the first half of the twentieth century.

The economic upturn that commenced after 1843 encouraged additional Worcester machinists to begin manufacturing machine tools. Samuel Coombs, a machinist in the local loom manufacturing firm of Phelps and Bickford, led the formation of the partnership of S. C. Coombs and Company. This firm and its successors, which continued after 1860, manufactured lathes and planers. Pierson Cowie started the manufacture of iron planing machines, and within about a year the firm of Woodburn, Light and Company had taken over the business. By the mid-1850s the successor firm employed seventy mechanics, and, along with diversified machinery production, it made an array of powered machine tools, including lathes, planers, slabbing machines, upright drills, and boring mills.

Around 1845 Alexander and Sewall Thayer started a shop to manufacture engine lathes, employing about ten mechanics; Alexander had innovated improvements in the existing machine. In the mid-1850s the firm became Thayer, Houghton and Company, and by the early 1860s it had about 150 employees. Along with diversified machine shop work, they made a wide array of powered machine tools, including lathes, planers, bolt-cutting machines, boring machines, gear cutters, and punching machines. These firms and others propelled Worcester into a prominent hub position in machine tool networks, and their sheer number and shifting ownership patterns confirms that the city had developed a vibrant community of practice in machine tool building.

Providence and Vicinity

Since 1800 the firms of Providence and vicinity ranked among the leaders in the key sectors of textile machinery and steam engines, making the area one of the greatest hubs of machinists. Because many of the firms built their own machine tools, they also could produce them for sale. They acquired knowledge about the latest technical improvements in machine tools from firms throughout New England, the New York City region, and Philadelphia. The sophisticated metalworking capabilities of so many businesses, however, provided few opportunities to sell machine tools to nearby companies before 1860. Arguably, the James S. Brown Machine Works in Pawtucket, whose renown dated from the early 1820s, when it operated as the partnership of Pitcher and Gay, ranked as the premier machine tool manufacturer during the last two decades of the antebellum.

As early as 1820, Brown designed his first improvement in machine tools, the addition of a slide rest to the turning lathe, and a decade later he designed a gear cutter for bevel gears. After Pitcher retired in 1842, Brown acquired full control of the firm. During the economic expansion he enlarged it, and textile machinery seems to have been the primary product line. In 1847 he added an iron foundry, and two years later he built a large machine shop. These changes advertised that the firm had become a significant participant in machinery markets. Brown supervised the construction of most of the shop's machine tools and contributed improvements to turning lathes and fluting engines, among others. Although textile equipment constituted most of its sales, the James S. Brown Machine Works probably also sold small numbers of machine tools. Along with steam engine works and other textile machinery firms, however, the firm faced a limited market for machine tools in Providence and vicinity. Yet it possessed the capacity to enlarge its production of machine tools for sale quickly. Shortly after the Civil War commenced in 1861, Brown's company shifted much of its productive capacity to making machine tools for gun manufacturers.

Fales, Jenks and Sons in Central Falls was also a major textile machinery manufacturer. Formed in 1830, it was manufacturing rotary pumps three years later and added a foundry in 1859 as well as other buildings after that. Like the Brown Machine Works, Fales, Jenks and Sons probably made a few machine tools for sale, but the firm used most of its tool production. In 1861 it built a large number of milling machines for gun makers and sewing machine manufacturers and, over the next two years, added a large machine shop. In contrast, Providence's W. T. Nicholson mainly manufactured small equipment—perhaps specialized lathes or milling machines, among others—for jewelers and silversmiths, and the firm turned out other machinists' equipment, such as vises and levels. After the start of the Civil War, Nicholson shifted quickly to making machine tools for rifling and milling which he sold to gun makers.

Firms in other industries besides textile machinery also produced modest quantities of machine tools for sale. The Thurston, Gardner and Company's most important product line consisted of steam engines, although, like other engine works, it carried on a diversified foundry business. It probably sold small amounts of heavy machine tools, which likely consisted of lathes—the chief machine tool in foundries and steam engine works—as well as planers, when these tools were perfected after 1840. Starting in the mid-1840s, it expanded to meet the sharply rising demands for industrial machinery. Like firms in that sector, it purchased machinery in Europe such as the Nasmyth steam hammers and heavy lathes and planers. These purchases upgraded the heavy machine tool expertise of Thurston and Gardner because some of this equipment, which was not patented, could be copied and sold. By the mid-1850s machine tools constituted a product line that the firm advertised alongside its steam engines, boilers, and sugar mills, which included steam mills.

By the late 1840s the foundry and steam engine works of Corliss, Nightingale and Company ranked among the leaders in employing sophisticated machine tools to build heavy machinery. Like other foundries, the firm built lathes and planing machines for its own use and for sale, and it also cut gear teeth and turned and cut bevel wheels. The Corliss company's attention to making and employing the latest machine tools to build steam engines and its efforts to sell these tools constituted its notable feature. In 1849, besides Corliss's patent for an automatic variable cut-off engine, he also received a patent for a gear-shaping machine. As the firm's annual production of steam engines rose, it continued to contribute machine tool improvements.

Although the Providence Tool Company sold few, if any, machine tools, this firm and its machinists operated as a hub in machine tool networks. The proprietors started the company in 1845 to manufacture small tools for carpenters and mechanics, and, initially, it employed forty to fifty workers. It steadily enlarged its buildings and built new ones, and at some point in the early 1850s it received a large infusion of capital from the Borden family in Fall River. The Bordens had a sizable investment in the Fall River Iron Works, one of New England's most important foundries outside of Boston's immediate environs, and they had interests or full ownership in other Fall River businesses, such as textile firms, a steamship line, a bank, and a gas works. The Fall River and Providence investors embarked on a large expansion of the Providence Tool Company, including building a foundry. They added heavy machine tools, including a twenty-five-ton punch press and a one hundred–horsepower steam engine to run the equipment. The product line—consisting of diverse hardware such as nuts, washers, chains, plate hinges, bolts, and other items used by locomotive, car, and shipbuilders as well as by machinists in many other industries—widened too.

The Providence Tool Company used sophisticated, specialized machine tools to turn out its diverse product line. Given that a firm such as this made most of its own equipment, one can assume that it employed an unusually talented groups of machinists. In 1861 the owners enticed Frederick Howe to be superintendent by offering him a rich financial package, appropriate for one of the nation's greatest machine tool innovators (see map 8.1). The timing of his arrival at the start of the Civil War enabled the firm to build a huge armory for manufacturing armaments, including sabers, bayonets, and muskets, and, with its large foundry capacity, it supplied gun barrels to other firearms makers. Utilizing Howe's expertise, the company made many of its machine tools for armament manufacturing, and its machinists contributed to further innovations in milling machines, an area in which Howe had gained fame. The same year he arrived in Providence, he asked Joseph Brown of the firm of Brown and

Sharpe to design the equipment that would become the Universal Milling Machine, and evidence suggests that Howe provided ideas about its design. Within two years the Providence Tool Company built a simpler, improved plain milling machine, most likely designed by Howe, and the firm licensed Brown and Sharpe to manufacture it. Thus, for about a decade following the mid-1850s the Providence Tool Company constituted a hub in machine tool networks.

Philadelphia

The migration of talented machinists—Edward Bancroft and William Sellers from Providence around 1848 and William Bement from Lowell in 1851—to Philadelphia in order to build machine tools for sale indirectly confirms the limited opportunities that existed to sell machine tools in New England. Many iron foundries, textile machinery firms, and firearms manufacturers possessed in-house capacity to build machine tools for their own use, and the large number that could build this equipment for others limited most firms to a few sales. In contrast, the Philadelphia area possessed few textile machinery firms and firearms manufacturers; thus, local machinists had little experience building lighter machine tools such as milling machines. By the early 1840s the heavy capital equipment firms (iron foundries and locomotive works) still used a small number of machine tools such as lathes, planers, and borers, and the leading machinists in most of these firms could make the tools. The swelling output of heavy capital equipment from local firms during the economic upturn following 1843, however, opened opportunities for building machine tools for sale.

The choice of Bancroft and Sellers to relocate from Providence to Philadelphia and to found their firm in 1848, therefore, exhibited brilliant strategic timing. They must have utilized their social and machinist networks to assess that the time was propitious to shift their base of operations. The ongoing mobility of machinists, which dated from at least 1810, kept the Providence machine shops in close contact with Philadelphia. Bancroft possessed strong ties to John Poole, his brother-in-law and traveling companion to New England metalworking establishments during the 1830s. Poole owned a prominent machinery firm near Wilmington, and firms in that foundry and machinery center had close connections with Philadelphia's metalworking firms. Sellers's numerous ties to Philadelphia reached through the extended Sellers family, which included members of the social and business elite of the city and surrounding area. The family's distinguished members had been active in engineering in Philadelphia for decades. The firm of Bancroft and Sellers, therefore, rested on sturdy foundations. The partners were brilliant machinists, and, as they expanded the firm, they drew on their network bridges to the top machine shops in the Philadelphia area and elsewhere in the East, becoming pivotal members of the community of practice of machine tool building which existed by the late 1840s and early 1850s.

In 1835 William Bement started out as an eighteen-year-old apprentice in the textile machinery firm of Moore and Colby in Peterborough, New Hampshire. The move placed him in the highly networked textile machinists of that area, which included numerous small mills and shops and the large textile complexes at Manchester and Nashua. Within two years he had risen to the position of foreman of the machine shop, and he soon became a partner when another one departed. In 1840 the twenty-three-year-old took the position of foreman and contractor in the machine shop of the Amoskeag Manufacturing Company, working under Superintendent William Burke, who would leave within five years to take over the Lowell Machine Shop. This network tie led Bement to a career opportunity that in time propelled him to the top rank of the nation's machinists. He stayed at Amoskeag two years, when he was offered a job as head of a textile machinery shop in Mishawaka, a budding industrial center near South Bend, Indiana; his relationship with Burke probably had something to do with it. The firm soon fell into difficulties, but Bement moved thirty-five miles northwest to a factory city in Michigan, and quickly became head of the machine shop of the St. Joseph Iron Foundry.

Burke must have kept track of Bement because, as soon as Burke became superintendent of the Lowell Machine Shop, he offered Bement a position as designer of the new line of machine tools that the Lowell shop would start selling. The extraordinary reputation of the shop presented a unique opportunity that would vault this twenty-eight-year-old to the top of his profession. He supervised the machine tool designing for three years and continued at Lowell until 1851, when Elijah Marshall, owner of a machine shop in Philadelphia which engraved rolls for calico printing, offered him a position. This job offer probably passed through several channels. Bement's leadership in designing the machine tool line would have been communicated to Philadelphia by machinists, many of whom had worked at the Lowell Machine Shop at some point. Burke's prominent hub position in the East's machinist networks gave him access to information about jobs. Because the Lowell mills engaged in calico printing, he knew which firms specialized in engraving for that printing process.

Marshall offered the thirty-four-year-old Bement a partnership, he accepted, and in 1851 the firm of Marshall, Bement and Colby commenced the manufacture of machine tools, operating in the midst of one of the most rapidly growing markets in the East, the burgeoning capital goods sector of the Philadelphia area. Shortly thereafter, Marshall and Colby retired, and Bement joined with an expert iron founder named Dougherty. During Bement's eleven years of work, from Amoskeag to his arrival in Philadelphia, his rapid upward career mobility had been lubricated by William Burke's behavior as a mentor and virtuous bourgeois.

The arrival of Bancroft, Sellers, and Bement in Philadelphia between 1848 and 1851 directly transferred technical skills from the Providence and Lowell areas, leading hubs of machine tool manufacturing, to that city. These hubs also incorporated the expertise in machine tool building from the communities of practice in Worcester and in the Connecticut Valley because these centers were connected. The New England nests of builders possessed skills in making heavy machine tools (such as lathes, planers, and boring machines) and lighter ones (such as drills, bolt-cutting machines, and milling machines). Although heavy metalworking equipment would be in greater demand in the Philadelphia area, the firms of Bancroft and Sellers—which became William Sellers and Company after Bancroft died in 1855—and of Bement and Dougherty also had the capacity to build light machine tools.

Bancroft, Sellers, and Bement thus served as network bridges between Philadelphia and the New England hubs, becoming central actors in enhancing the community of practice of machine tool building in Philadelphia. During the 1850s Philadelphia's locomotive works, such as Baldwin, became markets for their machine tool firms, and the large number of iron foundry and machinery works also generated business. After 1857 Sellers commenced a prolific decade of innovating machine tools which solidified his firm's stature. At the end of that period the Philadelphia booster Edwin Freedley could justly proclaim that the city's machine tool builders had achieved fame. Yet even the premier firms of William Sellers and Company and of Bement and Dougherty did not specialize in machine tools. By 1860 that equipment accounted for less than half of their total sales, and most of their output consisted of various machinery.

Firms Outside the Centers

Aside from firms in Worcester, Providence, Philadelphia, and the Connecticut Valley, other firms also produced machine tools for sale. John Gage began manufacturing machine tools in Nashua in 1837. He operated near textile machinery firms such as the Lowell Machine Shop and the Gay and Silver shop, both of which built their own machine tools and made some for sale. Like most firms, Gage's shop turned out various products besides machine tools. In 1851 he entered into the partnership of Gage, Warner and Whitney. The new partners must have contributed substantial capital because the firm built a shop the following year and began to manufacture stationary steam engines. Nonetheless, the machine tool business grew significantly, and by the mid-1850s the firm employed about seventy-five workers. Its product line included iron planers, light to heavy lathes, chucking lathes, bolt-cutting machines, drills, slabbing machines, gear-cutting machines, and light to heavy punching machines. The company's experience as a producer of both light and heavy machine tools, like that of some of its regional peers, provides a cautionary note about dividing production into a light machine tool sector in New England, and a heavy one in Philadelphia. Similarly, the Meriden Machine Company in Connecticut began as a small shop in 1839. Following various ownership changes, the firm expanded its business around the mid-1840s, as an economic expansion was under way. Over the next decade it enlarged machinery manufacturing and increased machine tool production for sale. The surrounding subregion of Hartford, Middletown, and New Haven was a hub of light machine tool innovation and manufacture, including the production of milling machines for the firearms industry. The Meriden company built a business spanning light and heavy industrial machine tools. By the mid-1850s its substantial machine shop and iron foundry employed 120 workers, and it sold $120,000 worth of machinery and machine tools annually. Many firms, including firearms manufacturers such as Whitney, Colt, and Sharps, built their own tools; thus, it would be reasonable to assume that most of the Meriden company's sales consisted of machinery rather than machine tools.[6]

SETTING THE BASE

Machine tool networks did not occupy a unique status in machinist networks. Instead, they constituted one more component that overlapped with the other networks of iron foundries, locomotive works, textile machinery firms, general machine shops, and firearms manufacturers. This complementarity was possible because most of the larger firms in these metalworking industries built their own machine tools. Although this equipment had unique design characteristics, in fact, machine tools constituted variations on the larger category of machinery, and know-how trading through machinist networks spread the technical skills necessary to build this equipment.

Even with the expanded use of machine tools which accompanied the economic upturn following 1843 and with the substantial jump in machine tool production during the next decade, a specialized machine tool industry had not emerged by 1860. A few small firms specialized in machine tools, and Worcester housed a number of them, but, if orders declined, they shifted to general machine shop work. For at least three decades after 1860 only a small number of industries—such as sewing machines, typewriters, and bicycles—had firms that submitted sizable orders for machine tools; thus, no large firm could specialize in them and survive. Even the machine tool sales of the Brown and Sharpe Company, one of the greatest producers of that equipment, did not exceed half of its total sales until at least 1900.

The absence of specialized machine tool firms did not indicate a primitive stage of development. Before 1860 a highly networked group of machinists achieved significant advances in machine tool building, often drawing on British designs. They built sophisticated machine tools, and their networks transmitted technical skills to the leading capital equipment producers. The centers of metal goods manufacturing—from iron foundries, steam engine works, and locomotive plants to textile machine shops and firearms factories—confronted related problems of shaping and cutting metal. Machine tool builders in these centers therefore faced a broad set of similar demands, what the economic historian Nathan Rosenberg called "technological convergence." The rapid diffusion of technical skills through personal connections and mechanic mobility contributed to the substantial advances in machine tool building, and this know-how trading forged a community of practice in that sector which spanned the leading centers of machine tool manufacturing. Thus, even though many firms disappeared and mechanics sometimes had short careers, advances in machine tool technology continued because the necessary technological knowledge and skills were embedded in networks; they were not solely the property of mechanics and firms.[7]

CHAPTER NINE

Machinists' Networks Forge the Pivotal Producer Durables Industry

> The contriving and making of machinery has become so common in this country, and so many heads and hands are at work with extraordinary energy. . . . As regards the class of machinery usually employed by engineers and machine makers, they are upon the whole behind those of England, but in the adaptation of special apparatus to a single operation in almost all branches of industry, the Americans display an amount of ingenuity, combined with undaunted energy.
>
> *"Report of the Committee on the Machinery of the United States of America," The American System of Manufactures*

The British experts who visited the United States during the early 1850s to evaluate its firearms manufacturing acquired ideas and machinery that would assist in establishing Britain's Enfield Armory, but, more important, they offered perceptive insights about the state of technology in the East, where they concentrated their visits. They witnessed the widespread use of machinery and learned about the sector's rapid expansion; nevertheless, near the end of the antebellum the East could not be characterized as fully industrialized. The share of equipment in the nation's domestic capital stock and the amount of equipment per capita had risen by the 1840s, but, on each measure, the level reached paled next to its value in 1900 (see fig. I.1). These astute observers saw a dynamism—they called it "energy," as workers busily transformed industrial practices and built machinery. Although most of the machine tools outside the firearms industry did not measure up to British standards, they found them in use in various industries. They were most astounded, however, by the deployment of an amazing variety of machinery and machine tools designed for highly specialized operations.[1]

In a range of metalworking sectors—iron foundries and steam engine works, locomotive works, textile machinery firms, general machine shops, firearms manufacturers, and machine tools—machinists operated in networks structured around hubs of machinists and firms. Within each sector these hubs connected to one another and to lesser concentrations of machinists, and many of the networks intersected across sectors. To be sure, some technical knowledge was not shared or control over patents restricted its dissemination, but even patents could spread technology through licensing, a lucrative practice of patentees. These machinist networks made possible the energetic building of machinery and machine tools.

In all of the sectors leading machinists were connected to one another through friendship, family, and job experiences, and the widespread practice of these top mechanics to move among machine shops early in their careers forged network ties that they employed for the remainder of their work lives. The top machinists possessed bourgeois virtue. Senior machinists supported one another's career moves by passing along information about job openings and by providing recommendations. They mentored promising apprentices and junior machinists by offering them positions and then recommending them for jobs elsewhere which offered opportunities for career advancement. The mechanism of the artifact-activity couple structured machinists' learning environment. This close bond between the artifacts (machines) with which they engaged and the skills and knowledge that they brought to bear when they worked on the machines and communicated with one another about that effort required on-the-job training and visits in order to acquire technical skills and innovations. Experienced machinists willingly shared their knowledge, and they opened their shops to visiting mechanics. Licensing agreements transmitted patented inventions, but even in these cases machinists engaged in extensive contacts to implement their inventions.

During the antebellum the machinist profession attracted young men from the upper classes of American society and the professional groups. The bustling metalworking firms operated at the leading edge of technological change, and participants confronted intellectual and technical challenges. The social, political, and economic networks of the East's elite provided conduits for the transmission of technical skills and innovations. The profession of machinist also attracted the upwardly mobile residents of prosperous farms, villages, and small towns because machinists received high wages. Along with their elite peers, the excitement and challenges of the new metalworking industries enamored them.

Young, aspiring machinists often learned technical skills while working as apprentices in machine shops near their homes, but, after completing their training, typically between the ages of eighteen and twenty-one, many moved to distant shops to find work. Machinist networks provided the information about these job opportunities, and the high mobility of machinists meant that new or expanding businesses that required machinists could attract them from distant places; they offered high wages to entice them to come. Machinists ranked among labor's industrial elite. The best machinists often became foremen of divisions of large machine shops by their early-to mid-twenties, and they reached the position of superintendent in charge of the entire machine shop between their late twenties and early thirties. During the 1820–60 period their annual pay packages, which sometimes included free housing, ranged from one thousand to three thousand dollars, and in a few cases they earned as much as five thousand dollars, making them among the highest-paid employees in the nation. At times the superintendent's position came with an equity stake of 20 to 50 percent in the firm, moving them into the ranks of the well-to-do.

COMMUNITIES OF PRACTICE

The pervasive know-how trading of technical skills and innovations among machinists and firms contributed to the emergence of communities of practice within each industrial sector—iron foundries and steam engine works, locomotive works, textile machinery firms, general machine shops, firearms manufacturers, and machine tools. Larger practicing communities encompassed multiple industrial sectors as some machinists moved across them during their careers. Likewise, their professional networks based on previous experiences in the same firms; on referrals from relatives, friends, and acquaintances; and on job mobility gave them access to diverse knowledge regarding technical skills and innovations, thus contributing to the creation of wider communities of practice. Some firms became so prominent that they acted as hubs of machinists networks, but the emergence of major hubs of multiple firms had greater consequences. Some of these hubs consisted of firms from one industrial sector, whereas others had firms from several sectors. The extraordinary level of interaction among machinists through their networks at the subregional, regional, and interregional scales (which encompassed the East) during the antebellum meant that the spatial scale of these networks often exceeded the market areas of the firms for which the machinists worked. This suggests that the network structure and behavior of the machinists facilitated the rapid transformation of the pivotal producer durables during the antebellum.

For most of the twentieth century and continuing into the twenty-first century the iron foundry has been viewed as a relic of a bygone era, and its activities are considered so opaque that its importance has been dismissed. Yet in the antebellum foundries arguably served as the earliest pivots of machinist networks and the incubators of some of the most important industries of the nineteenth century, especially steam engine building (and closely related marine engine works), locomotive works, and heavy industrial machinery. They contributed to the development of the machine tool industry, principally through their work on heavy industrial lathes, planers, and boring machines. They also built light machine tools such as milling machines, but they were not leaders in designing and building this equipment. The foundry processes were so material to the producer durables industries that major firms in sectors such as textile machinery and firearms added foundries to their production processes.

The East's iron foundries were concentrated in the metropolises of Boston, New York, and Philadelphia and in their industrial satellites. The involvement of the social, political, and economic elite (merchant wholesalers and financiers) in setting up foundries made these firms pivots of knowledge networks about technology. They paid high wages to attract the best machinists, and their employees developed extensive networks among themselves through which they engaged in know-how trading. Each of the metropolises and their satellites developed a community of practice of foundry work, yet the broadly similar capacities of these firms to build equipment suggests that the more important practicing community extended across the East. Some foundries achieved more success than others, and only a few made a profitable transition to locomotive manufacturing. Nevertheless, the major, as well as quite a few smaller, foundries could build a wide range of industrial machinery. This confirms the effectiveness of the machinist networks that bound foundries, and these networks acted as conduits—through job mobility, visits, and personal communications among machinists—for know-how trading of technical skills and innovations.

From the start iron foundries served a broad array of industrial sectors, whereas the textile machinery industry, which developed in general machine shops or in cotton mills, possessed a narrower focus. Nonetheless, even machine shops seldom specialized solely in textile machinery, and the larger, as well as some of the smaller, producers of textile equipment created linkages to other industrial sectors. The early, close association of textile machine shops with the establishment of cotton mills set the framework for textile machinery networks by the first two decades of the nineteenth century, and these networks persisted, in remarkable fashion, for the rest of the antebellum. The Providence hearth of textile machinists which originated with Samuel Slater and the Wilkinson family of machinists spawned local networks of textile machinists in New England and New York state, mostly through the mobility of mechanics. Once these local networks of textile machinists formed, they expanded nearby through the transfer of technical skills by means of apprenticeships and know-how trading. Secondary clusters of cotton textile machinists emerged near New York City and in the Philadelphia area. The personal relations among machinists and their mobility created and maintained bridges across these local networks, and this generated a larger community of practice of textile machinists. During the 1820s cotton textile boom the principal nests of textile machinist networks solidified in Providence and vicinity, Boston's environs, New York City's environs, and the Philadelphia area.

These nests constituted the primary training grounds for textile machinists, but their networks within each nest were even more significant. Much of the skills in building textile machinery came from general machinist skills, and most of the machine shops produced other types of machinery. Textile machine shops, therefore, trained many machinists who left to work in other industrial sectors. Know-how trading among the nests of textile machinists contributed strongly to the pervasive network links among the major machinist centers. Textile machinist bridges connecting these centers served as paths along which machinists crossed industrial sectors, such as moving from the Lowell Machine Shop to a Paterson locomotive firm or to a Philadelphia iron foundry.

The rise of locomotive builders reveals the broad intertwining of machinist networks and their intersectoral character. Early locomotive builders emerged, especially, out of foundries and textile machinery firms. As with most of the machine-building sectors, firms did not specialize in one line, such as locomotives. Engine building drew on machinist skills across a range of industrial sectors, and the networks that bound firms in different sectors acted as conduits for the transfer of technical skills and innovations in building locomotives. Besides the large locomotive producers, many foundries and machinery firms trained locomotive machinists. The Lowell Machine Shop, the Matteawan Company, and the Providence machine shops were prominent trainers, even though they (with the exception of the Lowell Shop) built few locomotives. As with other machine sectors, a few hubs—Philadelphia, Paterson, and the Boston region—came to dominate locomotive building, and their machinists and firms created local communities of practice. The more important practicing community of locomotive builders, however, encompassed machinist networks throughout the East, because leading machinists in railway repair shops and the chief engineers of railroads networked with their peers in locomotive firms. Furthermore, the Lowell Machine Shop, the Matteawan Company, and the Providence machine shops continued to provide skilled machinists for locomotive firms even after the three engine hubs dominated the industry.

The machinists in firearms networks developed their own community of practice around gun making, a narrow product line compared with most machine-building sectors. The Ordnance Department of the federal government significantly influenced the structure of firearms machinist networks, primarily after its reorganization in 1815. It ran the Springfield and the Harpers Ferry armories, and the senior officers in Washington used their control over contracts to private armories to shape the development of firearms manufacturing. These officers pursued a system of uniformity—a broad program to improve the organization and manufacture of armaments; it was not a quixotic effort to achieve perfect interchangeability.

The department exerted strong leadership and worked with the superintendents of the federal armories to facilitate the exchange of machinists among the armories and between them and the private arms contractors, most of whom operated in Massachusetts, Connecticut, and Rhode Island. Harpers Ferry had limited network ties with machinists in nearby Pennsylvania and Maryland, and most of its relations focused on the Springfield Armory. It was tightly integrated with private armories in New England through visits of top machinists to observe technical practices and machinery innovations, through temporary exchanges of machinists, and through job-hopping mechanics. Springfield, similar to the private armories, retained network ties through personal contacts and machinist job mobility to textile machine shops, general machine shops, and iron foundries. Consequently, even though firearms constituted a narrow product line, the networks interwove with larger machinist networks.

The machinist networks that bound the federal armories and the private firearms contractors created a community of practice in building light machine tools such as milling machines, and the Ordnance Department offered strong encouragement for this effort to improve metalworking equipment. John Hall at Harpers Ferry and the leading machinists at the Springfield Armory made noteworthy contributions to advances in machine tools which they shared among the federal and private armories. Machinists at private armories such as Simeon North's in Middletown (Connecticut) also added improvements in machine tools. By the late 1840s, however, private firearms manufacturers such as Robbins and Lawrence, as well as Colt, increasingly took the lead in machine tool advances. Their leading machinists had trained at manufacturing firms across a range of industrial sectors, including general machine work and textile machinery, as well as at private firearms manufacturers and the Springfield Armory.

Innovations in heavy machine tools such as large lathes, planers, and boring machines also came from iron foundries, steam engine works, locomotive builders, and textile machinery firms. Machinists who worked on machine tools moved among all these industries. As machine tool output increased during the end of the antebellum, firms that became leading manufacturers of this equipment (though not specialists yet) networked through their leading machinists with most of the industrial sectors. Machine tool manufacturing, therefore, became an integral part of the late antebellum industrial economy.

The communities of practice within each industrial sector—iron foundries and steam engine works, locomotive works, textile machinery firms, general machine shops, firearms manufacturers, and machine tools—intertwined across sets of sectors and even across all sectors, as in the case of machine tools. Networks often spanned larger territories than the sales areas of the products because personal relations and mechanic mobility structured the machinist networks. At the largest scale we can speak of a community of practice of machinists in the East.

THE UNIMPORTANCE OF INTERCHANGEABILITY

Perfect interchangeability in manufacturing, defined as the capacity to exchange the same component randomly among products without adjusting or fitting, had been considered a hallmark of the antebellum years, but that achievement has been thoroughly discredited. Even the standard of interchangeability kept changing as gauging and measurement became more precise. The United States military led the quest for interchangeability, but it aimed its program more toward a system of uniformity. Between 1815 and 1849 the Springfield Armory made substantial progress toward the goal of uniformity, and by mid-century it appears to have achieved practical interchangeability; arms parts could be interchanged with only modest adjustments. Nevertheless, this costly approach entailed an elaborate mechanized production system with careful and precise manufacture of each part through the use of a rational jig and fixture system and a rigid gauging system to check dimensions of the components. The innumerable gauges themselves were costly to manufacture, and any change in parts required new gauges.

From the 1820s to the 1850s only the United States military had the resources to implement this approach to interchangeable manufacturing fully. Leading private arms makers that produced for consumer markets did not follow the federal armory practice of interchangeability because it was not cost-effective. These firms extensively used sophisticated machine tools to make high-quality arms, but, rather than making each part identical, they made it as uniform as possible and focused their skilled work on the final fitting process. For all the acclaim heaped on Colt as the world's leading mechanized producer of revolvers, with state-of-the-art machine tools in his 1850s Hartford factory, the parts of the Colt revolvers did not interchange. The high fixed cost of Colt's armory practice limited its utility in the late antebellum to firms that had complex metal products and the potential for large production runs to cover the high fixed costs. Sewing machine manufacture, one of the few late-antebellum industries to employ armory practice, did not follow the strict form such as that implemented at the Springfield Armory, where they kept final filing and fitting to a low level by the 1850s. Instead, the highly mechanized armory form as practiced at Colt's Hartford factory, which allowed for more final filing and fitting, provided the model.

The effort to use special machines and machine tools, as well as to achieve uniformity in production of components, constituted a broad effort before 1860. The Baldwin Locomotive Works, which became the nation's premier engine builder, exemplifies this effort. Its attempts to introduce interchangeability in production began around 1839, when the firm tried to respond to the demands of customers for replacement parts that they could use without extensive machining. These efforts continued erratically into the early 1850s, but they involved only a few components; simple gauges sufficed as measuring devices. Progress to achieve interchangeability ratcheted higher after Charles Parry became superintendent of the Baldwin works in 1854, and within a decade a number of parts in locomotives were interchangeable with only modest machining. The institution of a rationalized production system capable of turning out large numbers of locomotives was more significant, however, than this small gain in interchangeability. That production system consisted of new locomotive designs that allowed for specialized piecework and included standard parts. Instead of perfect interchangeability, a prohibitively costly task for the six thousand parts in locomotives, they focused on an extensive gauging system to produce components with a built-in tolerance for adjustment. During the 1850s Baldwin added power tools such as lathes, borers, and planers to machine these components better; it expanded its investment in these tools even more during the next decade.[2]

THE MACHINIST NETWORKS CONTINUE

The antebellum machinist networks supported a broad-based development of the pivotal producer durables—metal fabricating, machinery, and machine tools. These networks set the foundation for the sharp rise in the deployment of equipment in manufacturing and in the economy as a whole between 1860 and 1900. Equipment's share of the domestic capital stock went from 9 percent to 28 percent, and the index of the real value of the amount of equipment per capita grew sixfold (see fig. I.1). These changes supported the large-scale industrialization over that period as the manufacturing share of commodity output (also including agriculture and mining) rose from 32 percent to 53 percent and the real value added in manufacturing jumped over sevenfold.

The industrial development of the Midwest commenced in the antebellum based on an agricultural-industrial growth process. It mirrored, in important ways, the East's earlier development, and machinists, likewise, were important to midwestern industrialization. Some leading eastern mechanics moved to the Midwest during the late antebellum, and this pattern of mobility continued afterward. They participated in founding some of the leading machinery and machine tool firms in cities such as Cincinnati, Cleveland, Detroit, and Chicago. Consequently, machinist networks of the East and the Midwest became integrated into a larger network that spanned the

American manufacturing belt. Even as machinist networks started to integrate across this territory in the late antebellum, after 1840 selected licensing linkages among iron foundries and machine shops across the belt, such as in the manufacture of harvesting machinery, were established. Over the next four decades input-output supply linkages in the pivotal producer durables likewise began to grow in importance in some sectors.

In the mid-1860s the visitor registry of William Sellers and Company in Philadelphia, a leading manufacturer of machine tools, foreshadowed the increasing significance of these linkages between the East and the Midwest. As expected, the share of local visitors from Philadelphia firms ranked high (13 percent), over one-third came from elsewhere in the Middle Atlantic, and another 9 percent came from New England. Yet the share from the Midwest amounted to 12 percent, and these visitors probably came to obtain ideas for making machine tools and to purchase some of them.[3]

The continued advances in metalworking technology and the capacity of old and new industrial centers to stay abreast of the latest improvements rested on the fundamental characteristic of machinist networks dating from the late eighteenth century. Technical skills and innovations were embedded in networks of machinists, firms, and clusters (individuals and firms), and these skills and innovations were not the sole property of any of the constituents. Individuals could have short or long careers, and companies could survive for a brief time or last for decades. Nonetheless, the networks retained knowledge of the technical skills and innovations, which became embedded in communities of practice. Much has been made of late-twentieth- and early twenty-first-century technology and its sophisticated job-hopping, youthful entrepreneurs. Yet machinists in the antebellum East anticipated modern behavior by over one hundred and fifty years.

Abbreviations

ASM — Nathan Rosenberg, ed., *The American System of Manufactures* (Edinburgh: Edinburgh University Press, 1969).

BAG — William R. Bagnall, *The Textile Industries of the United States* (Cambridge, Mass.: Riverside Press, 1893).

BISH — J. Leander Bishop, *A History of American Manufactures from 1608 to 1860*, 3 vols. (Philadelphia: Edward Young and Co., 1866).

DAB — *Dictionary of American Biography* (New York: Scribner's, various years).

DEY — Felicia J. Deyrup, "Arms Makers of the Connecticut Valley: A Regional Study of the Economic Development of the Small Arms Industry, 1798–1870," *Smith College Studies in History* 33 (1948).

GAZET — John C. Pease and John M. Niles, *The Gazeteer of the States of Connecticut and Rhode Island* (Hartford, Conn., 1819).

GIBB — George S. Gibb, *The Saco-Lowell Shops: Textile Machinery Building in New England, 1813–1949* (Cambridge: Harvard University Press, 1950).

HOUN — David A. Hounshell, *From the American System to Mass Production, 1800–1932: The Development of Manufacturing Technology in the United States* (Baltimore: Johns Hopkins University Press, 1984).

HUB — Guy Hubbard, "Development of Machine Tools in New England," *American Machinist* 59–61 (1923–24), series of twenty-three articles.

HUNT — Louis C. Hunter, *A History of Industrial Power in the United States, 1780–1930*, vol. 2: *Steam Power* (Charlottesville: University Press of Virginia, 1985).

JERE — David J. Jeremy, *Transatlantic Industrial Revolution: The Diffusion of Textile Technologies between Britain and America, 1790–1830s* (Cambridge, Mass.: MIT Press, 1981).

LEAD — Edwin T. Freedley, *Leading Pursuits and Leading Men: A Treatise on the Principal Trades and Manufactures of the United States* (Philadelphia: Edward Young, 1856).

LOZ — John W. Lozier, *Taunton and Mason: Cotton Machinery and Locomotive Manufacture in Taunton, Massachusetts, 1811–1861* (New York: Garland, 1986).

MCLANE — Louis McLane, *Documents Relative to the Manufactures in the United States Collected and Transmitted to the House of Representatives, 1832, by the Secretary of the Treasury*, House Doc. No. 308, 22nd Cong., 1st sess. (Washington, D.C.: Duff Green, 1833).

MRS — Merritt R. Smith, *Harpers Ferry Armory and the New Technology: The Challenge of Change* (Ithaca, N.Y.: Cornell University Press, 1977).

PATER L. R. Trumbull, *A History of Industrial Paterson* (Paterson, N.J.: Carleton M. Herrick, 1882).

RABER Michael S. Raber, Patrick M. Malone, Robert B. Gordon, and Carolyn C. Cooper, *Conservative Innovators and Military Small Arms: An Industrial History of the Springfield Armory, 1794–1968,* Report to the U.S. Department of the Interior, National Park Service (South Glastonbury, Conn.: Raber Associates, 1989).

ROE Joseph W. Roe, *English and American Tool Builders* (New York: McGraw-Hill, 1916).

ROOTS David R. Meyer, *The Roots of American Industrialization* (Baltimore: Johns Hopkins University Press, 2003).

STAT. U.S. Bureau of the Census, *Historical Statistics of the United States, Colonial Times to 1970, Bicentennial Edition,* 2 pts. (Washington, D.C.: Government Printing Office, 1975).

STEAM Levi Woodbury, *Steam Engines,* Transmitted to the House of Representatives, 1838, by the Secretary of the Treasury, House Doc. No. 21, 25th Cong., 3rd sess., *New American State Papers, Science and Technology,* vol. 7: *Steam Engines* (Wilmington, Del.: Scholarly Resources, 1973), 11–482.

Notes

INTRODUCTION

Epigraph: Joseph Whitworth, "New York Industrial Exhibition," ASM, 331.

1. Manuel Castells, *The Information Age: Economy, Society and Culture,* 3 vols. (Oxford: Blackwell Publishers, 1996–98).

2. Castells, *Information Age,* vol. 1: *The Rise of the Network Society,* 52–56, 151–200; Martin Kenney, ed., *Understanding Silicon Valley: The Anatomy of an Entrepreneurial Region* (Stanford: Stanford University Press, 2000); Chong-Moon Lee, William F. Miller, Marguerite G. Hancock, and Henry S. Rowen, eds., *The Silicon Valley Edge: A Habitat for Innovation and Entrepreneurship* (Stanford: Stanford University Press, 2000); AnnaLee Saxenian, *Regional Advantage: Culture and Competition in Silicon Valley and Route 128* (Cambridge: Harvard University Press, 1994); AnnaLee Saxenian, "Transnational Communities and the Evolution of Global Production Networks: The Cases of Taiwan, China, and India," *Industry and Innovation* 9, no. 3 (December 2002): 183–202.

3. Alexander J. Field, "On the Unimportance of Machinery," *Explorations in Economic History* 22 (October 1985): 378–401; Charles H. Fitch, "Report on the Manufactures of Interchangeable Mechanism" and "Report on the Manufacture of Hardware, Cutlery, and Edge-Tools; also Saws and Files," both in U.S. Bureau of the Census, *Report on the Manufactures of the United States at the Tenth Census, 1880* (Washington, D.C.: Government Printing Office, 1883); Albert W. Niemi Jr., *State and Regional Patterns in American Manufacturing, 1860–1900* (Westport, Conn.: Greenwood Press, 1974), 7, table 1; Kenneth L. Sokoloff, "Investment in Fixed and Working Capital during Early Industrialization: Evidence from U.S. Manufacturing Firms," *Journal of Economic History* 44 (June 1984): 545–56; U.S. Bureau of the Census, *Report on Power and Machinery Employed in Manufactures, Tenth Census of the United States, 1880,* vol. 22 (Washington, D.C.: Government Printing Office, 1888). For quote, see chapter epigraph.

4. BISH, vol. 2; Eugene S. Ferguson, ed., *Early Engineering Reminiscences (1815–40) of George Escol Sellers* (Washington, D.C.: Smithsonian Institution, 1965); HUB; Deirdre N. McCloskey, "Bourgeois Virtue and the History of P and S," *Journal of Economic History* 58 (June 1998): 297–317; ROE.

5. Meric S. Gertler, "Tacit Knowledge and the Economic Geography of Context, or the Undefinable Tacitness of Being (There)," *Journal of Economic Geography* 3 (2003): 75–99; Jeremy R. L. Howells, "Tacit Knowledge, Innovation and Economic Geography," *Urban Studies* 39, nos. 5–6 (2002): 871–84; Jan G. Lambooy, "Knowledge and Urban Economic Development:

An Evolutionary Perspective," *Urban Studies* 39, nos. 5–6 (2002): 1019–35; Peter Maskell and Anders Malmberg, "Localized Learning and Industrial Competitiveness," *Cambridge Journal of Economics* 23 (1999): 167–85; Steven Pinch, Nick Henry, Mark Jenkins, and Stephen Tallman, "From 'Industrial Districts' to 'Knowledge Clusters': A Model of Knowledge Dissemination and Competitive Advantage in Industrial Agglomerations," *Journal of Economic Geography* 3, no. 4 (2003): 373–88.

6. For the theory and empirical evidence supporting this discussion of network concepts, see Yanjie Bian, "Bringing Strong Ties Back In: Indirect Ties, Network Bridges, and Job Searches in China," *American Sociological Review* 62 (June 1997): 366–85; Ronald S. Burt, *Structural Holes: The Social Structure of Competition* (Cambridge: Harvard University Press, 1992); Linton C. Freeman, "Centrality in Social Networks: Conceptual Clarification," *Social Networks* 1 (1978–79): 215–39; Mark S. Granovetter, "The Strength of Weak Ties," *American Journal of Sociology* 78 (May 1973): 1360–80; Alejandro Portes, ed., *The Economic Sociology of Immigration: Essays on Networks, Ethnicity, and Entrepreneurship* (New York: Russell Sage Foundation, 1995); Stanley Wasserman and Katherine Faust, *Social Network Analysis: Methods and Applications* (Cambridge: Cambridge University Press, 1994).

7. Consider a local network of four machine firms in different towns, all within ten miles of one another. Some ties between firms will be shorter paths than others because of the types of contacts that evolved among the firms. We will not consider the shortest path in this local network as a bridge because longer paths are easily covered. A bridge is a tie that either is the only link between two contacts or is far more efficient as a link.

8. Diane Lindstrom, *Economic Development in the Philadelphia Region, 1810–1850* (New York: Columbia University Press, 1978); Glenn Porter and Harold C. Livesay, *Merchants and Manufacturers: Studies in the Changing Structure of Nineteenth-Century Marketing* (Baltimore: Johns Hopkins University Press, 1971); Allan R. Pred, *Urban Growth and the Circulation of Information: The United States System of Cities, 1790–1840* (Cambridge: Harvard University Press, 1973); Allan R. Pred, *Urban Growth and City-Systems in the United States, 1840–1860* (Cambridge: Harvard University Press, 1980); ROOTS; STAT, ser. Q321, Q329.

9. James Fleck, "Artefact—Activity: The Coevolution of Artefacts, Knowledge and Organization in Technological Innovation," *Technological Innovation as an Evolutionary Process*, ed. John Ziman (Cambridge: Cambridge University Press, 2000), 248–66; Robert B. Gordon, "Who Turned the Mechanical Ideal into Mechanical Reality?" *Technology and Culture* 29 (October 1988): 744–78; Joel Mokyr, "Evolutionary Phenomena in Technological Change," *Technological Innovation as an Evolutionary Process*, ed. John Ziman (Cambridge: Cambridge University Press, 2000), 52–65; W. J. Rorabaugh, *The Craft Apprentice: From Franklin to the Machine Age in America* (New York: Oxford University Press, 1986); Edward W. Stevens Jr., *The Grammar of the Machine: Technical Literacy and Early Industrial Expansion in the United States* (New Haven, Conn.: Yale University Press, 1995); Eric Von Hippel, *The Sources of Innovation* (New York: Oxford University Press, 1988), 3–10; Anthony F. C. Wallace, *Rockdale: The Growth of an American Village in the Early Industrial Revolution* (New York: Alfred A. Knopf, 1978), 237–39; Etienne Wenger, *Communities of Practice: Learning, Meaning, and Identity* (Cambridge: Cambridge University Press, 1998).

10. Robert C. Allen, "Collective Invention," *Journal of Economic Behavior and Organization* 4 (March 1983): 1–24; Joel Mokyr, "Demand vs. Supply in the Industrial Revolution," *Journal of Economic History* 37 (December 1977): 981–1008; Jacob Schmookler, *Invention and Eco-*

nomic Growth (Cambridge: Harvard University Press, 1966); Ross Thomson, "Introduction," *Learning and Technological Change*, ed. Ross Thomson (London: Macmillan, 1993), 1–5.

11. Monte A. Calvert, *The Mechanical Engineer in America, 1830–1910* (Baltimore: Johns Hopkins University Press, 1967), 3–27; Alfred D. Chandler Jr., *The Visible Hand: The Managerial Revolution in American Business* (Cambridge: Belknap Press of Harvard University Press, 1977), 81–187; Stevens, *Grammar of the Machine;* Walter G. Vincenti, *What Engineers Know and How They Know It* (Baltimore: Johns Hopkins University Press, 1990); Von Hippel, *Sources of Innovation;* Wallace, *Rockdale*, 211–19; Wenger, *Communities of Practice.*

12. John S. Brown and Paul Duguid, "Organizational Learning and Communities-of-Practice: Toward a Unified View of Working, Learning, and Innovation," *Organization Science* 2 (February 1991): 40–57; Edward W. Constant III, "The Social Locus of Technological Practice: Community, System, or Organization," in *The Social Construction of Technological Systems: New Directions in the Sociology and History of Technology*, ed. Wiebe E. Bijker, Thomas P. Hughes, and Trevor J. Pinch (Cambridge, Mass.: MIT Press, 1987), 223–42; Gertler, "Tacit Knowledge and the Economic Geography of Context"; Ross Thomson, "Crossover Inventors and Technological Linkages: American Shoemaking and the Broader Economy, 1848–1901," *Technology and Culture* 32 (October 1991): 1018–46; Steven W. Usselman, "Patents, Engineering Professionals, and the Pipelines of Innovation: The Internalization of Technical Discovery by Nineteenth-Century American Railroads," in *Learning by Doing in Markets, Firms, and Countries*, ed. Naomi R. Lamoreaux, Daniel M. G. Raff, and Peter Temin (Chicago: University of Chicago Press, 1999), 61–91; Vincenti, *What Engineers Know;* Von Hippel, *Sources of Innovation.*

13. Bernard G. Dennis Jr., Robert J. Kapsch, Robert J. LoConte, Bruce W. Mattheiss, and Steven M. Pennington, eds., *American Civil Engineering: The Pioneering Years* (Reston, Va: American Society of Civil Engineers, 2003); Richard R. Nelson and Gavin Wright, "The Rise and Fall of American Technological Leadership: The Postwar Era in Historical Perspective," *Journal of Economic Literature* 30 (December 1992): 1931–64; Nathan Rosenberg, "Technological Change in the Machine Tool Industry, 1840–1910," *Journal of Economic History* 23 (December 1963): 414–43; Michael Storper and Richard Walker, *The Capitalist Imperative: Territory, Technology, and Industrial Growth* (New York: Basil Blackwell, 1989); Gavin Wright, "Can a Nation Learn? American Technology as a Network Phenomenon," in *Learning by Doing in Markets, Firms, and Countries*, ed. Naomi R. Lamoreaux, Daniel M. G. Raff, and Peter Temin (Chicago: University of Chicago Press, 1999), 295–326.

CHAPTER 1: IRON FOUNDRIES BECOME EARLY HUBS OF MACHINIST NETWORKS

Epigraph: Advertisement that appeared in the *Aurora*, June 2, 1807; reprinted in Greville Bathe and Dorothy Bathe, *Oliver Evans: A Chronicle of Early American Engineering* (Philadelphia: Historical Society of Pennsylvania, 1935), 139–40.

1. For the bases of this section's discussion of the iron industry, see Gerald G. Eggert, *The Iron Industry in Pennsylvania*, Pennsylvania History Studies No. 25 (Harrisburg: Pennsylvania Historical Association, 1994), 1–14; Robert B. Gordon, *American Iron, 1607–1900* (Baltimore: Johns Hopkins University Press, 1996), 55–135; Paul F. Paskoff, *Industrial Evolution: Organization, Structure, and Growth of the Pennsylvania Iron Industry, 1750–1860* (Baltimore: Johns Hop-

kins University Press, 1983), 73, table 18; Arthur D. Pierce, *Iron in the Pines: The Story of New Jersey's Ghost Towns and Bog Iron* (New Brunswick, N.J.: Rutgers University Press, 1957); Glenn Porter and Harold C. Livesay, *Merchants and Manufacturers: Studies in the Changing Structure of Nineteenth-Century Marketing* (Baltimore: Johns Hopkins University Press, 1971), 37–45; James M. Ransom, *Vanishing Ironworks of the Ramapos: The Story of the Forges, Furnaces, and Mines of the New Jersey–New York Border Area* (New Brunswick, N.J.: Rutgers University Press, 1966); Joseph E. Walker, *Hopewell Village: A Social and Economic History of an Iron-Making Community* (Philadelphia: University of Pennsylvania Press, 1966).

2. Tench Coxe, "Digest of Manufactures, 1810," Communicated to the Senate, January 5, 1814, 13th Cong., 2nd sess., *New American State Papers, Manufactures* (Wilmington, Del.: Scholarly Resources, 1972), 1:201; Albert Gallatin, "Manufactures," Communicated to the House of Representatives, April 19, 1810, 11th Cong., 2nd sess., *New American State Papers, Manufactures* (Wilmington, Del.: Scholarly Resources, 1972), 1:128; GAZET; GIBB; David R. Meyer, "A Dynamic Model of the Integration of Frontier Urban Places into the United States System of Cities," *Economic Geography* 56 (April 1980): 120–40; Timothy Pitkin, *A Statistical View of the Commerce of the United States of America* (New Haven, Conn.: Durrie and Peck, 1835), 494; Porter and Livesay, *Merchants and Manufacturers*, 42–43; Allan R. Pred, *Urban Growth and the Circulation of Information: The United States System of Cities, 1790–1840* (Cambridge: Harvard University Press, 1973); Carroll W. Pursell Jr., *Early Stationary Steam Engines in America* (Washington, D.C.: Smithsonian Institution Press, 1969); ROOTS; Kenneth L. Sokoloff, "Inventive Activity in Early Industrial America: Evidence from Patent Records, 1790–1846," *Journal of Economic History* 48 (December 1988): 813–50; Kenneth L. Sokoloff and B. Zorina Khan, "The Democratization of Invention during Early Industrialization: Evidence from the United States, 1790–1846," *Journal of Economic History* 50 (June 1990): 363–78.

3. HUNT, 183–211.

4. Richard L. Hills, *Power from Steam: A History of the Stationary Steam Engine* (Cambridge: Cambridge University Press, 1989), 51–70, 75–81; Brooke Hindle, *Emulation and Invention* (New York: New York University Press, 1981), 25–45; HUNT, 12–13; Pursell, *Early Stationary Steam Engines in America*, 1–27; Robert H. Thurston, *A History of the Growth of the Steam-Engine*, centennial ed. (1878; rpt., Ithaca, N.Y.: Cornell University Press, 1939), 234–46.

5. David R. Meyer, "The Industrial Retardation of Southern Cities, 1860–80," *Explorations in Economic History* 25 (October 1988): 366–86; and "Midwestern Industrialization and the American Manufacturing Belt in the Nineteenth Century," *Journal of Economic History* 49 (December 1989): 921–37; Pred, *Urban Growth and the Circulation of Information*; ROOTS; Sokoloff, "Inventive Activity in Early Industrial America"; Sokoloff and Khan, "Democratization of Invention during Early Industrialization," 367, table 1; Gavin Wright, *Old South, New South: Revolutions in the Southern Economy since the Civil War* (New York: Basic Books, 1986).

6. Pursell, *Early Stationary Steam Engines in America*, 28–31; Archibald D. Turnbull, *John Stevens: An American Record* (New York: Century Co., 1928), 129–37.

7. Bathe and Bathe, *Oliver Evans*, 15–16, 33–74; Manfred Blake, *Water for the Cities: A History of the Urban Water Supply Problem in the United States* (Syracuse, N.Y.: Syracuse University Press, 1956), 24–43, 78–88; Talbot Hamlin, *Benjamin Henry Latrobe* (New York: Oxford University Press, 1955); HUNT, 33–36, 47–50, 177–79, 212–14; Pred, *Urban Growth and the Circulation of Information*; Pursell, *Early Stationary Steam Engines in America*, 43–47.

8. BISH, 2:99, 119; H. W. Dickinson, *Robert Fulton: Engineer and Artist* (London: John

Lane, 1913), 198–200; Hamlin, *Benjamin Henry Latrobe;* HUNT, 15–25, 178–79; John H. Morrison, *History of American Steam Navigation* (New York: W. F. Sametz and Co., 1903), 17–19, 41–42; Cynthia O. Philip, *Robert Fulton: A Biography* (New York: Franklin Watts, 1985); Pursell, *Early Stationary Steam Engines in America,* 50–51; George R. Taylor, *The Transportation Revolution, 1815–1860,* vol. 4: *The Economic History of the United States* (New York: Holt, Rinehart and Winston, 1951), 59; Thurston, *History of the Growth of the Steam-Engine,* 250–80; Turnbull, *John Stevens;* Thomas Weiss, "Economic Growth before 1860: Revised Conjectures," *American Economic Development in Historical Perspective,* ed. Thomas Weiss and Donald Schaefer (Stanford: Stanford University Press, 1994), 11–27.

9. See chapter epigraph.

10. Bathe and Bathe, *Oliver Evans;* BISH, 2:180, 3:178–81; Mark Granovetter, "Economic Action and Social Structure: The Problem of Embeddedness," *American Journal of Sociology* 91 (November 1985): 481–510; Hamlin, *Benjamin Henry Latrobe;* HUNT, 40–46, 177–78; Philip, *Robert Fulton;* Pursell, *Early Stationary Steam Engines in America,* 47–62; Thurston, *History of the Growth of the Steam-Engine,* 264–80; Turnbull, *John Stevens.*

11. Bathe and Bathe, *Oliver Evans;* Hamlin, *Benjamin Henry Latrobe;* HUNT, 128–44; Louis C. Hunter, *Steamboats on the Western Rivers* (Cambridge: Harvard University Press, 1949), 6–17; Philip, *Robert Fulton;* Daniel Preston, "William Thornton," in *American National Biography,* ed. John A. Garraty and Mark C. Carnes (New York: Oxford University Press, 1999), 21:609–11; Pursell, *Early Stationary Steam Engines in America;* Turnbull, *John Stevens;* John C. Van Horne, ed., *The Correspondence and Miscellaneous Papers of Benjamin Henry Latrobe,* 3 vols. (New Haven, Conn.: Yale University Press, 1984–88).

CHAPTER 2: A NETWORKED COMMUNITY BUILT BY COTTON TEXTILE MACHINISTS

Epigraph: Tench Coxe, "Digest of Manufactures, 1810," Communicated to the Senate, January 5, 1814, 13th Cong., 2nd sess., *New American State Papers, Manufactures* (Wilmington, Del.: Scholarly Resources, 1972), 1:163.

1. JERE, 83–86.

2. Flow production that is interrupted at successive stages differs from continuous flow production, such as the flour milling system of Oliver Evans.

3. JERE, 26–30, 76–82, 93–96, 187–88; Nathan Rosenberg, "The Direction of Technological Change: Inducement Mechanisms and Focusing Devices," *Economic Development and Cultural Change* 18 (October 1969): 1–24; Martin Van Buren, "Patents Granted by the United States," Secretary of State, Communicated to the House of Representatives, January 13, 1831, 21st Cong., 2nd sess., Doc. No. 50, *New American State Papers, Science and Technology, Patents* (Wilmington, Del.: Scholarly Resources, 1972), 4:185–88.

4. BAG; GAZET, 139–67, 202–28; JERE, 36–140; MCLANE, doc. 9; Thomas R. Navin, *The Whitin Machine Works since 1831: A Textile Machinery Company in an Industrial Village* (Cambridge: Harvard University Press, 1950), 16–20, app. 1, 479; ROOTS, 96–104; Caroline F. Ware, *The Early New England Cotton Manufacture: A Study in Industrial Beginnings* (Boston: Houghton Mifflin, 1931), 32; Robert B. Zevin, "The Growth of Cotton Textile Production after 1815," in *The Reinterpretation of American Economic History,* ed. Robert W. Fogel and Stanley L. Engerman (New York: Harper and Row, 1971), 122–47.

5. BAG, 368–75, 460–61, 474; Frances W. Gregory, *Nathan Appleton: Merchant and Entrepreneur, 1779–1861* (Charlottesville: University Press of Virginia, 1975), 5–18; MCLANE, docs. 4, 5, 10 (counties include Columbia, Rensselaer, Saratoga, Schenectady, and Washington); Lois K. Mathews, *The Expansion of New England* (Boston: Houghton Mifflin, 1909), 153; Ware, *Early New England Cotton Manufacture*, 32.

6. BAG, 501–9; MCLANE, doc. 10 (counties include Cayuga, Chenango, Herkimer, Jefferson, Oneida, Onondaga, and Otsego); Mathews, *Expansion of New England*, 154–67.

7. BAG, 160–61, 236–44, 373–75; Greville Bathe and Dorothy Bathe, *Jacob Perkins: His Inventions, His Times, and His Contemporaries* (Philadelphia: Historical Society of Pennsylvania, 1943), 1–55, app. 1, 200; GIBB, 11–12; JERE, 86–88, 102; LOZ, 52–94; Christopher Roberts, *The Middlesex Canal, 1793–1860* (Cambridge: Harvard University Press, 1938), 92–94; ROE, 118–19; Kenneth L. Sokoloff, "Inventive Activity in Early Industrial America: Evidence from Patent Records, 1790–1846," *Journal of Economic History* 48 (December 1988): 813–50; Van Buren, "Patents Granted by the United States," 188–93; Robert S. Woodbury, *History of the Lathe to 1850* (Cambridge, Mass.: MIT Press, 1961), 89–92.

8. BAG, 183, 269–73, 492; Irwin Feller, "Determinants of the Composition of Urban Inventions," *Economic Geography* 49 (January 1973): 47–58; Zorina Khan and Kenneth L. Sokoloff, "'Schemes of Practical Utility': Entrepreneurship and Innovation among 'Great Inventors' in the United States, 1790–1865," *Journal of Economic History* 53 (June 1993): 289–307; PATER; Allan R. Pred, *The Spatial Dynamics of U.S. Urban-Industrial Growth, 1800–1914* (Cambridge, Mass.: MIT Press, 1966), 86–142; ROOTS, 88–96; Jacob Schmookler, *Invention and Economic Growth* (Cambridge: Harvard University Press, 1966); Zevin, "Growth of Cotton Textile Production after 1815."

9. BAG; James L. Conrad Jr., "'Drive That Branch': Samuel Slater, the Power Loom, and the Writing of America's Textile History," *Technology and Culture* 36 (January 1995): 1–28; GAZET, 344; Jonathan T. Lincoln, "The Beginnings of the Machine Age in New England: David Wilkinson of Pawtucket," *New England Quarterly* 6 (December 1933): 716–32; LOZ, 57–73; ROE, 123–24.

10. BAG, 222–27; BISH, 2:71–72, 100, 164, and 3:18; JERE, 90; Philip Scranton, *Proprietary Capitalism: The Textile Manufacture at Philadelphia, 1800–1885* (Cambridge: Cambridge University Press, 1983), 75–134; Cynthia J. Shelton, *The Mills of Manayunk: Industrialization and Social Conflict in the Philadelphia Region, 1787–1837* (Baltimore: Johns Hopkins University Press, 1986), 26–53; Norman B. Wilkinson, "Brandywine Borrowings from European Technology," *Technology and Culture* 4 (Winter 1963): 1–13.

11. BAG, 388–89; Robert F. Dalzell Jr., *Enterprising Elite: The Boston Associates and the World They Made* (Cambridge: Harvard University Press, 1987); GIBB, 14–48; JERE, 190–98; Vera Shlakman, "Economic History of a Factory Town: A Study of Chicopee, Massachusetts," *Smith College Studies in History* 20, nos. 1–4 (October 1934–July 1935).

12. Carolyn C. Cooper, *Shaping Invention: Thomas Blanchard's Machinery and Patent Management in Nineteenth-Century America* (New York: Columbia University Press, 1991); Carolyn C. Cooper, "Nineteen-Century American Patent Management as an Invisible College of Technology," in *Learning and Technological Change*, ed. Ross Thomson (New York: St. Martin's Press, 1993), 40–61; Shelton, *Mills of Manayunk*, 33–37; Ware, *Early New England Cotton Manufacture*. Because distinctions between hand and machine power were difficult to make, all patents for spinning and weaving were included in tables 2.1 and 2.2. The proliferation

of patents with the shift to machine power suggests that little error is introduced by the inability to distinguish between these forms of power.

13. Eugene S. Ferguson, ed., *Early Engineering Reminiscences (1815–40) of George Escol Sellers* (Washington, D.C.: Smithsonian Institution, 1965); James T. Lemon, *The Best Poor Man's Country: A Geographical Study of Early Southeastern Pennsylvania* (Baltimore: Johns Hopkins University Press, 1972).

14. The only exceptions in tables 2.1 and 2.2 of patentees in the outer hinterland of New York City who were not based in central New York state are those in Walton and Queensbury in New York state and in Attleboro, Pa.

15. John L. Hayes, *American Textile Machinery* (Cambridge, Mass.: University Press, J. Wilson and Son, 1879), 19.

CHAPTER 3: THE FEDERAL ARMORIES AND PRIVATE FIREARMS FIRMS OPERATE IN OPEN NETWORKS

Epigraph: Subscribers, Gun Manufacturers, Borough of Lancaster, Pennsylvania, "Memorial to the House of Representatives of the United States," Communicated to the House of Representatives, February 4, 1803, 7th Cong., 2nd sess., *New American State Papers, Manufactures* (Wilmington, Del.: Scholarly Resources, 1972), 1:93.

1. DEY, 36–37; MRS, 28–41; RABER, 86–88; Merritt R. Smith, "Army Ordnance and the 'American System' of Manufacturing, 1815–1861," in *Military Enterprise and Technological Change: Perspectives on the American Experience*, ed. Merritt R. Smith (Cambridge, Mass.: MIT Press, 1985), 39–86.

2. DEY, 72–83, 233, app. B, table 2, 240, 245, app. D, tables 1, 4; HOUN, 32–33; Michael S. Raber, "Conservative Innovators, Military Small Arms, and Industrial History at Springfield Armory, 1794–1918," *IA: The Journal of the Society for Industrial Archeology* 14, no. 1 (1988): 1–21; RABER. During the War of 1812 workers spent extensive time repairing muskets, leaving less time for turning out new ones.

3. Christopher Clark, *The Roots of Rural Capitalism: Western Massachusetts, 1780–1860* (Ithaca, N.Y.: Cornell University Press, 1990); DEY, 72–79; Sam B. Hilliard, *Atlas of Antebellum Southern Agriculture* (Baton Rouge: Louisiana State University Press, 1984), 28–31, maps 26–32; Margaret E. Martin, "Merchants and Trade of the Connecticut River Valley, 1750–1820," *Smith College Studies in History* 24, nos. 1–4 (October 1938–July 1939): 74–101, 131–69; David R. Meyer, "The Industrial Retardation of Southern Cities, 1860–1880," *Explorations in Economic History* 25 (October 1988): 366–86; MRS, 25–103; Richard J. Purcell, *Connecticut in Transition: 1775–1818* (Washington, D.C.: American Historical Association, 1918), 78–90; Gavin Wright, *Old South, New South: Revolutions in the Southern Economy since the Civil War* (New York: Basic Books, 1986), 17–50.

4. DEY, 37–67; Robert S. Woodbury, "The Legend of Eli Whitney and Interchangeable Parts," *Technology and Culture* 1 (Summer 1960): 235–53.

5. Joseph Anderson, ed., *Town and City of Waterbury*, 3 vols. (New Haven, Conn.: Price and Lee Co., 1896), 2:233–318; Shirley S. DeVoe, *The Tinsmiths of Connecticut* (Middletown, Conn.: Wesleyan University Press, 1968); DEY, 41, 48, 218–28, app. A, tables 3–4; C. Bancroft Gillespie, *A Century of Meriden: "The Silver City,"* 3 pts. (Meriden, Conn.: Journal Publishing Co., 1906), pt. 1, 346–61; Constance McL. Green, *Eli Whitney and the Birth of American*

Technology (Boston: Little, Brown, 1956); Jeannette Mirsky and Allan Nevins, *The World of Eli Whitney* (New York: Macmillan, 1952); S.N.D. North and Ralph H. North, *Simeon North: First Official Pistol Maker of the United States* (Concord, N.H.: Rumford Press, 1913); ROOTS, 71–86, 96–104; Smith, "Army Ordnance and the 'American System' of Manufacturing, 1815–1861." Also see chapter epigraph.

6. Green, *Eli Whitney and the Birth of American Technology;* HOUN, 31; Mirsky and Nevins, *The World of Eli Whitney;* RABER, 88–93, 206–8; Merritt R. Smith, "Eli Whitney and the American System of Manufacturing," in *Technology in America: A History of Individuals and Ideas,* ed. Carroll W. Pursell Jr., 2nd ed. (Cambridge, Mass.: MIT Press, 1990), 45–61; Woodbury, "Legend of Eli Whitney."

7. BAG, 236–44; Bruce Clouette and Matthew Roth, *Bristol, Connecticut: A Bicentennial History, 1785–1985* (Canaan, N.H.: Phoenix Publishing, 1984), 51, 57; DeVoe, *Tinsmiths of Connecticut,* 3–34, 66; DEY, 220–28, app. A, table 4, 240, app. D, table 1; Gillespie, *Century of Meriden,* pt. 1, 346–47; HOUN, 28–29, 32–33; John J. Murphy, "Entrepreneurship in the Establishment of the American Clock Industry," *Journal of Economic History* 26 (June 1966): 169–86; Catharine M. North, *History of Berlin, Connecticut* (New Haven, Conn.: Tuttle, Morehouse and Taylor Co., 1916), 288–90; North and North, *Simeon North* (quotes from reprinted letter dated November 7, 1808, and from reprinted contract of April 16, 1813, signed by Simeon North and Callender Irvine), 64–65, 80–82; RABER, 34–35, 89 n. 13, 122; Kenneth D. Roberts, *The Contributions of Joseph Ives to Connecticut Clock Technology, 1810–1862* (Bristol, Conn.: American Clock and Watch Museum, 1970), 17, 27, tables 7, 8; ROOTS, 71–86.

8. ASM; Carolyn C. Cooper, "'A Whole Battalion of Stockers': Thomas Blanchard's Production Line and Hand Labor at Springfield Armory," *IA: The Journal of the Society for Industrial Archeology* 14, no. 1 (1988): 37–57; Donald R. Hoke, *Ingenious Yankees: The Rise of the American System of Manufactures in the Private Sector* (New York: Columbia University Press, 1990), 52–99; HOUN, 29; MRS, 92, 109, 117; Murphy, "Entrepreneurship in the Establishment of the American Clock Industry"; North, *History of Berlin, Connecticut,* 289–90; North and North, *Simeon North* (quotes from reprinted report of James Stubblefield and Roswell Lee to Decius Wadsworth, March 20–22, 1816), 106; RABER, 96–97, 237, 244–46; Smith, "Army Ordnance and the 'American System' of Manufacturing, 1815–1861"; Merritt R. Smith, "John H. Hall, Simeon North, and the Milling Machine: The Nature of Innovation among Antebellum Arms Makers," *Technology and Culture* 14 (October 1973): 573–91; Kenneth L. Sokoloff, "Inventive Activity in Early Industrial America: Evidence from Patent Records, 1790–1846," *Journal of Economic History* 48 (December 1988): 832, fig. 2; STAT, ser. W99; Robert S. Woodbury, *History of the Lathe to 1850* (Cambridge, Mass.: MIT Press, 1961), 89–93; and *History of the Milling Machine* (Cambridge, Mass.: MIT Press, 1960).

9. Mirsky and Nevins, *World of Eli Whitney* (quote, 268); MRS, 83, 153; RABER, 86–87; Smith, "Army Ordnance and the 'American System' of Manufacturing, 1815–1861."

10. BAG; Richard D. Brown, *Massachusetts: A Bicentennial History* (New York: W. W. Norton, 1978), 145; Carolyn C. Cooper, *Shaping Invention: Thomas Blanchard's Machinery and Patent Management in Nineteenth-Century America* (New York: Columbia University Press, 1991), 17–18; DEY, 240–41, 245, 247, app. D, tables 1, 2, 4, 5; HOUN, 28–29; MRS; PATER; RABER; Smith, "Army Ordnance and the 'American System' of Manufacturing, 1815–1861."

11. Nathan Rosenberg, "Factors Affecting the Diffusion of Technology," *Explorations in*

Economic History 10 (Fall 1972): 3–33; Nathan Rosenberg, "Technological Change in the Machine Tool Industry, 1840–1910," *Journal of Economic History* 23 (December 1963): 414–43.

CHAPTER 4: IRON FOUNDRIES RULE THE HEAVY CAPITAL EQUIPMENT INDUSTRY

Epigraph: Jacob Abbott, "The Novelty Works," *Harper's New Monthly Magazine* 2 (May 1851): 724.

1. Alfred D. Chandler Jr., "Anthracite Coal and the Beginnings of the Industrial Revolution in the United States," *Business History Review* 46 (Summer 1972): 141–81; Robert E. Gallman, "Commodity Output, 1839–1899," *Trends in the American Economy in the Nineteenth Century*, Studies in Income and Wealth, vol. 24 (Princeton, N.J.: Princeton University Press, 1960), 43, table A-1 (variant A); STAT, ser. A7; Thomas Weiss, "U.S. Labor Force Estimates and Economic Growth, 1800–1860," in *American Economic Growth and Standards of Living before the Civil War*, ed. Robert E. Gallman and John J. Wallis (Chicago: University of Chicago Press, 1992), 27, table 1.2 (variant C). For the states included within the regional divisions in fig. 4.2, see Jeremy Atack, Fred Bateman, and Thomas Weiss, "The Regional Diffusion and Adoption of the Steam Engine in American Manufacturing," *Journal of Economic History* 40 (June 1980): 285, table 1.

2. Jeremy Atack, "Fact in Fiction? The Relative Costs of Steam and Water Power: A Simulation Approach," *Explorations in Economic History* 16 (October 1979): 409–37; Atack, Bateman, and Weiss, "Regional Diffusion and Adoption of the Steam Engine"; HUNT, 73.

3. ROOTS; Charles G. Washburn, *Industrial Worcester* (Worcester, Mass.: Davis Press, 1917), 111. Most of the evidence in this section is drawn from MCLANE.

4. BISH, 3:178–81; "Cornelius Henry Delamater," DAB, 5:211–12; "James Peter Allaire," DAB, 1:128; "James Peter Allaire," *National Cyclopedia of American Biography* (New York: James T. White and Co., 1935), 24:355–56; MCLANE, doc. 10, no. 48, doc. 11, nos. 15–16; "Manufactures of Dutchess County," *Hunt's Merchants' Magazine* 15 (October 1846): 370–76; "Manufacturing Industry of New York," *Hunt's Merchants' Magazine* 16 (January 1847): 92–94; Carroll W. Pursell Jr., *Early Stationary Steam Engines in America* (Washington, D.C.: Smithsonian Institution Press, 1969), 51; "Thomas Rogers," DAB, 16:112–13; Robert H. Thurston, *A History of the Growth of the Steam-Engine*, centennial edition (1878; rpt., Ithaca, N.Y.: Cornell University Press, 1939), 214–15, 273–80.

5. BISH, 3:18, 28; John K. Brown, *The Baldwin Locomotive Works, 1831–1915* (Baltimore: Johns Hopkins University Press, 1995), 3–7; Andrew Dawson, *Lives of the Philadelphia Engineers: Capital, Class and Revolution, 1830–1890* (Hants, Eng.: Ashgate, 2004), 21; Eugene S. Ferguson, ed., *Early Engineering Reminiscences (1815–40) of George Escol Sellers* (Washington, D.C.: Smithsonian Institution, 1965); LEAD, 288; MCLANE, doc. 13, no. 4, doc. 16, no. 2; "Matthias W. Baldwin," DAB, 1:541–42; "Samuel V. Merrick," DAB, 12:557–58; Bruce Sinclair, *Philadelphia's Philosopher Mechanics: A History of the Franklin Institute, 1824–1865* (Baltimore: Johns Hopkins University Press, 1974).

6. "Cyrus Alger," DAB, 1:177–78; MCLANE, doc. 3; Pursell, *Early Stationary Steam Engines in America*, 102–3.

7. STEAM.

8. HUNT, 74, 76, 78–79, tables 4, 6, 8; David R. Meyer, "The Industrial Retardation of Southern Cities, 1860–1880," *Explorations in Economic History* 25 (October 1988): 366–86.

9. Each of the metropolis names as market outlets in table 4.4 refers to both the city and its environs within a radius of about fifty miles. In all of the cases most of the sales actually were in the city itself.

10. Joseph Anderson, ed., *Town and City of Waterbury*, 3 vols. (New Haven, Conn.: Price and Lee Co., 1896), 2:275–406; "Charles W. Copeland," DAB, 4:423; Margaret E. Martin, "Merchants and Trade of the Connecticut River Valley, 1750–1820," *Smith College Studies in History* 24, nos. 1–4 (October 1938–July 1939): 18–73; Pursell, *Early Stationary Steam Engines in America*, 103; ROOTS, 41–46, 71–87, 266–78; STEAM.

11. BISH, 3:110–12, 208; Victor S. Clark, *History of Manufactures in the United States, 1607–1860* (Washington, D.C.: Carnegie Institution of Washington, 1916), 377–78, 502–4; Edwin T. Freedley, *Philadelphia and its Manufactures* (Philadelphia: Edward Young, 1858), 290–91; LEAD, 272–80; "Manufactures, Albany, New York," *Hunt's Merchants' Magazine* 21 (July 1849): 56–57; Peter Temin, *Iron and Steel in Nineteenth-Century America* (Cambridge, Mass.: MIT Press, 1964), 36–39.

12. Victor S. Clark, *History of Manufactures in the United States, 1860–1914* (Washington, D.C.: Carnegie Institution of Washington, 1928), 185, 296, 685. The price levels for "all commodities" and for the subset of "metals and metal products" were similar in 1850 and 1860; therefore, the decadal changes in capitalization and value added per employee measure real dollar amounts quite well. See STAT, ser. E52, E58.

13. Abbott, "Novelty Works"; "American Machinery—Matteawan," *Scientific American* 4 April 27, 1850, 253; BISH, 3:121–24, 126–28; "Charles Morgan," DAB, 13:164–65; Alexander J. Field, "Land Abundance, Interest/Profit Rates, and Nineteenth-Century American and British Technology," *Journal of Economic History* 43 (June 1983): 405–31; "George W. Quintard," DAB, 15:314–15; John G. B. Hutchins, *The American Maritime Industries and Public Policy, 1789–1914* (Cambridge: Harvard University Press, 1941), 325–68; John A. James and Jonathan S. Skinner, "The Resolution of the Labor-Scarcity Paradox," *Journal of Economic History* 45 (September 1985): 513–40; LEAD, 290–91; "Manufacture of Iron—The Novelty Works," *Hunt's Merchants' Magazine* 23 (October 1850): 463–65; "Manufactures in the City of New York," *Hunt's Merchants' Magazine* 37 (September 1857): 380–85; "Manufactures of Dutchess County," *Hunt's Merchants' Magazine*; "Manufacturing Industry of New York," *Hunt's Merchants' Magazine*; Peter Temin, "Labor Scarcity and the Problem of American Industrial Efficiency in the 1850s," *Journal of Economic History* 26 (September 1966): 277–98; "The West Point Foundry at Cold Spring," *Hunt's Merchants' Magazine* 16 (April 1847): 421. Also see the chapter epigraph.

14. BISH, 3:244–49; "Boston, Massachusetts," *Hunt's Merchants' Magazine* 39 (July 1858): 45; Freedley, *Philadelphia and its Manufactures*, 434–36; LEAD, 287–97; "Robert L. Thurston," DAB, 18:520–21; "Statistics of Providence, R.I., Manufactures," *Hunt's Merchants' Magazine* 23 (August 1850): 247–48.

15. BAG, 373–75; BISH, 3:32; Monte A. Calvert, *The Mechanical Engineer in America, 1830–1910* (Baltimore: Johns Hopkins Press, 1967), 8–9; "George H. Corliss," DAB, 4:441; HUNT, 251–300; LEAD, 307–8; LOZ, 398–401; W. J. Rorabaugh, *The Craft Apprentice: From Franklin to the Machine Age in America* (New York: Oxford University Press, 1986), 63–64; Anthony F. C. Wallace, *Rockdale: The Growth of an American Village in the Early Industrial Revolution* (New York: Alfred A. Knopf, 1978); "William Sellers," DAB, 16:576–77.

16. STAT, ser. E58; George Wallis, "New York Industrial Exhibition," ASM, 260–61; Joseph Whitworth, "New York Industrial Exhibition," ASM, 331–33, 337 (quote). The crude machinery measures for 1840 and 1860 provide a reasonable comparative approximation of the iron foundry and machinery sector. In 1840 iron castings are reported separately; therefore, foundries, which mostly produced castings along with a small amount of machinery, are not included. At the same time, textile machinery is included in the machinery category. In 1860 textile machinery and locomotives are separated from the machinery category.

CHAPTER 5: NETWORKED MACHINISTS BUILD LOCOMOTIVES

Epigraph: "The Transportation of Passengers and Wares, a Visit to the Norris Locomotive Works," *United States Magazine* 2 (October 1855): 151, 167.

1. John K. Brown, *The Baldwin Locomotive Works, 1831–1915* (Baltimore: Johns Hopkins University Press, 1995), 14, 173, 243–46, app. B; Albert Fishlow, *American Railroads and the Transformation of the Ante-Bellum Economy* (Cambridge: Harvard University Press, 1965), 4–6, 149–53; U.S. Bureau of the Census, *Manufactures of the United States in 1860, Eighth Census* (Washington, D.C.: Government Printing Office, 1865); John H. White, *American Locomotives: An Engineering History, 1830–1880*, rev. and exp. ed. (1968; rpt., Baltimore: Johns Hopkins University Press, 1997), 4–13. Also see chapter epigraph.

2. Christopher T. Baer, *Canals and Railroads of the Mid-Atlantic States, 1800–1860* (Wilmington, Del.: Regional Economic History Research Center, Eleutherian Mills–Hagley Foundation, 1981); James D. Dilts, *The Great Road: The Building of the Baltimore and Ohio, The Nation's First Railroad, 1828–1853* (Stanford: Stanford University Press, 1993), 196–99, 237–38, 257–58; Fishlow, *American Railroads and the Transformation of the Ante-Bellum Economy*, 399, table 54; Edward C. Kirkland, *Men, Cities, and Transportation: A Study in New England History, 1820–1900*, 2 vols. (Cambridge: Harvard University Press, 1948); Milton Reizenstein, *The Economic History of the Baltimore and Ohio Railroad, 1827–1853*, Johns Hopkins University Studies in Historical and Political Science, ser. 15, nos. 7–8 (Baltimore: Johns Hopkins University Press, 1897), 35; ROOTS, chap. 5; George R. Taylor, *The Transportation Revolution, 1815–1860*, vol. 4: *The Economic History of the United States* (New York: Holt, Rinehart and Winston, 1951); White, *American Locomotives*, 19. Regional definitions for fig. 5.2: New England (Maine, New Hampshire, Vermont, Massachusetts, Rhode Island, and Connecticut); Middle Atlantic (New York, Pennsylvania, New Jersey, Maryland, and Delaware); South (Virginia, West Virginia, North Carolina, South Carolina, Georgia, Florida, Alabama, Mississippi, Louisiana, Texas, Kentucky, and Tennessee); and Midwest (Ohio, Indiana, Illinois, Michigan, Wisconsin, Iowa, Missouri, and California).

3. BISH, 3:178–81; Brown, *Baldwin Locomotive Works*, xxvi–xxx, 3–4, 13, 31–39; Angus Sinclair, *Development of the Locomotive Engine*, annotated edition, ed John H. White Jr. (1907; rpt., Cambridge, Mass.: MIT Press, 1970), 249–50; "Thomas Rogers," DAB, 16:112–13; Robert H. Thurston, *A History of the Growth of the Steam-Engine* (1878; rpt., Ithaca, N.Y.: Cornell University Press, 1939), 210–14; White, *American Locomotives*, 13–24.

4. "American Machinery—Matteawan," *Scientific American* 5, April 27, 1850, 253; BISH, 2:208, 3:32–35; Brown, *Baldwin Locomotive Works*, 169; GIBB, 63–103, 183–95; LEAD, 313–14; LOZ; "Manufactures of Dutchess County," *Hunt's Merchants' Magazine* 15 (October 1846): 370–76;

PATER, 72–148 (quote, 119); Sinclair, *Development of the Locomotive Engine*, 200–209; STEAM, E10; White, *American Locomotives*.

5. "Asa Whitney," DAB, 20:155–56; Jeremy Atack, "Comment," in *Learning by Doing in Markets, Firms, and Countries*, ed. Naomi R. Lamoreaux, Daniel M. G. Raff, and Peter Temin (Chicago: University of Chicago Press, 1999), 96, table 2C.1; BISH, 3:32–35; Brown, *Baldwin Locomotive Works*, 6–17, 37, 61–63, 241–42, app. A, 255; Sinclair, *Development of the Locomotive Engine*, 282–94, 345–46; Bruce Sinclair, *Philadelphia's Philosopher Mechanics: A History of the Franklin Institute, 1824–1865* (Baltimore: Johns Hopkins University Press, 1974); "The Transportation of Passengers and Wares, a Visit to the Norris Locomotive Works," *United States Magazine*, 151–67; White, *American Locomotives*, 57–62, 543; "William Norris," DAB, 13:555–56.

6. "Charles Danforth," DAB, 5:65–66; PATER; "Thomas Rogers," DAB, 112–13; White, *American Locomotives*, 541.

7. George W. Browne, *The Amoskeag Manufacturing Co. of Manchester, New Hampshire* (Manchester, N.H.: Amoskeag Manufacturing Co., 1915); Robert F. Dalzell Jr., *Enterprising Elite: The Boston Associates and the World They Made* (Cambridge: Harvard University Press, 1987); DEY; GIBB, 92–97, 195, 641, app. 6; Arthur M. Johnson and Barry E. Supple, *Boston Capitalists and Western Railroads: A Study in the Nineteenth-Century Railroad Investment Process* (Cambridge: Harvard University Press, 1967); LEAD, 305–15; LOZ; Sinclair, *Development of the Locomotive Engine*, 200–209, 250–51; Sinclair, *Philadelphia's Philosopher Mechanics;* Martha Taber, "A History of the Cutlery Industry in the Connecticut Valley," *Smith College Studies in History* 41 (1955); White, *American Locomotives*, 14, 538, 540.

8. Brown, *Baldwin Locomotive Works*, 20–21, 31; GIBB, 195; LOZ, 490–520; PATER; White, *American Locomotives*, 15.

CHAPTER 6: RESILIENT COTTON TEXTILE MACHINIST NETWORKS

Epigraphs: "Statistics of Lowell Manufactures," *Hunt's Merchants' Magazine* 16 (April 1847): 423; "American Machinery—Matteawan," *Scientific American* 5, April 27, 1850, 253.

1. JERE, 276, table D.1, 279, table D.5; U.S. Bureau of the Census, *Report on the Manufactures of the United States at the Tenth Census, 1880* (Washington, D.C.: Government Printing Office, 1883).

2. GIBB, tables 1, 4, 9, 47, 101, 188, app. 1, 631; Paul F. McGouldrick, *New England Textiles in the Nineteenth Century: Profits and Investment* (Cambridge: Harvard University Press, 1968), 18–20, 146–47; Robert S. Woodbury, *Studies in the History of Machine Tools* (Cambridge, Mass.: MIT Press, 1972); Robert B. Zevin, "The Growth of Cotton Textile Production after 1815," in *The Reinterpretation of American Economic History*, ed. Robert W. Fogel and Stanley L. Engerman (New York: Harper and Row, 1971), 122–47. Textile machinery cost is measured in current dollars in fig. 6.3. There is no index of wholesale prices of machinery for 1827–60, but the Warren-Pearson "all commodity" index suggests that real prices stayed within a modest range between index values of 75 and 111, and prices in the late 1820s and late 1850s were virtually the same. See STAT, ser. E52.

3. Samuel Batchelder, *Introduction and Early Progress of the Cotton Manufacture in the United States* (Boston: Little, Brown, 1863); George W. Browne, *The Amoskeag Manufacturing Co. of Manchester, New Hampshire* (Manchester, N.H.: Amoskeag Manufacturing Co., 1915); GIBB; John L. Hayes, *American Textile Machinery* (Cambridge, Mass.: University Press, John

Wilson and Son, 1879); LOZ; McGouldrick, *New England Textiles in the Nineteenth Century*, 18–20, 268; Vera Shlakman, "Economic History of a Factory Town: A Study of Chicopee, Massachusetts," *Smith College Studies in History* 20, nos. 1–4 (October 1934–July 1935); "Statistics of Lowell Manufactures," 422–23 (also see chap. epigraph).

4. BAG, 407–11; BISH, 3:319–23; "David Wilkinson," DAB, 20:222; Henry H. Earl, *A Centennial History of Fall River, Mass.* (New York: Atlantic Publishing and Engraving Co., 1877); LOZ; Barbara M. Tucker, *Samuel Slater and the Origins of the American Textile Industry, 1790–1860* (Ithaca, N.Y.: Cornell University Press, 1984).

5. BAG, 524–32; BISH, 2:283; Browne, *Amoskeag Manufacturing Co. of Manchester, New Hampshire*, 42–60; GIBB; LOZ, 47–49, 255–59; MCLANE, docs. 3–5; Thomas R. Navin, *The Whitin Machine Works since 1831: A Textile Machinery Company in an Industrial Village* (Cambridge: Harvard University Press, 1950); ROE, 124, 217; Bruce Sinclair, *Philadelphia's Philosopher Mechanics: A History of the Franklin Institute, 1824–1865* (Baltimore: Johns Hopkins University Press, 1974), 195–216; Caroline F. Ware, *The Early New England Cotton Manufacture: A Study in Industrial Beginnings* (Boston: Houghton Mifflin, 1931), 81, 89; Charles G. Washburn, *Industrial Worcester* (Worcester, Mass.: Davis Press, 1917).

6. "American Machinery—Matteawan," *Scientific American* (see chap. epigraph); "Charles Danforth," DAB, 5:65–66; GIBB, 78; LOZ, 32–34, 160–70, 209–15; McGouldrick, *New England Textiles in the Nineteenth Century*, 248–50, app. D, table 48; MCLANE, doc. 3, no. 135, doc. 11, no. 23; "Manufactures of Dutchess County," *Hunt's Merchants' Magazine* 15 (October 1846): 370–76; PATER; George S. White, *Memoir of Samuel Slater*, 2nd ed. (Philadelphia, 1836), 256.

7. LEAD, 334–38; U.S. Bureau of the Census, *Compendium of the Sixth Census, 1840* (Washington, D.C.: Blair and Rives, 1841). The evidence for 1832 (MCLANE) may underestimate the Philadelphia area's textile machinery industry, but its small scale is consistent with evidence for later in the antebellum. See discussion at the end of this chapter.

8. Judith A. McGaw, *Most Wonderful Machine: Mechanization and Social Change in Berkshire Paper Making, 1801–1885* (Princeton, N.J.: Princeton University Press, 1987); MCLANE, doc. 3, doc. 10, no. 20, doc. 14; Margaret E. Martin, "Merchants and Trade of the Connecticut River Valley, 1750–1820," *Smith College Studies in History* 24, nos. 1–4 (October 1938–July 1939); Shlakman, "Economic History of a Factory Town"; U.S. Bureau of the Census, *Compendium of the Sixth Census, 1840*.

9. Clive Day, "The Early Development of the American Cotton Manufacture," *Quarterly Journal of Economics* 39 (May 1925): 450–68; U.S. Bureau of the Census, *Compendium of the Sixth Census, 1840*; U.S. Bureau of the Census, *Report on the Manufactures of the United States at the Tenth Census, 1880*. The U.S. Census of 1820 and MCLANE have incomplete tallies of the textile industry. The discussion of changes in the market for textile machinery may therefore have some errors in the details. Nonetheless, the overall changes are consistent with diverse evidence about the growth of the cotton textile industry and the patterns that emerged by the 1850s.

10. BAG, 501–9; BISH, 3:296–97; Browne, *Amoskeag Manufacturing Co. of Manchester, New Hampshire*, 42–111; "Charles T. James," DAB, 9:572–73; GIBB; LEAD, 308–9, 311–13, 339–40; LOZ; McGouldrick, *New England Textiles in the Nineteenth Century*, 248–50, app. D, table 48; William Mass, "Mechanical and Organizational Innovation: The Drapers and the Automatic Loom," *Business History Review* 63 (Winter 1989): 876–929; Navin, *The Whitin Machine Works since 1831*; ROE, 167–70, 252–56; Daniel P. Tyler, *Statistics of the Condition and Products of Cer-*

tain Branches of Industry in Connecticut, for the Year Ending October 1, 1845 (Hartford, Conn.: John L. Boswell, 1846); U.S. Bureau of the Census, *Manufactures of the United States in 1860, Eighth Census* (Washington, D.C.: Government Printing Office, 1865); Washburn, *Industrial Worcester*, 86–89; "William Mason," DAB, 12:377–78.

11. U.S. Bureau of the Census, *Manufactures of the United States in 1860, Eighth Census.*

CHAPTER 7: THE CRADLES OF THE METALWORKING MACHINERY INDUSTRY

Epigraph: Jacob Abbott, "The Armory at Springfield," *Harper's New Monthly Magazine* 5 (July 1852): 146

1. Arthur F. Burns and Wesley C. Mitchell, *Measuring Business Cycles* (New York: National Bureau of Economic Research, 1946), 102, table 23; Charles H. Fitch, "Report on the Manufactures of Interchangeable Mechanism" and "Report on the Manufacture of Hardware, Cutlery, and Edge-Tools; also Saws and Files," both in U.S. Bureau of the Census, *Report on the Manufactures of the United States at the Tenth Census, 1880* (Washington, D.C.: Government Printing Office, 1883); LEAD, 327–33; Duncan M. McDougall, "Machine Tool Output, 1861–1910," *Output, Employment, and Productivity in the United States after 1800*, Studies in Income and Wealth, vol. 30 (New York: National Bureau of Economic Research, 1966), 497–519; Ross M. Robertson, "Changing Production of Metalworking Machinery, 1860–1920," *Output, Employment, and Productivity in the United States after 1800*, Studies in Income and Wealth, vol. 30 (New York: National Bureau of Economic Research, 1966), 479–95; Nathan Rosenberg, "Technological Change in the Machine Tool Industry, 1840–1910," *Journal of Economic History* 23 (December 1963): 414–43; W. Steeds, *A History of Machine Tools, 1700–1910* (Oxford: Clarendon Press, 1969); U.S. Bureau of the Census, *Report on Power and Machinery Employed in Manufactures, Tenth Census of the United States, 1880*, vol. 22 (Washington, D.C.: Government Printing Office, 1888).

2. Jacob Abbott, "The Novelty Works," *Harper's New Monthly Magazine* 2 (May 1851): 721–34; "George H. Corliss," DAB, 4:441; HUNT, 251–300; LEAD, 287–88, 290, 434–36; "Manufacture of Iron—The Novelty Works," *Hunt's Merchants' Magazine* 23 (October 1850): 463–65; Carroll W. Pursell Jr., *Early Stationary Steam Engines in America* (Washington, D.C.: Smithsonian Institution Press, 1969), 51, 100; "Report of the Committee on the Machinery of the United States of America," ASM, 87–197; Steeds, *History of Machine Tools*, 26–45; "The West Point Foundry at Cold Spring," *Hunt's Merchants' Magazine* 16 (April 1847): 421; Joseph Whitworth, "New York Industrial Exhibition," ASM, 327–89; Robert S. Woodbury, *History of the Lathe to 1850* (Cambridge, Mass.: MIT Press, 1961).

3. John K. Brown, *The Baldwin Locomotive Works, 1831–1915* (Baltimore: Johns Hopkins University Press, 1995), 14–21, 167–69; LOZ, 401–7, 461–68; PATER, 114–24, 128–44; Angus Sinclair, *Development of the Locomotive Engine*, annotated edition, ed. John H. White Jr. (1907; rpt., Cambridge, Mass.: MIT Press, 1970), 232–40, 282–83, 345–46; "Thomas Rogers," DAB, 16:112–13; "The Transportation of Passengers and Wares, a Visit to the Norris Locomotive Works," *United States Magazine* 2 (October 1855): 151–67; John H. White, *American Locomotives: An Engineering History, 1830–1880*, rev. and expanded ed. (1968; rpt., Baltimore: Johns Hopkins University Press, 1997), 16, 20–21, tables 1–2.

4. GIBB; LEAD, 308–9, 327, 339–40; LOZ, 34, 49, 345–48.

5. DEY; MRS; RABER; Merritt R. Smith, "Army Ordnance and the 'American System' of Manufacturing, 1815–1861," in *Military Enterprise and Technological Change: Perspectives on the American Experience*, ed. Merritt R. Smith (Cambridge, Mass.: MIT Press, 1985), 39–86.

6. Abbott, "The Armory at Springfield" (also, see chapter epigraph); Brown, *Baldwin Locomotive Works, 1831–1915*, 243–46, app. B; Carolyn C. Cooper, *Shaping Invention: Thomas Blanchard's Machinery and Patent Management in Nineteenth-Century America* (New York: Columbia University Press, 1991), 85–121; and "'A Whole Battalion of Stockers': Thomas Blanchard's Production Line and Hand Labor at Springfield Armory," *IA: The Journal of the Society for Industrial Archeology* 14, no. 1 (1988): 37–57; "Cyrus Buckland," DAB, 3:229; DEY, 144–74; Fitch, "Report on the Manufactures of Interchangeable Mechanism," 25; MRS, 127–36; RABER; Paul J. Uselding, "Technical Progress at the Springfield Armory, 1820–1850," *Explorations in Economic History* 9 (Spring 1972): 291–316.

7. MRS; S.N.D. North and Ralph H. North, *Simeon North: First Official Pistol Maker of the United States* (Concord, N.H.: Rumford Press, 1913), 158–74; RABER, 243–45; Merritt R. Smith, "John H. Hall, Simeon North, and the Milling Machine: The Nature of Innovation among Antebellum Arms Makers," *Technology and Culture* 14 (October 1973): 573–91.

CHAPTER 8: MACHINE TOOL NETWORKS

Epigraph: Edwin T. Freedley, *Philadelphia and its Manufactures* (Philadelphia: Edward Young, 1858), 131.

1. DEY, 55–67, 117–32; MRS; ROE; Merritt R. Smith, "Army Ordnance and the 'American System' of Manufacturing, 1815–1861," in *Military Enterprise and Technological Change: Perspectives on the American Experience*, ed. Merritt R. Smith (Cambridge, Mass.: MIT Press, 1985), 39–86.

2. "Cyrus Buckland," DAB, 3:229; GIBB, 73, 216; "James Tyler Ames," DAB, 1:248; MRS, 288–90; "Nathan Peabody Ames," DAB, 1:249–50; "Report of the Committee on the Machinery of the United States of America," ASM, 97–117, 180–81, 187–91 (quote, 101); Nathan Rosenberg, "Introduction," ASM, 1–86; Vera Shlakman, "Economic History of a Factory Town: A Study of Chicopee, Massachusetts," *Smith College Studies in History* 20, nos. 1–4 (October 1934–July 1935): 24–34, 81–89; Martha Taber, "A History of the Cutlery Industry in the Connecticut Valley," *Smith College Studies in History* 41 (1955): 14–18; George Wallis, "New York Industrial Exhibition," ASM, 284–85 (quote).

3. DEY, 122–23; "Frederick W. Howe," DAB, 9:286; HOUN, 69–71; HUB, 59, December 20, 1923, 919–22; HUB, 60, February 14, 1924, 255–58; HUB, 60, March 20, 1924, 437–41; HUB, 60, June 12, 1924, 875–78; "Report of the Committee on the Machinery of the United States of America," ASM, 103–4 (quote), 107, 115–16, 181–84; ROE; Rosenberg, "Introduction," ASM, 17; Merritt R. Smith, "The American Precision Museum," *Technology and Culture* 15 (July 1974): 413–37; Robert S. Woodbury, *History of the Milling Machine* (Cambridge, Mass.: MIT Press, 1960), 31–41.

4. "Christopher M. Spencer," DAB, 17:446–47; "A Day at the Armory of 'Colt's Patent Fire-arms Manufacturing Company,'" *United States Magazine* 4 (March 1857): 221–49 (quote, 233); "Elisha K. Root," DAB, 16:144–45; "Elisha K. Root," *National Cyclopedia of American Biography* (New York: James T. White and Co., 1922), 18:313; Charles H. Fitch, "Report on the Manufactures of Interchangeable Mechanism," U.S. Bureau of the Census, *Report on the Manufactures of the United States at the Tenth Census, 1880* (Washington, D.C.: Government Printing Office, 1883), 21–22, 27; "Francis A. Pratt," DAB, 15:172–73; Donald R. Hoke, *Ingenious Yankees: The Rise*

of the American System of Manufactures in the Private Sector (New York: Columbia University Press, 1990), 101–30; William Hosley, *Colt: The Making of an American Legend* (Amherst: University of Massachusetts Press, 1996), 26, 59–60; HOUN; Robert A. Howard, "Interchangeable Parts Reexamined: The Private Sector of the American Arms Industry on the Eve of the Civil War," *Technology and Culture* 19 (October 1978): 633–49; HUB, 59, July 5, 1923, 1–4; HUB, 60, April 24, 1924, 617–20; MRS, 288, 290; "Report of the Committee on the Machinery of the United States of America," ASM, 100–104, 115–18; ROE, 137, 164–85; Paul J. Uselding, "Elisha K. Root, Forging, and the 'American System,'" *Technology and Culture* 15 (October 1974): 543–68; Woodbury, *History of the Milling Machine*, 33–35.

5. BISH, 3:227, 315, 321; Freedley, *Philadelphia and its Manufactures*, 436–37; Alden Hatch, *Remington Arms* (New York: Rinehart and Co., 1956), 34–82; Howard, "Interchangeable Parts Reexamined"; HUB, 60, April 24, 1924, 617–20; ROE, 175–76.

6. BISH, vol. 3; John K. Brown, *The Baldwin Locomotive Works, 1831–1915* (Baltimore: Johns Hopkins University Press, 1995), 169; "Coleman Sellers," DAB, 16:574–75; Henry H. Earl, *A Centennial History of Fall River, Mass.* (New York: Atlantic Publishing and Engraving Co., 1877), 41–50; "George H. Corliss," DAB, 4:441; HUNT, 251–300; LEAD; LOZ, 398–407; Ross M. Robertson, "Changing Production of Metalworking Machinery, 1860–1920," in *Output, Employment, and Productivity in the United States after 1800*, Studies in Income and Wealth vol. 30 (New York: National Bureau of Economic Research, 1966) 479–95; ROE; Charles G. Washburn, *Industrial Worcester* (Worcester, Mass.: Davis Press, 1917), 111–23; "William Sellers," DAB, 16:576–77; Woodbury, *History of the Milling Machine*, 38–42. Also see chapter epigraph.

7. Duncan M. McDougall, "Machine Tool Output, 1861–1910," in *Output, Employment, and Productivity in the United States after 1800*, Studies in Income and Wealth, vol. 30 (New York: National Bureau of Economic Research, 1966), 497–519; Robertson, "Changing Production of Metalworking Machinery"; Nathan Rosenberg, "Technological Change in the Machine Tool Industry, 1840–1910," *Journal of Economic History* 23 (December 1963): 414–43.

CHAPTER 9: MACHINISTS' NETWORKS FORGE THE PIVOTAL PRODUCER DURABLES INDUSTRY

Epigraph: "Report of the Committee on the Machinery of the United States of America," ASM, 128–29, 193.

1. ASM. Also see chapter epigraph.

2. John K. Brown, *The Baldwin Locomotive Works, 1831–1915* (Baltimore: Johns Hopkins University Press, 1995), 170–83; HOUN; Robert A. Howard, "Interchangeable Parts Reexamined: The Private Sector of the American Arms Industry on the Eve of the Civil War," *Technology and Culture* 19 (October 1978): 633–49; RABER, 86–94; Nathan Rosenberg, "Technological Change in the Machine Tool Industry, 1840–1910," *Journal of Economic History* 23 (December 1963): 414–43; Merritt R. Smith, "Army Ordnance and the 'American System' of Manufacturing, 1815–1861," in *Military Enterprise and Technological Change: Perspectives on the American Experience*, ed. Merritt R. Smith (Cambridge, Mass.: MIT Press, 1985), 39–86.

3. Robert E. Gallman, "Commodity Output, 1839–1899," *Trends in the American Economy in the Nineteenth Century*, Studies in Income and Wealth, vol. 24 (Princeton, N.J.: Princeton University Press, 1960), 26, table 4, and 43, table A-1; David R. Meyer, "Emergence of the American Manufacturing Belt: An Interpretation," *Journal of Historical Geography* 9 (April 1983):

145–74; David R. Meyer, "Midwestern Industrialization and the American Manufacturing Belt in the Nineteenth Century," *Journal of Economic History* 49 (December 1989): 921–37; ROE; ROOTS; Gordon M. Winder, "Before the Corporation and Mass Production: The Licensing Regime in the Manufacture of North American Harvesting Machinery, 1830–1910," *Annals of the Association of American Geographers* 85 (September 1995): 521–52; Gordon M. Winder, "The North American Manufacturing Belt in 1880: A Cluster of Regional Industrial Systems or One Large Industrial District," *Economic Geography* 75 (January 1999): 71–92.

Essay on Sources

PRIMARY SOURCES

The accuracy and details of U.S. government data covering the 1790–1860 period vary widely. Nonetheless, these sources are quite useful if employed judiciously and if a focus is kept on the broader patterns of the data. Sometimes generalizations about one point in time that has missing data can be corroborated by more detailed data at another point in time. Although patent data are specific to one view of industrial activities, the lists of patentees, their residences, and the types of patents for the 1791–1820 period contained in Martin Van Buren, "Patents Granted by the United States," Secretary of State, Communicated to the House of Representatives, January 13, 1831, 21st Cong., 2nd sess., Doc. No. 50, *New American State Papers, Science and Technology*, vol. 4: *Patents* (Wilmington, Del.: Scholarly Resources, 1972), 80–504, provide a long-term perspective for a variety of industries. For this study the main use was for the steam-related and the textile equipment patents.

For the time around 1810 the best sources, albeit unsystematic and with selective coverage, are Tench Coxe, "Digest of Manufactures, 1810," Communicated to the Senate, January 5, 1814, 13th Cong., 2nd sess., *New American State Papers*, vol. 1: *Manufactures* (Wilmington, Del.: Scholarly Resources, 1972), 160–410; and Albert Gallatin, "Manufactures," Communicated to the House of Representatives, April 19, 1810, 11th Cong., 2nd sess., *New American State Papers*, vol. 1: *Manufactures* (Wilmington, Del.: Scholarly Resources, 1972), 124–42. For the early 1830s, which picks up the economic changes of the extended period of 1820s growth, Louis McLane, *Documents Relative to the Manufactures in the United States Collected and Transmitted to the House of Representatives, 1832, by the Secretary of the Treasury*, House Doc. No. 308, 22nd Cong., 1st sess. (Washington, D.C.: Duff Green, 1833), is a treasure trove of evidence. Many areas of the East and particular industries have little or no coverage, but the tables and details that are available provide an excellent window on industry at that time.

The report on steam engines by Levi Woodbury, *Steam Engines*, transmitted to the House of Representatives, 1838, by the Secretary of the Treasury, House Doc. No. 21, 25th Cong., 3rd sess. *New American State Papers, Science and Technology*, vol. 7: *Steam Engines* (Wilmington, Del.: Scholarly Resources, 1973), 11–482, is unparalleled as a source for an industrial sector. Some major gaps exist, especially for New York state; nevertheless, the data on steam engine manufacturers and their buyers provide an exceptional view of the scale of industrial development as of the late 1830s.

Although often criticized for incompleteness and inaccuracies, the U.S. Bureau of the Census, *Compendium of the Sixth Census, 1840* (Washington, D.C.: Blair and Rives, 1841) possesses evidence for many individual manufactures by county and state. When used to identify overall patterns, which also can be linked to other censuses, this data can be employed to develop generalizations about industrial development. The manufacturing data for 1850, Secretary of the Interior, *Abstract of the Statistics of Manufactures, Seventh Census, 1850* (Washington, D.C.: Government Printing Office, 1859), unfortunately, is incomplete and only available at the state level for individual manufactures. Nevertheless, this evidence can be linked to data for other points in time to fill in some of the industrial changes. Without question the U.S. Bureau of the Census, *Manufactures of the United States in 1860, Eighth Census* (Washington, D.C.: Government Printing Office, 1865), offers the best data on manufacturing for the antebellum at the county, state, and national levels. With the extensive level of detail the data also can be used to confirm generalizations from previous incomplete censuses.

Contemporary magazines contain many nuggets of information that can be used in conjunction with other evidence to develop generalizations. The well-known *Hunt's Merchants' Magazine* provided useful facts about individuals, manufactures, and firms, and the vignettes of antebellum firms in *Harper's New Monthly Magazine, Scientific American,* and *United States Magazine* supplied rich texture.

SECONDARY SOURCES

The book's focus on the full range of metalworking manufactures for the antebellum East precluded the use of archival materials for individual firms. This placed the onus on finding a rich set of secondary materials, and the resources available are quite extraordinary. To use them effectively, however, required the integration of numerous, disparate sources from materials written during the antebellum, in a few cases, to the broader array from the late nineteenth century. Twentieth-century resources also are quite substantial because many scholars have completed fine studies that provided evidence and insights. This book's study of machinist networks required backgrounds on many individuals, most of whom did not have formal biographies written about them. Thus, the *Dictionary of American Biography* proved to be a valuable source of background material that would have been impossible to assemble otherwise.

Theory

The study of knowledge and its role in economic development is an active area of research in many disciplines, and some good ways to enter this literature are Meric S. Gertler, "Tacit Knowledge and the Economic Geography of Context, or the Undefinable Tacitness of Being (There)," *Journal of Economic Geography* 3 (2003): 75–99, Jeremy R. L. Howells, "Tacit Knowledge, Innovation and Economic Geography," *Urban Studies* 39, nos. 5–6 (2002): 871–84; and Peter Maskell and Anders Malmberg, "Localised Learning and Industrial Competitiveness," *Cambridge Journal of Economics* 23 (1999): 167–85. A large literature discusses the theory of social networks and can be accessed through electronic library resources using the theory term (*social networks*). Sociologists developed many of the early contributions to theory and empirical work, and this research has now spread to other disciplines. Excellent introductions are available in Ronald S. Burt, *Structural Holes: The Social Structure of Competition* (Cambridge: Harvard University Press, 1992); and Stanley Wasserman and Katherine Faust, *Social Network Analysis: Methods and Applications* (Cambridge: Cambridge University Press, 1994).

The study of communities of practice is developed especially well in engineering, which is relevant to this book, and some good sources are Monte A. Calvert, *The Mechanical Engineer in America, 1830–1910* (Baltimore: Johns Hopkins Press, 1967); and Walter G. Vincenti, *What Engineers Know and How They Know It* (Baltimore: Johns Hopkins University Press, 1990). The concept has broader applications, including many types of organizations, which can be surveyed in Etienne Wenger, *Communities of Practice: Learning, Meaning, and Identity* (Cambridge: Cambridge University Press, 1998).

Antebellum Industrialization

The classic source, J. Leander Bishop, *A History of American Manufactures from 1608 to 1860*, 3 vols. (Philadelphia: Edward Young and Co., 1866), is essential for anyone doing research on the antebellum period. It is written by someone with access to materials and to people involved in manufacturing in the late antebellum. The second volume includes many details about industrial development and the third volume has vignettes on firms and cities. Another well-known source is Victor S. Clark, *History of Manufactures in the United States, 1607–1860* (Washington, D.C.: Carnegie Institution of Washington, 1916). David R. Meyer, *The Roots of American Industrialization* (Baltimore: Johns Hopkins University Press, 2003), presents an interpretation of the agricultural-industrial transformation of the East. Many other sources on industrialization during this period can be found in it, including the substantial literature in economic history.

Machinery

Good entrées to important questions about machinery are David A. Hounshell, *From the American System to Mass Production, 1800–1932: The Development of Manufacturing Technology in the United States* (Baltimore: Johns Hopkins University Press, 1984); and Donald R. Hoke, *Ingenious Yankees: The Rise of the American System of Manufactures in the Private Sector* (New York: Columbia University Press, 1990). The series of articles on interchangeable mechanism written by Charles H. Fitch, which appear in U.S. Bureau of the Census, *Report on the Manufactures of the United States at the Tenth Census, 1880* (Washington, D.C.: Government Printing Office, 1883), provide a sweeping view of the development of many important types of machinery.

Insightful interpretations of the relative importance of machinery during the antebellum are covered in Alexander J. Field, "On the Unimportance of Machinery," *Explorations in Economic History* 22 (October 1985): 378–401; and Kenneth L. Sokoloff, "Investment in Fixed and Working Capital during Early Industrialization: Evidence from U.S. Manufacturing Firms," *Journal of Economic History* 44 (June 1984): 545–56. A valuable way of conceptualizing how changes in machinery technology occur, which combines network ideas and community of practice, is Gavin Wright, "Can a Nation Learn? American Technology as a Network Phenomenon," in *Learning by Doing in Markets, Firms, and Countries*, ed. Naomi R. Lamoreaux, Daniel M. G. Raff, and Peter Temin (Chicago: University of Chicago Press, 1999), 295–326. One of the best contemporary sources on machinery and the leaders of many of the sectors covered in this book is Edwin T. Freedley, *Leading Pursuits and Leading Men: A Treatise on the Principal Trades and Manufactures of the United States* (Philadelphia: Edward Young, 1856).

Iron Foundries and Steam-Engine Works

Excellent starting points for iron foundries and steam engines are Robert B. Gordon, *American Iron, 1607–1900* (Baltimore: Johns Hopkins University Press, 1996), which gives the background for understanding iron foundries; Louis C. Hunter, *A History of Industrial Power in the United States, 1780–1930*, vol. 2: *Steam Power* (Charlottesville: University Press of Virginia, 1985), which details how iron foundries and the steam engine industry operated; and Carroll W. Pursell Jr., *Early Stationary Steam Engines in America* (Washington, D.C.: Smithsonian Institution Press, 1969), which offers a survey of the people and firms that built the steam engine industry. The classic work from one of the nineteenth-century leaders of the steam engine industry is Robert H. Thurston, *A History of the Growth of the Steam-Engine*, centennial ed. (1878; rpt., Ithaca, N.Y.: Cornell University Press, 1939). Fine biographies of some of the leaders of the steam engine industry are available, including Greville Bathe and Dorothy Bathe, *Oliver Evans: A Chronicle of Early American Engineering* (Philadelphia: Historical Society of Pennsylvania, 1935); Talbot Hamlin, *Benjamin Henry Latrobe* (New York: Oxford University Press, 1955); and Archibald D. Turnbull, *John Stevens: An American Record* (New York: Century Company, 1928).

Locomotives

The best way to start research on the locomotive industry is to examine the studies by Angus Sinclair, *Development of the Locomotive Engine*, annotated ed., ed. John H. White Jr. (1907; rpt., Cambridge, Mass.: MIT Press, 1970); and John H. White, *American Locomotives: An Engineering History, 1830–1880*, rev. and exp. ed. (1968; rpt., Baltimore: Johns Hopkins University Press, 1997). Studies of individual locomotive firms, which also provide insights into the antebellum industry, include John K. Brown, *The Baldwin Locomotive Works, 1831–1915* (Baltimore: Johns Hopkins University Press, 1995); and John W. Lozier, *Taunton and Mason: Cotton Machinery and Locomotive Manufacture in Taunton, Massachusetts, 1811–1861* (New York: Garland Publishing, 1986). A book rich in history and anecdotes about Paterson's locomotive industry, written by an individual who knew some of the antebellum legends, is L. R. Trumbull, *A History of Industrial Paterson* (Paterson, N.J.: Carleton M. Herrick, 1882).

Textile Machinery

The textile machinery industry has been researched more than any other equipment sector, in part because the cotton textile industry received so much attention. Many of the studies combine the manufacture of textiles and the equipment used. The classic fact book of individuals and firms which must be consulted is William R. Bagnall, *The Textile Industries of the United States* (Cambridge, Mass.: Riverside Press, 1893). A study that provides depth on the full sweep of the industry is David J. Jeremy, *Transatlantic Industrial Revolution: The Diffusion of Textile Technologies between Britain and America, 1790–1830s* (Cambridge, Mass.: MIT Press, 1981). Philadelphia has been the site of fine case studies, including Philip Scranton, *Proprietary Capitalism: The Textile Manufacture at Philadelphia, 1800–1885* (Cambridge: Cambridge University Press, 1983); and Cynthia J. Shelton, *The Mills of Manayunk: Industrialization and Social Conflict in the Philadelphia Region, 1787–1837* (Baltimore: Johns Hopkins University Press, 1986).

One of the most famous studies of textile machinery is John L. Hayes, *American Textile Machinery* (Cambridge, Mass.: University Press, John Wilson and Son, 1879). Excellent case studies of individual textile machinery firms include George S. Gibb, *The Saco-Lowell Shops: Textile Machinery Building in New England, 1813–1949* (Cambridge: Harvard University Press, 1950); Lozier, *Taunton and Mason*; William Mass, "Mechanical and Organizational Innovation: The Drapers and the Automatic Loom," *Business History Review* 63 (Winter 1989): 876–929; and

Thomas R. Navin, *The Whitin Machine Works since 1831: A Textile Machinery Company in an Industrial Village* (Cambridge: Harvard University Press, 1950). Works on Samuel Slater, the earliest, well-known textile machinery mechanic, include James L. Conrad Jr., "'Drive That Branch': Samuel Slater, the Power Loom, and the Writing of America's Textile History," *Technology and Culture* 36 (January 1995): 1–28; and George S. White, *Memoir of Samuel Slater*, 2nd ed. (Philadelphia, 1836). Many case examples of textile machinery firms are included in Trumbull, *History of Industrial Paterson;* and Charles G. Washburn, *Industrial Worcester* (Worcester, Mass.: Davis Press, 1917).

Firearms and Machine Tools

One of the early gems of research on the firearms industry, which includes machine tools, is Felicia J. Deyrup, "Arms Makers of the Connecticut Valley: A Regional Study of the Economic Development of the Small Arms Industry, 1798–1870," *Smith College Studies in History* 33 (1948). The federal government's role in promoting the firearm's industry is detailed in Merritt R. Smith, "Army Ordnance and the 'American System' of Manufacturing, 1815–1861," in *Military Enterprise and Technological Change: Perspectives on the American Experience*, ed. Merritt R. Smith (Cambridge, Mass.: MIT Press, 1985), 39–86. The leading federal armories and their use of machine tools are covered in major case studies: Michael S. Raber, Patrick M. Malone, Robert B. Gordon, and Carolyn C. Cooper, *Conservative Innovators and Military Small Arms: An Industrial History of the Springfield Armory, 1794–1968*, Report to the U.S. Department of the Interior, National Park Service (South Glastonbury, Conn.: Raber Associates, 1989); and Merritt R. Smith, *Harpers Ferry Armory and the New Technology: The Challenge of Change* (Ithaca, N.Y.: Cornell University Press, 1977).

Machine Tools

The literature on machine tools is quite disparate, ranging from metal products industries and machinery firms to case studies of manufacturing companies. Here the focus is on research dealing directly with machine tools. The classic book that remains salient is Joseph W. Roe, *English and American Tool Builders* (New York: McGraw-Hill, 1916). The series of twenty-three articles by Guy Hubbard that deal with machine tools in New England, published in the *American Machinist* 59–61 (1923–24), is unusual in its details about the people, the firms, and the industry, and it should be consulted by anyone doing research in this area. An excellent source on the development of important machine tools is Robert S. Woodbury, *Studies in the History of Machine Tools* (Cambridge, Mass.: MIT Press, 1972).

The article that posed important hypotheses about machine tools is Nathan Rosenberg, "Technological Change in the Machine Tool Industry, 1840–1910," *Journal of Economic History* 23 (December 1963): 414–43. The reports of British visitors to the United States factories during the 1850s contain many insights about the use of machine tools. These are collected, with an excellent introduction, in *The American System of Manufactures*, ed. Nathan Rosenberg (Edinburgh: Edinburgh University Press, 1969). An extensive literature examines the development of machine tools and their use in the arms industry, including Merritt R. Smith, "John H. Hall, Simeon North, and the Milling Machine: The Nature of Innovation among Antebellum Arms Makers," *Technology and Culture* 14 (October 1973): 573–91; and Robert S. Woodbury, "The Legend of Eli Whitney and Interchangeable Parts," *Technology and Culture* 1 (Summer 1960): 235–53.

Index